Systems, Cybernetics, Control, and Automation

Ontological, Epistemological, Societal, and Ethical Issues

RIVER PUBLISHERS SERIES IN AUTOMATION, CONTROL AND ROBOTICS

Indexing: All books published in this series are submitted to Thomson Reuters Book Citation Index (BkCI), CrossRef and to Google Scholar

The "River Publishers Series in Automation, Control and Robotics" is a series of comprehensive academic and professional books which focus on the theory and applications of automation, control and robotics. The series focuses on topics ranging from the theory and use of control systems, automation engineering, robotics and intelligent machines.

Books published in the series include research monographs, edited volumes, handbooks and textbooks. The books provide professionals, researchers, educators, and advanced students in the field with an invaluable insight into the latest research and developments.

Topics covered in the series include, but are by no means restricted to the following:

- Robots and Intelligent Machines
- Robotics
- Control Systems
- Control Theory
- Automation Engineering

For a list of other books in this series, visit www.riverpublishers.com

Systems, Cybernetics, Control, and Automation

Ontological, Epistemological, Societal, and Ethical Issues

Spyros G. Tzafestas
National Technical University of Athens
Athens, Greece

Published 2017 by River Publishers
River Publishers
Alsbjergvej 10, 9260 Gistrup, Denmark
www.riverpublishers.com

Distributed exclusively by Routledge
4 Park Square, Milton Park, Abingdon, Oxon OX14 4RN
605 Third Avenue, New York, NY 10017, USA

Systems, Cybernetics, Control, and Automation Ontological, Epistemological, Societal, and Ethical Issues / by Spyros G. Tzafestas.

Routledge is an imprint of the Taylor & Francis Group, an informa business

ISBN 978-87-93609-07-5 (print)

"Without the belief that Nature is subject to the laws
there can be no science".

"We changed our environment so radically
that now we have to change ourselves to live
in this new environment".

"Let us remember that the automatic machine
is the precise economic equivalent of slave labor.
Any labor which competes with slave labor must accept
the economic consequences of slave labor".

NORBERT WIENER (1894–1964)
Father of Cybernetics

GREEK INVENTORS OF MECHANICAL AND FEEDBACK MECHANISMS

Ancestors of Classical and Modern Systems and Control Engineers

ΑΡΧΙΜΗΔΗΣ/ARCHIMEDES

287–212 BC
SCREW LEVERS
COMPOUND PULLEYS
SCREW PUMP

ΚΤΗΣΙΒΙΟΣ Ο ΑΛΕΞΑΝΔΡΕΥΣ

KTESIBIOS OF ALEXANDRIA

285–222 BC
WATER CLOCK
SUCTION PUMP
HYDRAULIS

ΗΡΩΝ Ο ΑΛΛΕΞΑΝΔΡΕΥΣ

HERO OF ALEXANDRIA

10–85 AD
AUTOMATOPOIETICA
WIND POWER ORGAN
AEROPILE

Contents

Preface

Progress is only possible by passing from a state of undifferentiated wholeness to differentiation of parts.
Ludwig von Bertalanffy

The boundaries of 'our universe' are not the boundaries of the universe.
Ervin Laszlo

All human interactions are opportunities either to learn or to teach.
M. Scott Peck

The purpose of this book is to provide a consolidated overview of four closely interrelated and co-developed fields of science and technology, namely: *'Systems', 'Cybernetics', 'Control', and 'Automation'*. It is devoted to the study of fundamental ontological, epistemological, social impact, and ethical issues of the above fields.

'General Systems theory' **(GST)** was initiated by *Ludwig von Bertalanffy* emphasizing the fact that natural biological systems are 'open' and interact with their environment. He revealed the 'isomorphism' exhibited by natural systems which he used to study systems in a unified way.

'System Dynamics' **(SD)** was coined by *Jay Forrester* as a methodology to investigate and understand the dynamic performance of systems through the development and computer simulation of appropriate models.

Cybernetics, which overlaps considerably with GST was initiated by *Norbert Wiener* as the scientific/engineering field that studies control and communication in the animal and the machine, in a unified way, also including social systems.

Control, in its current form, started its development during World War II (focusing on military applications) and has since then been remarkably developed (under the name *'modern control theory'*) reaching a high mature state.

Automation, from the ancient Greek **automa**tiza**tion** (*αυτοματοποι ητική/automatopoietike*) and the American **automa**tic organiza**tion,** uses feedback control and information technology to automate industrial, office, and managerial processes. Automation is implemented at several levels from full machine control to full human control.

These fields exhibit several influences on society, and rise, during their use and application, many ethical concerns and dilemmas.

This book studies the above fields, and their societal, ethical, and philosophical issues in a single volume for the first time, at a depth and width deemed sufficient for the reader to see the scientific/technological and societal/ethical aspects of the four fields. The book can be used in engineering courses as a unified reference providing fundamental material on ontology, epistemology, social impact, ethics, and general philosophy of the four fields. The book can also be used for independent reading.

January 2017 **Spyros G. Tzafestas**

List of Figures

List of Tables

List of Abbreviations

A	Adaptive Automation
AC	Adaptive Control
ACT	Automation and Control Technology
AI	Artificial Intelligence
AIC	Algorithmic Information Content
AOM	Academy of Management
AR	Augmented Reality
AT	Assistive Technology
ATC	Air Traffic Controller
BIBO	Bounded-Input Bounded-Output
BP	Back Propagation
CAD	Computer-Aided Design
CAS	Complex Adaptive System
CCT	Classical Control Theory
CIM	Computer-Integrated Manufacturing
CLD	Causal Loop Diagram
CNS	Communication Networks System
CybOrg	Cybernetic Organism
DCS	Distributed Control System
DSS	Decision Support System
ECS	Economic Control System
FBC	Feedback Controller
FFC	Feedforward Controller
FG	Fuzzy-Genetic
FIM	Fuzzy Inference Mechanism
FIR	Flight Information Region
FLC	Fuzzy Logic Control
FMS	Flight Management System
GA	Genetic Algorithm
GAC	Genetic Algorithm-based Control
GST	General Systems Theory
HCA	Human-Centered Automation
HF	Human Factors
HITL	Human-In-The-Loop
HJB	Hamilton-Jacobi-Bellman

HMI	Human-Machine Interface
IA	Industrial Automation
IATA	International Air Transportation Association
ICT	Information and Communication Technology
IFR	Instrument Flight Rules
ICAO	International Civil Aviation Organization
IP	Internet Protocol
JIT	Just-in-Time
KB	Knowledge Base
KBS	Knowledge Based System
LAN	Local Area Protocol
LOA	Level of Automation
LIS	Library and Information Science
MAM	Mobile Autonomous Manipulator
MAN	Metropolitan Area Protocol
MAS	Multi-Agent System
MBPC	Model-Based Predictive Control
MCS	Management Control System
MCT	Modern Control Theory
MIMO	Multi-Input Multi-Output
MIS	Management Information System
MLP	Multilayer Perceptron
MR	Mixed Reality
NCS	Networked Control System
NF	Neurofuzzy
NG	Neurogenetic
NN	Neural Network
NSPE	National Society of Professional Engineers
OA	Office Automation
OC	Organizational Cybernetics
OR	Operational Research
PAT	Process Automation Technology
PMS	Power Management System
PNS	Principle of Natural Selection
PID	Proportional plus Integral plus Derivative
PLC	Programmable Logic Controller
PT	Philosophy of Technology
PwSN	People with Special Needs
QoL	Quality of Life
QoS	Quality of Service
RBF	Radial Basis Function

SA	Situation Awareness
SAE	Society of Automotive Engineers
SCADA	Supervisory Control And Data Acquisition
SCCA	Systems, Cybernetics, Control, and Automation
SD	System Dynamics
SE	Systems Engineering
SFD	Stock and Flow Diagram
SIP	Stages of Information Processing
S-R	Stimulus-Response
SISO	Single-Input Single-Output
SITL	Society-In-The-Loop
SoSE	System of Systems Engineering
SP	Systems Philosophy
STC	Self-Tuning Control
TCP	Transmission Control Protocol
TPS	Transaction Processing System
VSM	Viable System Model
VR	Virtual Reality
WAN	Wide Area Protocol
WHO	World Health Organization
WMR	Wheelchair Mounted Robot
WMS	Workflow Management System

Names and Achievements of SCCA Pioneers

LUDVIG VON BERTANLAFFY (1901-1972) General Systems Theory, Bertanlaffy model

ERVIN LASZLO (1932-) General evolution Theory, Systems philosophy

JAY W. FORRESTER (1918-2016) Urban/World dynamics, System dynamics

RUSSEL L. ACKOFF (1919-2009) Purposeful systems theory, f-Laws

NORBERT WIENER (1894-1964) Cybernetics, Wiener filter

HEINZ VON FOERSTER (1911-2002) Second-order cybernetics, Constructivism

GREGORY BATESON (1904-1980) Socio-cybernetics, Ecology of mind

W. ROSS ASHBY (1903-1972) Complex systems, Requisite variety

GORDON PASK (1928-1996) New cybernetics

HUMBERTO MATURANA (1928-) Autopoiesis/Family therapy

STAFFORD BEER (1926-2002) Management cybernetics

ILYA PRIGOZINE (1917-2003) Dissipative structure

WALTERR EVANS (1920-1999) Root locus method

HARRY NYQUIST (1889-1976) Nyquist stability criterion

H. WADE BODE (1905-1982) Bode plots

GENE FRANKLIN (1927-2012) Sampled data control

RICHARD BELLMAN (1920-1984) Dynamic programming

RUDOLF KALMAN (1930-2016) Optimal filtering/control

ALEXANDR LYAPUNOV (1857-1918) Lyapunov stability

LEV PONTRYAGIN (1908-1988) Maximum principle

THOMAS SHERIDAN (1929-) Supervised control

LOFTI ZADEH (1921-) Fuzzylogic/control

Abstracts

Chapter 1: Introductory Concepts and Outline of the Book

This chapter serves as an introduction to the book, providing preliminary conceptual material about ontological, epistemological, societal, and ethical issues of 'Systems, Cybernetics, Control. and Automation' **(SCCA)**. In addition, the chapter provides preparatory information concerning systems philosophy, control philosophy, and cybernetics philosophy.

Chapter 2: Basics of Philosophy, Philosophy of Science, and Philosophy of Technology

Philosophy (love of wisdom) provides the means via which man can comprehend the world, discovering 'what is true' and 'what is being', and using his mind to support his life. This chapter provides a brief discussion of the fundamental elements of philosophy (basic concepts, historical evolution, and branches), philosophy of science (what is science, what are experimental theories and observations, and questions about scientific data and the law of science), and philosophy of technology focusing on the related philosophical questions.

Chapter 3: Background Concepts: Systems Ontology, History, and Taxonomy

System theory is concerned with the relations between the parts of a system (e.g., the organs, the muscles, tissue, bone, and cells of the human body). The properties of a system depend on the way its parts interact to work as a whole. This chapter presents background material on the system concept including ontological elements of systems (environment, boundary, closed/open systems, etc.), a brief historical review, and a multi-dimensional taxonomy of systems.

Chapter 4: General systems Theory and Systems Dynamics

General systems theory **(GST)** was originated by Ludwig von Bertalanffy as the science or methodology of *'open systems',* and is connected to the science of *'wholeness'*. System dynamics **(SD)** was initiated by Jay Forrester as a methodology to understand the dynamic behavior of systems by constructing dynamic models and simulating them on the computer. This chapter provides the fundamental ontological and epistemological elements of GST and SD. Basic ontological questions addressed include 'what is GST?', What is SD?', What is isomorphism in GST?', and 'What is systems thinking?'. Epistemological elements studied include: Bertalanffy's view of GST, Forrester's causal loop diagrams, a scheme for performing SD, the system thinking continuum, and the axioms of GST.

Chapter 5: Cybernetics

Cybernetics was coined by Norbert Wiener in 1948 as the science that studies in a unified way control and communication in the animal and the machine. Cybernetics and GST essentially deal with the study of the same problem, namely the *'organization'* problem. This chapter discusses fundamental ontological and epistemological aspects of cybernetics namely: definition(s), history, first-order and second-order cybernetics highlighting the difference between them, social systems, and sociocybernetics including the inherent processes of autopoiesis, autocatalysis, and self-organization.

Chapter 6: Control

Control is based on the concept of feedback and aims to improve crucial characteristics of systems such as stability, accuracy, speed of response, and to design controllers that optimize energy-like performance criteria. This chapter provides a review of the control theory field starting with answers to the questions 'what is feedback?' and 'what is control?', and a brief historical survey of control from ancient times to present. Then, it presents the fundamental concepts and techniques of classical control (closed-loop system, stability, performance specifications, root-locus, frequency response methods, phase-lag, phase lead and phase lag–lead compensations, PID control, and discrete time control). Next, the chapter outlines the basic elements of modern control theory, namely: state space modeling, Lyapunov stability, state-feedback control, optimal and stochastic control, and model-free/ behavioral control. Finally, the chapter presents the fundamental ontological elements of networked control systems.

Chapter 7: Complex and Nonlinear Systems

Complex and nonlinear systems constitute two major areas of systems with very important theoretical and practical results. Complexity crosses both natural/physical sciences and humanity sciences (sociology, psychology, management, economics, etc.). This chapter studies the ontological and epistemological issues of complex and nonlinear systems (bifurcations, strange attractors, chaos, fractals, emergence, and complex adaptive systems), and discusses the 'complexity' concept focusing on the issue of complexity measures. Next, the chapter presents the concepts of 'adaptation' and 'self-organization' in natural, technological, and societal systems, with emphasis on the mechanisms via which they can be achieved, including representative examples.

Chapter 8: Automation

Automation exploits feedback control and information and communication technology **(ICT)** to automate the operation of technological and societal systems. A primary concern in modern automation is the study of the physical, mental, and psychological features of humans at work, which can help in the development of human-centered or human-minding automation. This chapter is devoted to the ontology and epistemology of industrial and office automation systems, including selected applications. It also discusses fundamental issues of human-machine interfaces **(HMIs)** used in human-automation interaction and cooperation, and virtual reality in automation and medical robotics, and investigates crucial human factors that must be considered in the design and operation of automation systems.

Chapter 9: Societal Issues

All four fields studied in this book have strong overlapping and complementary implications on society. This chapter provides an overview of these implications. Specifically, the chapter discusses the role of GST in society development, investigates the relation of cybernetics and society, examines the general impact of control on industrial, management and economic systems, and presents key issues of the impact of automation on society (e.g., productivity, employment, and working conditions) including the impact of robotics and office automation.

Chapter 10: Ethical and Philosophical Issues

Systems, cybernetics, control, and automation can be used for the good and the bad, and over the years have created crucial ethical concerns and dilemmas. This chapter investigates these ethical concerns. Starting with a discussion of the question 'What is ethics?' The chapter examines the ethical issues and concerns of 'systems engineering' and 'systems thinking'. Then, it discusses the ethics of cybernetics focusing on the ethics of 'cybernetic organisms' *(cyborgs)* and 'implants', and the ethics of control and automation including robot ethics *(roboethics)* and management control ethics. The chapter ends by giving an overview of systems philosophy, control philosophy, and cybernetics philosophy.

1

Introductory Concepts and Outline of the Book

Everything must be made as simple as possible, but not simpler.
Albert Einstein

The famous balance of nature is the most extraordinary of all cybernetic systems. Left to itself, it is always self-regulated.
Joseph Wood Krutch

It is always easier to destroy a complex system than to selectively alter it.
Roby James

1.1 Introduction

Systems, cybernetics, control, and automation **(SCCA)** are four scientific and technological fields, which are now at a highly mature state and have contributed substantially to the development, growth, and progress of human society. Collectively, their real-life applications range over a very large gamma of man-made or biological systems, including transportations, power generation, chemical industry, robotics, manufacturing, cybernetics organisms (cyborgs), aviation, economic systems, enterprise, systems, medical/health systems, environmental applications, and so on. A large number of models, methods, and tools were developed to assure the high efficiency of SCCA applied in practical situations.

The present chapter serves as an introduction to the SCCA's ontological, epistemological, societal, and ethical issues discussed in the book. Preliminary preparatory material is provided for all four fields:

- General systems theory/system dynamics.
- Cybernetics (first-and-second order).

- Feedback control (classical and modern).
- Automation (industrial and non-industrial).

In addition, general introductory information is provided in the chapter about systems philosophy, control philosophy, and cybernetics philosophy.

1.2 Systems Theory, System Dynamics, and Cybernetics: A Preliminary Look

1.2.1 Systems Theory and System Dynamics

The concept of *General Systems Theory* **(GST)** was first introduced by *Ludwig von Bertalanffy*, orally in the 1930s, and in his various publications after World War II. In his historical survey, von Bertalanffy points out that 'in a certain sense it can be said that the notion of system is as old as European philosophy' [1]. He says that "if we try to define the central motif in the birth of philosophical-scientific thinking with the Ionian pre-Socratics of the 6th Century B.C., one way to spell it on would be as follows. Man in early culture, and even primitives of the day, experience themselves as being 'thrown' into a hostile world, governed by chaotic and incomprehensible demonic forces which, at best, may be propitiated or influenced by way of magical practices".

Von Bertalanffy gave a mathematical description of system properties, namely wholeness, sum, growth, competition, allometry, mechanization, centralization, finality, and equifinality, derived from the system modeling by simultaneous differential equations. In his works, he was primarily interested in developing the biology-inspired theory of 'open systems', i.e., of systems exchanging energy and matter with the environment as all living systems do, and he established as a general system model the concept of '*open system*' which interacts with its environment both in feed-forward and feedback modes (Figure 1.1).

Actually, it would be more appropriate to speak of '*system theories*' in plural instead of '*system theory*' in singular, since a wide set of scientific and cultural theories can be called '*system theories*'. These theories include:

- *General systems theory* (Ludwig von Bertalanffy, Niklas Luhman, and others).
- *Cybernetics* (Norbert Wiener) which is the theory of regulation and interaction of man-machine systems.
- *Information theory* (Claude Shannon) which defined information technically via the probability and signal-to-noise ratio concepts.
- *Meta-mathematics* (Kurt Godel) which is, among others, concerned with the concept of recursion used e.g., in AI.

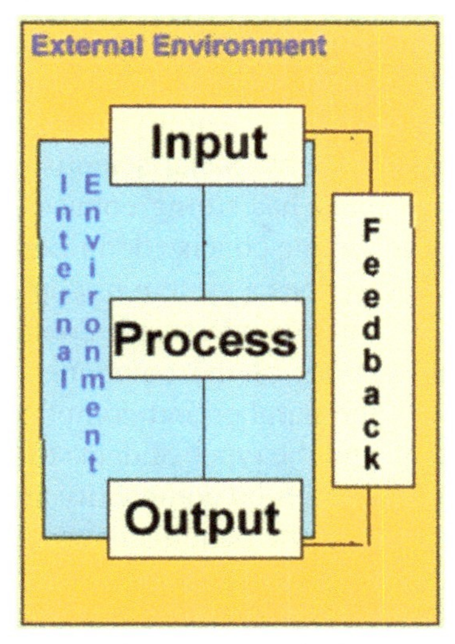

Figure 1.1 Bertalanffy's concept of 'open system' with input and feedback from the system's environment (Feedback in open systems has inspired many thinkers to examine their application to living systems).

Source: www.freshbrainz.blogspot.gr (/2009_01_achive.html).

All these systems theories interact, and are historical technological products of the 1940's and 1950's. General systems theory claims to be the umbrella of all systems theories. General system theory was viewed by von Bertalanffy as a method of organizing the interactions between component parts of a larger organism. It is this 'organizing information' view that made GST easily adaptable to many scientific fields including psychology, sociology, and social work. About the interaction of components von Bertalanffy writes [2]:

> '*A system is an entity which maintains its existence through the mutual interaction of its parts*'.

The *complete information* about a dynamical system at a certain time is determined by its state at that time, which typically is multi-dimensional determined by many state quantities. According to the view of Laplace, similar causes effectively determine similar outcomes: Therefore, in the phase space, trajectories that start close to each other also remain close to each other over time. Nonlinear dynamical systems with *deterministic chaos* show an exponential dependence on initial conditions for bounded orbits, i.e., the separation of trajectories with close initial states increases exponentially.

A general class of dynamic systems is the class of *complex systems* which include systems of many areas and societal applications. The concept of complexity is considered to be part of a unifying framework of science. Complexity has particularly studied by social scientists or philosophers of social science. Actually, complexity has strong connection with probability and information, but there is no unique concise definition of complexity. In a wide sense, the term *complex* describes a system or component of which the design or function or both is difficult to understand. Complexity is determined by various factors, e.g., the number of the components and the intricacy of the interface between them, the number and complication of conditional branches, the degree of nesting, and the types of data structures [3]. As *Brian Artur* [4] states 'Common to all studies on complexity are systems with multiple elements that adapt or react to the patterns created by these elements.' A special class of complex systems involves the so-called **CAS** introduced by *Holland* [5, 6]. A CAS is a large collection of various parts (components, agents) interconnected in a hierarchical way such that organization persists or grows over time without centralized control through a dynamical, continuously unfolding process, individual agents within the system actively (but imperfectly) gather information from neighboring agents and from the external environmental. Within the CAS, competition operates to maintain or strengthen certain properties while constraining or eliminating others. As a result, the CAS is characterized by potential change and adaptation either through modification of its rules, connections, and responses, or through alteration of the external environment [7]. A simple schematic illustration of CAS is shown in Figure 1.2. The local agents interact and lead to emergence of overall (*holistic*) system patterns which in turn, by feedback, influence the behaviors of the agents.

System Dynamics (SD) was founded by *Jay Forrester* as an aspect of system theory aiming to understand the dynamic behavior of complex systems through *stocks, flows, time delays*, and *internal feedback loops*. The basic concepts of SD are system dynamics' simulation, feedback, nonlinearity, and loop dominance. GST and SD are supported by the so-called *systems thinking* (or *thinking systematically*) which enables system scientists to understand the world and the physical/social phenomena. Systems-thinking is actually a 'new way of thinking to understand and manage complex systems'. Systems thinking is a soft methodology and qualitative analysis based on '*causal diagrams*'. System dynamics is a hard methodology and quantitative analysis which uses stock and flow diagrams, and is grounded on control and modern nonlinear systems theory. It is designed to be a convenient practical tool for

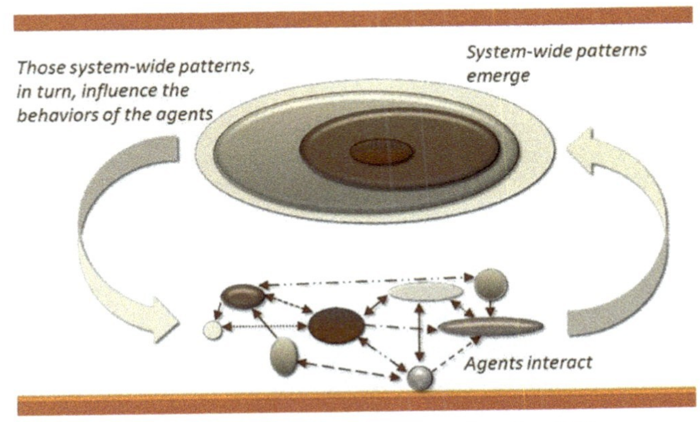

Figure 1.2 Schematic representation of a CAS.

Source: www.slideplayer.com/slide/6399997 (Arch EE Workshop 17–18 June 2015).

organization policy-planners and other decision makers that enables them to solve crucial and demanding problems of their organizations.

Figure 1.3 shows the three key aspects which should be involved in any 'systems thinking' definition, namely [8]:

- *Purpose, goal or function* (described in a clear understandable way, and related to everyday life).
- *Elements* (representing the features of systems thinking).
- *Interconnections* (describing the way the elements or characteristics feed into and relate to each other).

Systems thinking involves and is carried out at the following top-down levels:

- *Strategy or feel level* (people, purpose).
- *Tactical or think level* (preparation, performance).
- *Operational or process level* (performing production of goods and services).

According to *Klir* [9] 'systems thinking focuses on those properties of systems and associated problems that emanate from the general notion of *system hood*, while the divisions of the classical science have been done largely on properties of *thinking hood*'.

Figure 1.4 shows the '*triangular-pyramid*' ('tetradian') representation of systems thinking where the four vertices correspond to (i) person-to-person relationships, (ii) conversations (information exchange), (iii) product and service transactions, and (iv) aspirational shared purpose (vision and values).

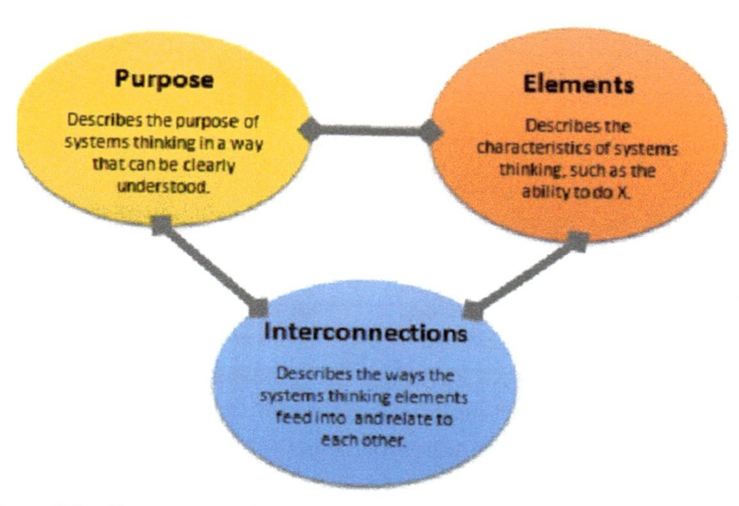

Figure 1.3 Components of system test required for a system thinking definition.
Source: www.researchgate.net [8].

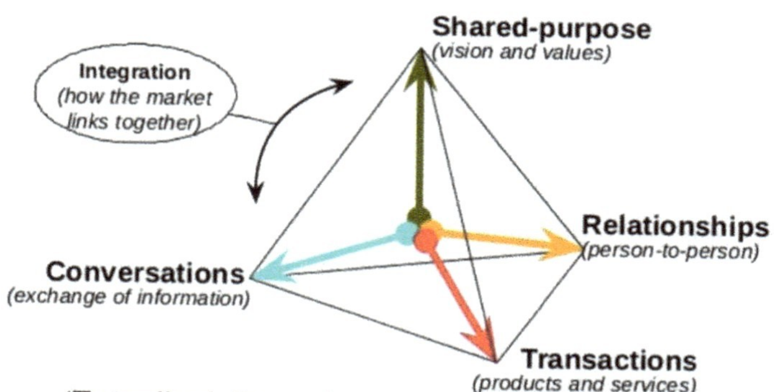

Figure 1.4 Triangular pyramid illustration of systems-thinking.
Source: www.slideshare.net (/tetradian/bridging-enterprisesearcharchitecture-and-systemsthinking, T. Graves, Principal Consultant, www.tetradian.com).

1.2.2 Cybernetics

Cybernetics is the study of human-machine interaction employing the principles of feedback, control and communication. Cybernetics as a concept in society has its origin in Plato's work that referred to it as 'government'. Formally, the field of cybernetics was founded by *Norbert Wiener* with his 1948 book 'Cybernetics: or Control and Communication in the Animal and the Machine' which articulated the marriage of control and communication inspiring numerous scientists and engineers of later generations. In the 1960s and 70s, cybernetics generated systems approaches that entered social sciences from management and social control to economics. In 1960 *J.R.C. Licklider* (MIT) publishes a seminal paper entitled: '*Man-computer symbiosis*' that exploited cybernetics ideas to produce a roadmap for the future development of interactive computing [10]. Biological cybernetics (*biocybernetics*) is the application of cybernetics and systems theory to further understanding of systems biology. Cybernetic systems operate through transfer of information (physical, chemical, electrical, electromagnetic). Systems theory advocates that everything affects everything and so organized living systems never operate in an open-loop (non-feedback) way. We know that physics is the general theory of matter and energy relationships. Similarly, cybernetics is a general theory of information processing, feedback control, and decision making. One can view cybernetics as the foundation of social sciences in about the same way that physics is the foundation of the engineering principles.

In general, cybernetics provides coherent and non-vitalistic explanations for continually ordered natural and biological phenomena (e.g., cognition, regularized behavior, etc.) which have historically been studied by scientific and analytical thinkers relying on vitalistic functions.

- *Paul Pangaro states:* 'Cybernetics is simultaneously the most important science of the age and the least recognized and understood [....]. It has as much to say about human interactions as it does about machine intelligence'.
- *Peter Corning writes:* 'The single most important property of a cybernetic system is that it is controlled by the relationship between endogenous goals and external environment. Cybernetics is about purposiveness, goals, information flows, decision-making, control processes and feedback, properly defined, at all levels of living systems'.

The three basic assumptions of cybernetics are the following:

- **Systemism** *(Reality is to be apprehended in a systemic manner. Cybernetics focuses on structure, not on the matter).*
- **Analogisation of systems** *(Isomorphisms).*
- **Energetic and material nature of interactions** '*(energy and matter principle)*'.

About *System Analogies*, *Stefan Banach* says: 'A mathematician is one who sees analogies between theorems; a better mathematician is one who can see analogies between proofs, and the best mathematician can find analogies between theories. The ultimate mathematician would see analogies between analogies'. In cybernetics one must indeed see *analogies between theories.*

Cybernetics differs from AI. Cybernetics is based on a '*constructivist*' view of the world whereas AI is based on the '*realist*' view that knowledge is a commodity which can be stored in a machine, and was defined by John McCarthy (1956) as 'the discipline of building machines that are as intelligent as human". According to Marvin Minsky the application of such stored knowledge to the real world constitutes 'intelligence' manifested, for example, by rule-based expert (software) systems. Paul Pangaro argues that 'the differences between AI and cybernetics are not merely semantic but determine fundamentally the direction of research performed from a cybernetic versus AI stance' [11]. Pangaro has coined the diagram of Figure 1.5 to illustrate the different ways of interpreting and treating the aspects of:

- Representation,
- Memory,
- Reality,
- Epistemology,

by AI and cybernetics.

Cybernetics as coined by Wiener is called 'first-order' cybernetics or '*cybernetics of observed systems*'. Heinz von Foerster has developed the 2nd-order cybernetics (cybernetics of cybernetics) which includes the observer in the system study. Von Foerster argued that 2nd-order cybernetics is actually the '*cybernetics of observing systems*' [12]. Francis Heylingen argues that 'second-order cybernetics studies the role of the (human) observer in the construction of models of systems and other observers'

The key features of first-order and second-order cybernetics that determine their differences are shown in Table 1.1.

Generalizing the ideas of second-order cybernetics, scientists and thinkers have attempted to develop a *second-order science* by expanding (first-order) science through the addition of the observer to what is observed, and study

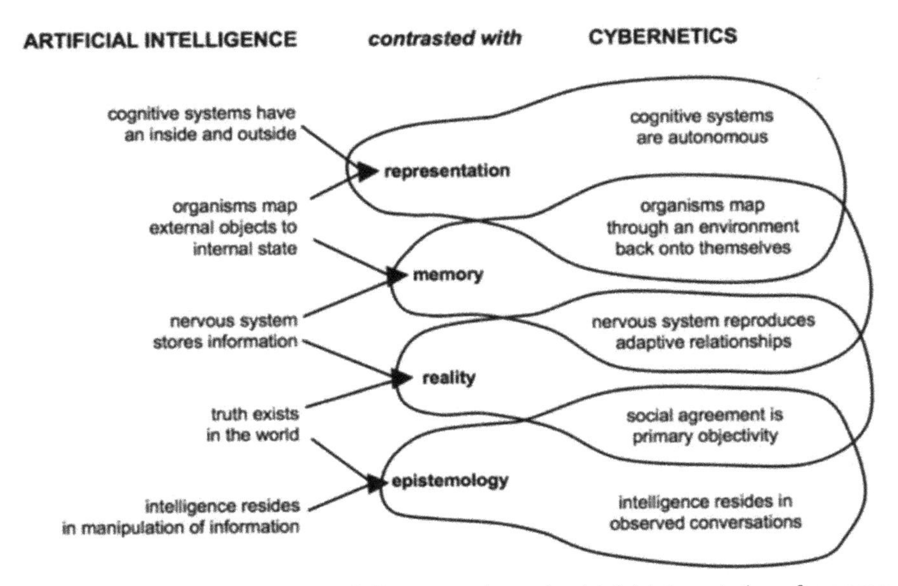

Figure 1.5 Artificial intelligence (*left*) versus cybernetics (*right*) interpretation of representation, memory, reality, and epistemology.

Source: www.pangaro.com/definition-cybernetics.html

Table 1.1 Key features and differences of first- and second-order cybernetics

First-Order Cybernetics Deals with:	Second-Order Cybernetics Deals with:
Observed system	Observing system
Model purpose	Modeler purpose
Controlled systems	Autonomous systems
System variables' interaction	'Observer/observed' interaction
Social systems theories	Theories of 'ideas/society' interactions

the coevolution of theories and society [13, 14]. We would move our thinking from considering 'science as creating descriptions of systems' to 'science as active part of social systems'. First-order science explores the world. Second-order science is a reflexive form of first-order science (dealing with the reflection on explorations). Louis H. Kaufman has pointed out that for every well-developed first-order scientific field, a corresponding second-order field or discipline can be constructed. For example, second-order science may include and study the following:

- Sociology of sociology.
- Philosophy of philosophy.

- Functions of functions.
- Tests of tests.
- Patterns of patterns.
- Computation of computation.
- Cybernetics of cybernetics.
- Economics of economics.
- Linguistics of linguistics.
- Logic of logic.
- Geometry of geometry.
- Mathematics of mathematics, etc.

1.3 Control and Automation: A Preliminary Look

1.3.1 Control

Automatic or feedback control plays a fundamental role in modern society and lies at the core of automation. The engineering use of control is very much older than the theory and can be traced back to the Greek engineer *Ktesibios* (285–222 BC). Working for the king Ptolemaeos II he has designed and constructed the so-called 'water clock'. After about two centuries *Heron of Alexandria* (Greek mathematician and engineer ~100 BC) has designed several regulating mechanisms (e.g., the famous mechanism for the automatic opening of Temple doors, and the automatic distribution of wine) [15]. A big step forward in the development of control engineering was made during the industrial revolution, with the major finding being *James Watt's* fly-ball governor [16], a mechanism that was able to regulate the speed of steam engine by throttling the flow of steam. The key results of control theory were developed around the period of the Second World by *Nyquist, Bode, Nichols* and *Evans*, and are now called the '*classical control theory*' (**CCT**). In the 1960s the state–space approach to control was developed which allowed multivariable and time varying systems to be treated in a unified way. Particular results derived using this approach include the optimal state-estimator (*Kalman* filter), the linear pole (eigenvalue) placement controller, and the linear-quadratic controller (deterministic and stochastic). All these results are collectively called '*modern control theory*' (**MCT**).

Adaptive, hierarchical, and decentralized control theory was then followed, and in the 1980s the development of 'robust control theory'$_2$ was made (H_2, H_{infinity}, l_1, and μ-theory control). In parallel with the above theories, intensive work was done in the analysis and design of nonlinear, distributed-parameter, coordinated/decentralized, and model free/intelligent (fuzzy, neural, neurofuzzy, expert, and behavior-based) controllers.

The central concept in all the above control methods is the notion of *feedback*, through which feedback control systems or closed-loop systems are realized. Feedback is distinguished in:

- Positive feedback, and
- Negative feedback.

Positive feedback causes a change in the same direction as the input (stimulus), and thereby augments the change (error) moving the system state further from the equilibrium point. In other words, positive feedback abruptly displaces a system away from a set point and enhances the stimuli thereby leading to greater departure of the output from the desired value. In the human body, positive feedback usually leads to instability and is tolerated by the body only for a short time. Examples of biological positive feedback are: blood clotting (stop blood flow), uterus contraction during childbirth (parturition), labor contractions, hormone increase, etc. In man-made systems, positive feedback is typically purposely used to achieve desired and controllable oscillatory performance.

Negative feedback (Figure 1.6) is used to regulate the output of the system such that to be maintained within a small interval around the desired output (controlled variable) value or response, and assures the following features:

- Gain desensitization (i.e., gain less sensitive to system parameter variations).
- Reduction of nonlinear effects (e.g., nonlinear distortion effects).
- Reduction of disturbance and noise effects.
- Increase of system bandwidth leading to faster response.
- Increase of accuracy of tracing the desired steady state response.

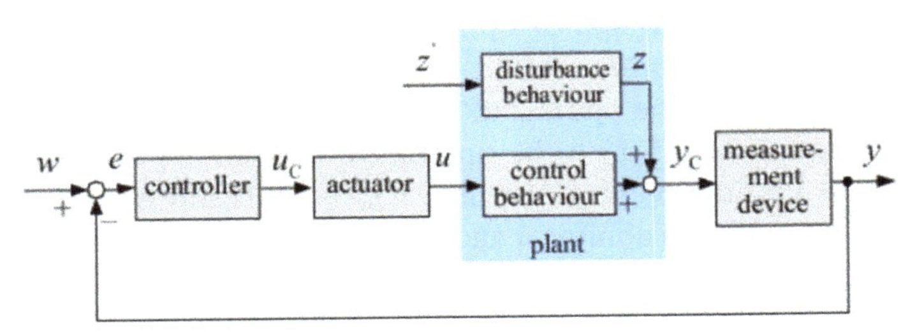

Figure 1.6 General structure of closed-loop system with negative feedback (w: set point/command/input signal, e: error signal, u_C: control signal, u: manipulated signal, z: disturbance signal, y: controlled signal/variable).

In living organisms negative feedback counteracts any changes and brings the body back to *homeostasis* (normal steady-state condition). Examples of homeostasis are temperature control, glucose control, blood pressure control, body water regulation, hydrogen ion (pH) regulation, etc. The feedback in humans is achieved by the cooperation of receptors (nerves), control or coordinating center (brain), and effectors (muscles, glands). Actually, chemical, endocrine, thermal, and nervous systems work together to achieve and maintain homeostasis. For example, hormones responsible for the control of the glucose level are (i) insulin, and (ii) glucagon. The levels of the glucose in the blood are sensed by the chemoreceptors of the pancreas which releases either insulin (by beta cells) or glucagon (by alpha cells) that are directed to the liver in amounts that depend on the glucose concentration. If the glucose concentration level increases, then less glucagon and more insulin is released and (secreted) by the pancreas directed to the liver. If the glucose level decreases, the pancreas releases less insulin and more glucagon targeting the liver. The above negative feedback process in blood glucose regulation is illustrated in Figure 1.7(a). In contrast, Figure 1.7(b) shows that positive feedback in blood sugar leads to further increase of the sugar level (i.e., high blood glucose is not regulated).

A unified classical controller used over the years for controlling process and other industrial plants is the so-called *PID-or 3-term-controller* which involves three additive controller components, namely *proportional* (P) controller, *integral* **(I)** controller, and *derivative* **(D)** controller. Proportional controllers exert control analogous to the present system error, integral controllers have memory and work with the past error values summed up over a given time interval of the past, and derivative control work with predicted values of the error some time ahead in the future. It is noted that most controllers (linear or nonlinear) have implicit proportional, integral and derivative control terms.

A primary tool for the design of optimal controllers in man-made systems is the theory of *optimization* (static and dynamic). *Static optimization* employs the standard function optimization by equating to zero the gradient (derivative) of the function $\mathbf{f}(x)$ to be optimized, and solving the equation $\partial\mathbf{f}(\mathbf{x})/\partial\mathbf{x} = \mathbf{0}$ to find the point $\mathbf{x}$ at which $\mathbf{f}(\mathbf{x})$ is optimum (maximum or minimum). If this equation is nonlinear and cannot be solved by linear algebra we use numerical techniques *(Gradient technique:* 1st order, and *Newton-like techniques:* 2nd order). For solving dynamic optimization problems (control problems, etc.), we use the *dynamic programming method (Bellman's principle of optimality)*, and the *minimum/maximum principle (Pontryagin).*

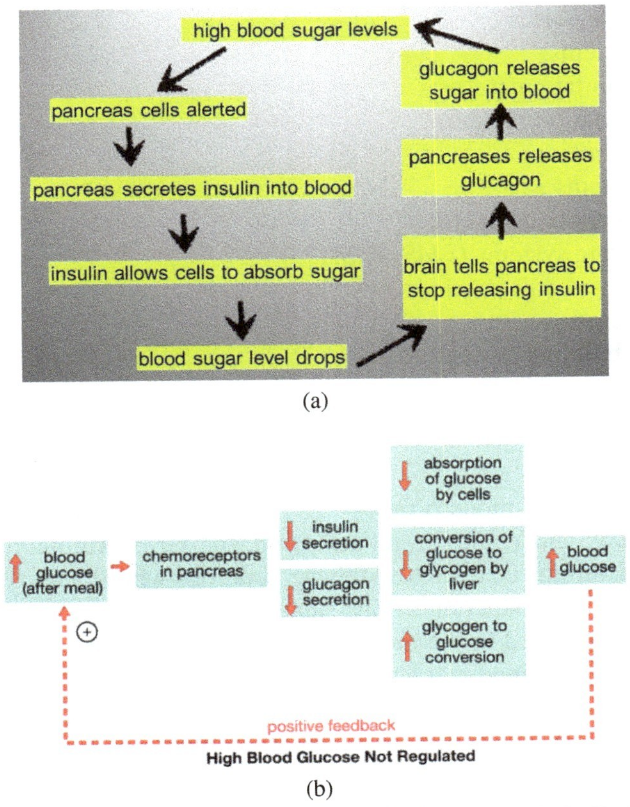

Figure 1.7 (a) Sugar level regulation by negative feedback control, (b) Positive feedback implies that high sugar level in the blood is increased further.

Source: (a) www.slideplayer.com (/26/8406888/slides/slide_5.jpg), and (b) www.moodle2.rockyview.ab.ca (Alberta Education: Module2–The endocrine system).

Richard Bellman pointed out the need to adopt mathematical control theory and models to face the undesired problems of the environment. He says: 'Mankind's history has been a struggle against a hostile environment. We finally reached a point where we can begin to dominate our environment and cease being victims of the vagaries of nature in the form of fire, flood, famine, and pestilence [...]. As soon as we understand this fact, our mathematical interests necessarily shift in many areas from descriptive analysis to control theory' (In: *Some Vistas of Modern Mathematics*, University of Kentucky Press, 1968).

The applications of control theory and engineering coincide in their majority with the applications of automation. Some of them are the following:

- Automotive systems.
- Railway systems.
- Land transportation systems.
- Sea transportation systems.
- Aviation systems.
- Robotic systems.
- Continuous process plants.
- Manufacturing systems (FMS, CIM).
- Enterprises and organizations.
- Office systems.
- Intelligent building systems.
- Electrical and mechanical energy systems.

1.3.2 Automation

The term *automation* in modern times was coined in 1952 by *D.S. Harder* of Ford Company, and comes from the composite term '**auto***matic organiza***tion**'. It was introduced to involve the methodology which analyzes, organizes and control the production means such that all material, machine, and human resources are used in the most effective way [17, 18]. In the history of technology it is described that actually the first who used the term *automatization* was *Heron of Alexandria in* his work *περί Αυτοματοποιητικής* (Peri automatopoitikes, meaning 'about automatization') [15]. A generally accepted definition of automation is: 'Automation is the use of controlled systems such as computers to control industrial machinery and processes, replacing human operators' (Wikipedia: Automation). In general, the term automation refers to any process or function which is self-driven and reduces, then eventually eliminates the need for human intervention. As in the above definition of automation, it is stated, that the automated operation of the systems is achieved by using. The feedback is closed through suitable-measurement and sensing devices, and the control action is exerted by suitable actuators (motors and other prime movers or human executives) through proper interfaces and displays [19, 20]. According to [21], 'automation that is appropriate for application in realistically complex sociotechnical domains should be based on an integrated understanding of the technical, human, organizational, economic, and cultural attributes of the application, and can be enhanced by development and use of particular types of the application'.

As it happens in all technological domains, automation must satisfy a certain set of requirements to justify its application. At minimum these criteria should include [21]:

- Adequacy of the means and funds employed.
- Acceptability of the elaborated technical solution.

Other possible criteria and problem attributes to be considered in automation efforts include [21, 22]:

- *Social dimension of automation* (Machines must always be selected in relation to the subjects of the work, i.e., humans).
- *Humanization of technology* (As the field of human-machine systems design is progressing, more people become able to perform evaluative and analytical investigations. Thus humanization of technology is needed).
- *Human centered (or human minding) design* methodology (A design methodology is needed to enable designers to achieve goals in a way compatible with the human-minding philosophy of the design).

Human-factors (physical, psychological, cognitive) contribute to the human-centered methodology, and play a dominant role for the successful, safe, and efficient operation of automation systems. Three key human factors that should be considered in the design of automation systems (particularly in flight decks systems) are:

- *Allocation of function* (i.e., what specific task is assigned to human or to automation).
- *Stimulus-response compatibility* (which is concerned with the relationship-geometric or conceptual – between a *stimulus*, e.g., a display, and a *response* e.g., a control action).
- *Internal model of the operator* (i.e., the operator's internal representation of the controlled system, including his/her conceptual understanding about the system components, processes, and input/output quantities).

Other human factors that should be taken into account are:

- *Psychological factors* (job stress, job satisfaction).
- *Physical stress factors* (job design, muscular strength).
- *Human bias* (weight bias, speed bias, height bias, horizontal distance bias, temperature bias).

- *Human error* (observation error, error types characterized by their cause, e.g., slips or failures of execution, or mistakes that occur when the plan is wrong itself although the actions go as planned).
- *Knowledge-based mistakes* (which occur as the result of the hypothesis testing procedure or interference in functional reasoning due to wrong analogies).

Industrial, process, and manufacturing automation belongs to one of three main types (shown in Figure 1.8), which have been evolved over the years, and are briefly described as follows [23]:

- *Static automation* This is technology-centered automation that cannot face or regulate large variations in workload.
- *Flexible automation* This is a form of Adaptive Automation **(AA)** which reduces the drawbacks of static automation by dynamically shifting tasks between operator and automation. It merely makes use of task and user models without taking external events into account (i.e., it is just user-centered).
- *Intelligent AA* This interaction-based kind of automation goes one step beyond flexible (adaptive) automation by explicitly including world models, and therefore interactively taking into account external events.

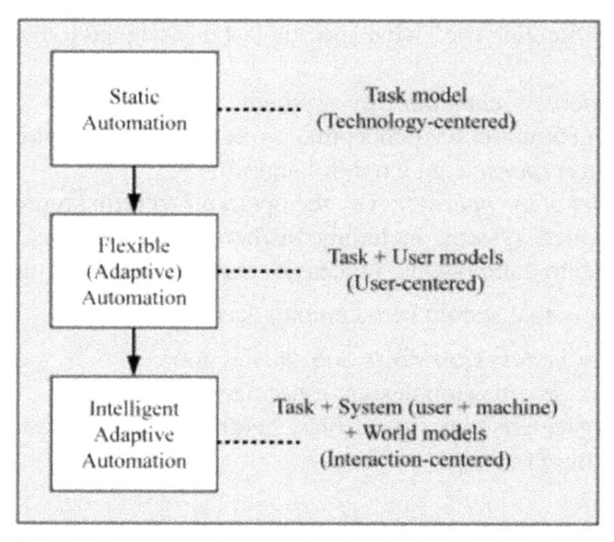

Figure 1.8 The three types of automation (static, flexible, intelligent adaptive).

Source: https://yokogawa.com/blog/smart-workload-balancing

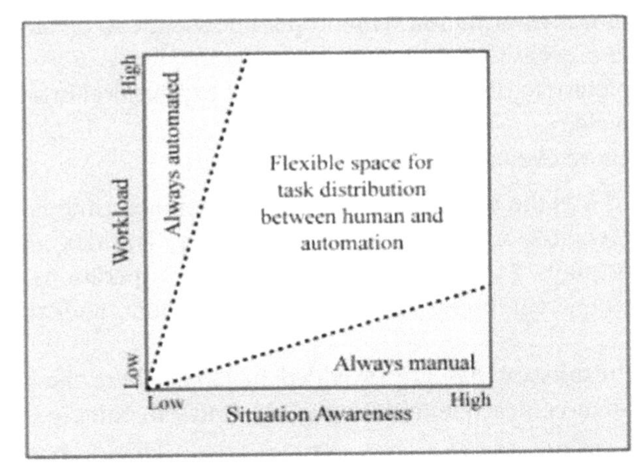

Figure 1.9 Flexible space of intelligent AA (low workload with high situation awareness suggest manual operation, and high workload with low situation awareness suggest maximum automation).

Source: https://yokogawa.com/blog/smart-workload-balancing

Figure 1.9 shows the flexible space of task distribution between operator and automation in the work load/situation awareness plane. The rule here is that tasks which need high workload but provide little situation awareness should always be automated. But tasks that impose low workload and provide high situation awareness should preferably always be manually performed. In practice, automation of tasks lies between these two extremes (upper/lower bounds of 'level of automation') which are shown in Figure 1.9 by the dotted lines. The criterion for selecting tasks that should be automated is the so-called '*cognitive workload value*' factor which is computed taking into account both workload and situation awareness [24].

Situation awareness (**SA**) refers to being aware of your surroundings, [25], and involves:

- The perception of the elements of the environment within a time-space domain.
- Comprehension of the meaning of these elements.
- Projection of their state in the near future.

According to Dominguez, in automation SA should include the following specific aspects [26]:

- Information extraction from the environment.

- Integration of this information with proper knowledge to create a mental picture of the present situation.
- Use of this picture to direct further perceptual exploration in a continual perceptual cycle.
- Anticipate future events.

A crucial part of SA is the ability to understand how much time is available until some event occurs or intervention is necessary. Besides automation and aviation, SA plays a dominant role in military operations, medical calls, vehicle driving, search and rescue, law enforcement, and many other applications.

Fundamental results on AA are provided in [27], where the two main approaches to human-centered automation are elaborated, namely:

- *Intermediate levels of automation* (**LOA**s) for maintaining operator involvement in complex automation systems.
- **AA** for managing operator workload via dynamic control allocations between the human and machine over time.

The authors of [27] have coined the 3-dimensional diagram of Figure 1.10 for illustrating and providing possible answers or thoughts to the questions of *what, how, and when to automate*. This diagram is based on the use of the concepts of **LOA**, **AA**, and *stages of information processing* (**SIP**), which are the variables along the three axes.

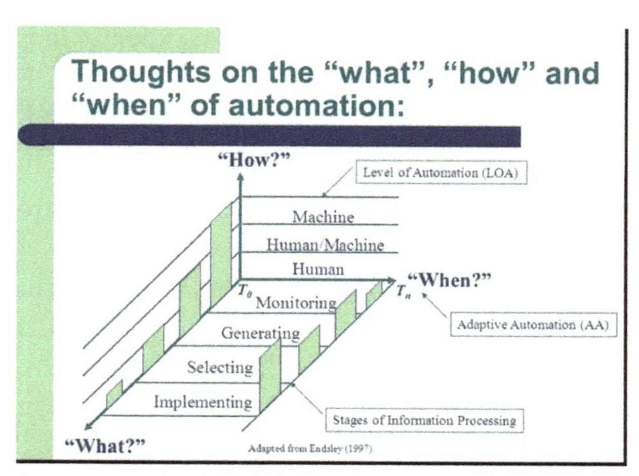

Figure 1.10 Three dimensional self-explained diagram of the 'what, how, and when to automate' questions.

Source: www.slideplayer.com (slide 6943170) [28].

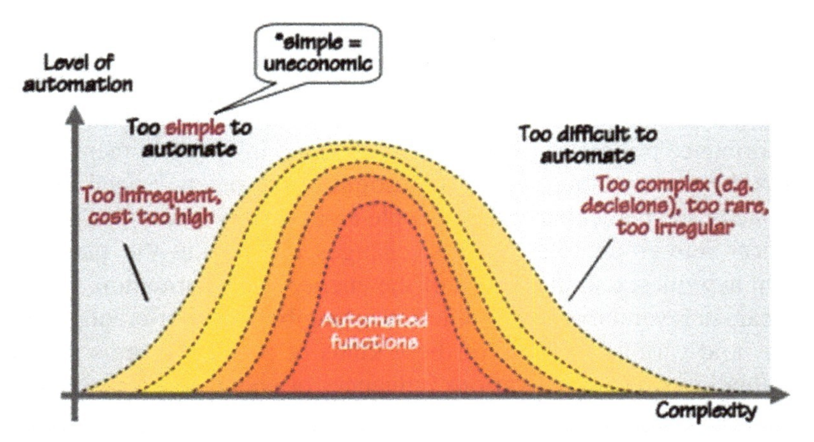

Figure 1.11 An Illustration of the 'automation left-over principle' on the 'complexity/LOA' plane. Functions that are automated belong to the internal bell area which is constrained by a lower and an upper bound of complexity.

Source: www.kitchensoap.com (2013/08/20/a-mature-role-for-automation-part ii/).

Another useful concept concerning automation is the '*automation left-over principle*' which is pictorially illustrated in Figure 1.11 on the 'complexity/level of automation' plane. Functions that cannot be assigned to machines are left for humans to perform. In other words, this principle says that 'everything that can be mechanized should be mechanized' as long as it is guaranteed that mechanization or automation will always work correctly and not suddenly require operator intervention or support. Therefore, full automation should be attempted only when it is possible to predict every possible condition and situation. Applying this principle the operator's life becomes more comfortable, not the opposite.

Human-Machine Interaction A key role in achieving successful automation (and human-centered automation) plays the *human-machine interaction* **(HMI)**, which is now a well-established field of computer systems and automation. HMI uses concepts, principles, and techniques of human factors engineering, cognitive and experimental psychology, and other closely related disciplines [29, 30]. Standard HMI devices include: keyboards and pointing devices (touch screens, light pens, graphic tablets, track balls, mice, and joysticks). More advanced HMIs that are used in automation include: graphical user interfaces **(GUIs)**, windowing systems, visual displays, and intelligent HMIs (e.g., natural language HMIs/NLI, multimodal HMIs, and force sensing/tactical HMIs).

1.4 Societal and Ethical Issues

Systems, cybernetics, control, and automation operate in our society, and so unavoidably they have society implications that call for the application of ethical performance principles by the humans that are involved in them.

Systems theory contributes to the development of society. Social development is one of the main concerns of people and includes issues like power and influence, intergroup relationships, changes concerning the planning development activities, and utilization of the energy and information. Social energy appears in several forms (tangible and intangible), and different people in a society hold various amounts of these kinds of power. Systems theory helps in evaluating whether the various influences of energy are beneficial or not, and understanding how to assist communities gain access to and regulate them. In general, systems theory contributes to society in many aspects, e.g. for:

- Selecting development goals.
- Planning strategy to reach goals.
- Performing activities to achieve goals.
- Evaluating progress.
- Using the evaluation results for developing improved planning strategies and activities.

Cybernetics helps to study societal processes, e.g., to identify, describe, and explain the structure and behavior of social systems, or to apply systems methodology (model construction, computer simulation, etc.) to the study of social organizational aspects and processes [31]. Behavioral cybernetics studies how to maintain internal and external balance of organismic systems (like society, business organizations. economic systems, etc.). For example, economic stability can be maintained by regulating (increasing or decreasing) productivity. Cybernetics is concerned with systems in conversation (i.e., systems interacting and talking to each other). In other words cybernetics is engaged in processes through which information is exchanged or communicated between each subsystem or each element in a particular system (e.g., a body or a society).

Control and Automation have overlapping impacts on society ranging from pure technological applications (process industries, transportations, energy systems, etc.), and biological applications to office automation and managerial, economic and environmental applications. All these applications contribute to the improvement of the quality of life (health, prosperity, human development, education, quality of working conditions, etc.). Of course, control and automation, and technology in general, have also several negative effects (e.g., increased environmental pollution, climatic change, increased consumption of earth resources, etc.) which need serious consideration and proper (local and global) measures [32, 33].

Ethics and morality is a primary human concern since the ancient times, having a philosophical, societal, religious and real-life dimension. Figure 1.12 gives a map of the origins of ethics that start from religion and philosophy going down to the family level, and through codes, regulations and laws arrive at the everyday ethical actions.

Naturally, over the years several philosophers have developed different theories of ethics, the dominant of which are the following:

- The deontological theory *(Kant).*
- Utilitarian theory *(Mill).*
- Social contract theory *(Hobbes).*
- Value-based theory *(Dewry).*
- Case-based theory *(Casuistry).*

Ethics regulates the behavior of humans in their relation and cooperation with other persons both in every-day life and in their work environment.

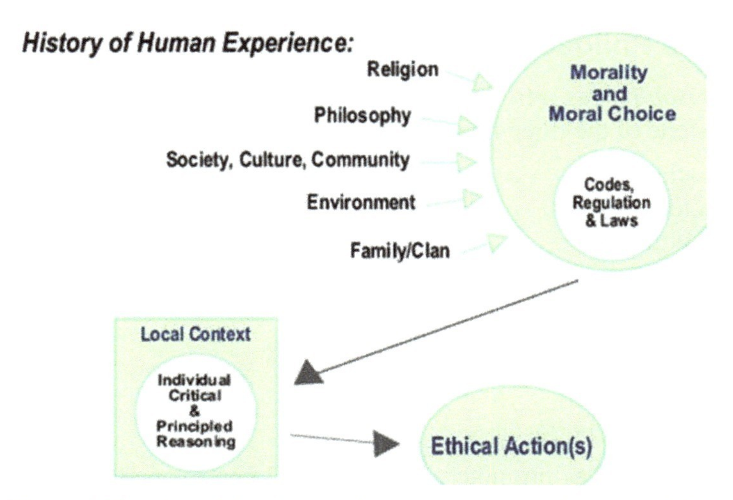

Figure 1.12 Map of the history of human experience referring to ethics.

Source: www.nursingworld.org/ClientResources/Images/OJIN/FIGURE.GIF

Scientists and engineers should behave ethically in performing their tasks adhering to the general and professional codes of ethics. Professional societies have formulated and released *ethics codes* which provide guidance for interaction between their members, such that they can serve both each other and the society as a whole. Some widely known engineering and management professional codes are:

- National Society of Professional Engineers (**NSPE**) code.
- Electrical and Electronic Engineers (**IEEE**) code.
- American Society for Mechanical Engineers (**ASME**) code.
- Academy of Management (**AOM**) code.
- *Accreditation Board of Engineering and Technology* (**ABET**) code.

Each field of science and engineering exhibits its own ethical issues that have to be carefully examined and faced over time.

In this book, we will be concerned with the following:

- Ethics of systems engineering.
- Ethics of systems thinking.
- Ethics of cybernetics.
- Ethics of control and automation.
- Ethics of management control.

1.5 Systems Philosophy

Systems Philosophy was introduced by Ervin Laszlo in 1972 [34]. Systems philosophy is one of the three fields studied within systemics, namely:

- System science.
- Systems technology.
- Systems philosophy.

The principal domains of systems philosophy are [35]:

- System ontology.
- Systems paradigms.
- System axiology.
- Applied systems philosophy.

Systems philosophy was promoted, besides Laszlo and von Bertalanffy, by the following thinkers (non-exhaustively) [36–39]:

- *Hassan Ozbekham* (Global Problematique).
- *Leo Apostel* (Integrated Worldview).

- *David Rousseau* (Value Realism).
- *Gerald Midgley* (System Intervention).

1.6 Control and Cybernetics Philosophy

Control philosophy is a system of generalizing philosophical considerations about the subject and methods of control, the position of control among other scientific fields, etc. Novikov [40] argues that cybernetics is a branch of control science that studies its most theoretical regularities. Philosophical issues of control that need deep consideration include:

- Control in life an society.
- Control principles.
- Control mechanisms.
- Control functions.
- Forms and types of control.
- Conditions of control.

Control methodology is the theory of activity organization. Activity is any purposeful human action.

Cybernetics philosophy is concerned with philosophical aspects of organization models, feedback, goads, and conversation. It tries to apprehend the capacity and limits of systems of all kinds, namely biological, man-made, and social systems. Contrary to that AI follows a realist way to what *intelligence* constitutes in the real world, cybernetics was evolved from a *constructivist* view of the world in which objectivity derives from shared agreement about meaning, and where intelligence (or information) is the product of interaction rather than commodity stored in a computer [40].

1.7 Outline of the Book

The book involves ten chapters.
Chapter 1 (the present chapter) entitled: '*Introductory Concepts and Outline of the Book*' presents preparatory material on systems, cybernetics, control, and automation for a quick look to the subject of the book.

Chapter 2 '*Basics of Philosophy, Philosophy of Science, and Philosophy of Technology*', discusses the basic concepts of philosophy, philosophy of science and philosophy of technology. The discussion starts with the historical evolution, and includes the principal branches, and a number of other early branches of philosophy. Then, it presents fundamental concepts of philosophy of science and philosophy of technology.

Chapter 3 '*Background Concepts: Systems Ontology, History, and Taxonomy*', provides a set of well-known definitions of the system concept, followed by a summary of the basic ontological elements of systems (open-closed systems, environment, boundary). Then, the chapter makes a short tour to the history of systems theory, and gives a multi-dimensional system's taxonomy.

Chapter 4 '*General Systems Theory and System Dynamics*', presents the fundamental ontological and epistemological issues of 'general systems theory' **(GST)** and 'system dynamics' **(SD)**. The questions: 'what is GST?', 'what is SD?', and 'what is system thinking?, and 'what is isomorphism?' are considered. The principal epistemological aspects of GST and SD (principles, causal loop diagrams, SD modeling and simulation) are then considered, including the axioms of GST.

Chapter 5 '*Cybernetics*', discusses the fundamental ontological and epistemological aspects of cybernetics. Specifically, it provides a review of certain answers to the question 'what is cybernetics' given by pioneers in the field, and a brief historical review of cybernetics. Then, the chapter discusses first-order versus second-order cybernetics, and outlines the basic features of social systems and socio cybernetics.

Chapter 6 '*Control*', presents a review of the control theory field. The chapter starts with the answers to the ontological questions 'what is feedback?' and 'what is control?', and then provides a short historical review from the prehistoric and early control period to the classical and modern control periods. Next, the chapter presents fundamental epistemological elements (concepts and methods) of *CCT* (stability, closed loop control, root locus method, frequency domain methods, compensator design, discrete-time control systems), and *modern control theory* (state space, controllability, observability, Lyapunov stability, state- feedback control, optimal and stochastic control, model-free control (fuzzy, neural, mixed, and expect control)). The chapter concludes with fundamental ontological aspects of networked control systems (structure, properties, advantages, disadvantages, vulnerability).

Chapter 7 '*Complex and Nonlinear Systems*', is devoted to the study of complexity, complex systems, and nonlinear systems. It addresses the questions: 'what is complexity?' and 'what is a complex system?' Then, it presents a number of complexity measures, and a list of stability and control methods of nonlinear systems. Next, the chapter discusses the concepts of 'bifurcations', 'chaos', 'strange attractors', 'fractals', and 'CASs'. Finally, a look at the concepts of 'adaptation' and 'self-organization' is made.

Chapter 8 '*Automation*', examines several existing answers to the ontological question 'what is automation?', and reviews a set of representative works on automation, selected from the literature. It discusses epistemologically the classes of industrial automation and office automation which differ in their nature and characteristics. Then, the chapter outlines four further classes of automation systems, namely robotic automation systems, aircraft automation systems, air traffic control systems, and automated driving vehicles, and discusses fundamental concepts/of human-machine interfaces (**HMIs**) in human-automation interaction. Finally, the chapter presents conceptual issues of virtual, augmented, and mixed reality including medical applications, and outlines three key human factors that must be considered in the design and operation of automation systems.

Chapter 9 '*Societal Issues*', studies the influence on society of systems, cybernetics, control, and automation. Specifically, after a quick look at the question 'what is society?' the chapter: (i) discusses the role of systems theory in the society development, the relation of cybernetics and society, and the impact of control on industrial and economic/management systems. The chapter closes with a discussion of the key issues of the impact of automation on productivity, capital formation, employment, and working conditions.

Chapter 10 '*Ethical and Philosophical Issues*', is devoted to the ethics of systems, cybernetics, control, and automation. A short discussion of the question: 'what is ethics?', and an outline of the major 'ethics theories' are first given, followed by the ethical issues and concerns of systems engineering and systems thinking. Then, the chapter examines the ethics of cybernetics, 'cybernetic organisms (cyborgs)' and 'human implants'. Finally, the chapter provides a general overview of the ethics of control and automation/robotics, including relevant codes of professional ethics, and presents the key concepts of systems philosophy and cybernetics philosophy.

In overall, the book gives a spherical picture of the interrelated fields of 'systems', 'cybernetics', 'control' and 'automation', covering the most fundamental ontological and epistemological aspects, concepts, and methods. The four fields, which were developed and evolved in parallel, are now at a very mature state with an enormous number of theoretical results and practical societal applications. The book involves sufficient material on the impact of the four fields on society, and the ethical issues arising in their design and use.

References

[1] von Bertalanffy, L. (1972). "The history and status of general systems theory," in *Trends in General Systems Theory*, ed. J. Klir (New York, NY: Wiley-Interscience).

[2] Davidson, M. (1983). *Uncommon Sense: The Life and Thought of Ludwig von Bertalanffy*. New York, NY: J.P. Tarcher, Inc.

[3] Weng, G., Bhalla, U. S., and Iyengar, R. (1999). Complexity in biological signaling systems. *Science* 284, 92–96.

[4] Brian Arthur, W. (1999). Complexity and the economy. *Science* 284, 107–109.

[5] Holland, J. H. (1992). *Adaptation in National and Artificial Systems*. Cambridge, MA: MIT Press.

[6] Holland, J. H. (1995). *Hidden Order: How Adaptation Builds Complexity*. Reading, MA: Addison Wesley.

[7] Eidelson, R. J. (1997). Complex adaptive systems in the behavioral and social sciences. *Rev. Gen. Psychol.* 1, 42–71.

[8] Arnold, R., and Wade, J. P. (2015). "A definition of systems thinking: a systems approach," in *Proceedings of 2015 Conference on Systems Engineering Research, December, 2015* (Science Direct, Procedia Computer Science).

[9] Klir, G. (2001). *Facet of Systems Science*. New York, NY: Kluwer/Plenum.

[10] Licklider, J. C. R. (1960). Man-Computer Symbiosis. *IRE Trans. Hum. Fact. Econ*. HFE-1, 4–11.

[11] Pangaro, P. (2001). *Origins of AI in Cybernetics, Lecture of the course 'Introduction to Cybernetics'*. Stanford, CA: Stanford University.

[12] Mueller, K. M., and Riegler, A. (2014). Second-order science: The revolution of scientific structures. *Constr. Found.* 10, 7–15.

[13] Umpleby, S. A. (2011). Second-order economics as an example of second-order cybernetics. *Cybern. Hum. Know*. 1893–1894, 172–177.

[14] von Foerster, H. (1979). *Cybernetics of Cybernetics, Understanding of Understanding, Keynote Address, Essay on Cybernetics and Cognition*. Berlin: Springer.

[15] Lazos, C. (1998). *Engineering and Technology in Ancient Greece*. Patras: University of Patras Press.

[16] Bennet, S. (1963). *History of Control Engineering 1800–1930*. Amsterdam: Peter Peregrinus.

[17] Bagut, I. L. (1965). *The Age of Automation, The New American Library of World Literature*. New York, NY: Mentor Books.

[18] Francois, W. (1964). *Automation, Industrialization Comes of Age*, New York, NY: Collier Books.

[19] Rasmussen, J. (1986). *Information Processing, in Human Machine Interaction*, Amsterdam: North Holland.

[20] Shell, R. L., and Hall, E. L. (2000). *Handbook of Industrial Automation*. New York, NY: Marcel Dekker.

[21] Martin, T., Kivinen, J., Rijnsdorf, J. E., Rodd, M. G., and Rouse, W. B. (1991). Appropriate Automation: Integrating Technical, Human, Organizational, Economic and Cultural Factors. *Automatica* 27, 901–917.

[22] Tzafestas, S. G. (2010). *Human and Nature Minding Automation: An Overview of Concepts, Methods, Tools and Applications*. Berlin: Springer.

[23] Hou, M., Banbury, S., and Burns, S. (2014). *Intelligent adaptive systems: An interaction-centered design perspective*. Boca Raton, FL: CRC Press.

[24] Parasuraman, R., Sheridan, T. B., and Wickens, C. D. (2008). *Situation awareness, mental workload and trust in automation: Viable, empirically supported cognitive engineering constructs. J. Cogn. Eng. Decis. Mak.* 2, 140–160.

[25] Endsley, M. R. (1998). "A comparative analysis SAGAT and SART for evaluations of situation awareness," in *Proceedings of the Human Factors and Ergonomic Society 42nd Annual Meeting*, Santa Monica, CA, 82–86.

[26] Dominguez, C., Vidulich, M., Vogel, M., and McMillan, G. (1994). *Situation Awareness: Papers and Annotated Bibliography*. Arlington, TX: Armstrong Laboratory, Human System Center (AL/CF-TR-1994-0085).

[27] Kaber, D. B., and Endsley, M. R. (2004). The effects of level of automation and AA on human performance, situation awareness, and workload in a dynamic control task. *Theor. Issues Ergon. Sci.* 5, 113–153.

[28] Kaber, D. B. (1999). *A Case for Theory-Based Research of Automation and Adaptive Automation*. Raleigh, NC: North Carolina State University.

[29] Backer, R. M., and Buxton, W. A. (1987). *Readings in Human-Computer Interaction: A Multidisciplinary Approach*, Los Altos, CA: Morgan Kaufman.

[30] Lajoic, S. P., and Derry, S. J. (eds) (1993). *Computers and Cognitive Tools*. Hillsdale, NJ: Erlbaum.

[31] Buckley, W. (1967). *Sociology and Modern Systems Theory*. Englewood Cliffs, NJ: Prentice Hall.

[32] Schafer, B. J. *The social problem of automation, From the Collection of the Computer History Museum*. Available at: www.computerhistory.org
[33] Tzafestas, S. G. (2017). *Energy, Information, Feedback, Adaptation, and Self-organization, The Fundamental Elements of Life and Society*. Berlin: Springer.
[34] Laszlo, E. (1972). *Introduction to Systems Philosophy: Toward a New Paradigm of Contemporary Thought*. Philadelphia, PA: Gordon and Breach Publishers, 1972.
[35] von Bertalanffy, L. (1976). *General Systems Theory*. New York, NY: Brazillier.
[36] Meadows, D. H., and Randers, J. (1972). *The Limits to Growth*. New York, NY: Universe Books.
[37] Aerts, D., Apostel, L., DeMoor, B., Hellemans, S., Maex, E., Van Belle, H., and Van der Veken, J. (1944). *Worldviews: From Fragmentation to Interaction*. Brussels: VUB Press.
[38] Rousseau, D. (2012). "Could spiritual intuitions map a scientifically plausible ontology?," in *Proccedings of Joint Conference of the Scientific and Medical Network and the Society for Scientific Explorations: Mapping Time, Mind and Space, October 2012*, Drogheda, 18–21.
[39] Midgley, G. (2000). *System Intervention: Philosophy, Methodology, and Practice*. Berlin: Springer.
[40] Novikov, D. A. (2016). "Chapter-2 Cybernetics," in *Springer Series, 'Studies in Systems, Decision, and Control*, Vol. 47, Berlin: Springer.

2

Basics of Philosophy, Philosophy of Science, and Philosophy of Technology

Only absolute true is that there are No absolute truths.
Socrates

Science is the creator of prosperity.
Plato

Knowing yourself is the beginning of all wisdom.
Aristotle

2.1 Introduction

Philosophy provides the means for understanding the world around us. The term *'philosophy'* originates from the Greek composite word *'φιλοσοφία'* (philosophia), where *'φιλῶ'* ('philo' means love) and *'σοφια'* ('sophia' means wisdom). Thus, *'philosophy'* is *'love of wisdom'*. A philosophy is a system of beliefs about reality, and is a natural product of man's rational mind and reasoning. The conclusions are possible because the world does exist in a particular way. Philosophy offers the premises through which man can comprehend the world, discovering what is true and using his mind to support his life [1, 2].

Problems in philosophy that are still under investigation include the following:

- Identity and otherness.
- Mind/body problem.
- Freedom and slavery.
- Ethical status of information.
- Ontology of the virtual.
- Language and thought.
- Political tolerance between different religions.

- The nature of the life of excellence.
- The ultimate worth of the goals you seek.
- The nature of the process of knowledge.
- Relation between medieval and modern concepts and ideas.
- What is the best ethical thinking?
- Is there life after death?
- Can we objectively know anything?
- Do we have free will?

To do philosophy we need to build any argument or theory as true a premise as possible following, e.g., Descartes' approach to develop such an example of philosophical thinking by continually turning down any belief which can be logically in doubt. Talking about philosophy we usually refer to any philosophy that follows the *'linguistic turn'* in philosophy, meaning the shift towards the use of semantic and structural analysis. By the late 1940s modern philosophy had philosophers like Heidegger investigating the nature of '*being*', and by the 1960s the problems of *being* had largely given way to problems of '*knowing*'.

The purpose of this chapter is to briefly discuss the fundamental elements of philosophy which are deemed to be useful for reading the ethical and philosophical aspects of the book. Specifically, the chapter:

- Discusses the ontological question 'what is philosophy', and the principal historical evolution stages of philosophy.
- Outlines the principal branches of philosophy, namely: logic, metaphysics (ontology), epistemology, ethics, and aesthetics.
- Discusses a number of further early branches of philosophy (e.g., philosophy of language, philosophy of mind, philosophy of mathematics, etc.)
- Presents the fundamental issues of philosophy of science and philosophy of technology, including a list of the big philosophers and their views.
- Provides a brief discussion of fundamental questions in philosophy of science and philosophy of technology.

2.2 What is Philosophy?

Philosophy (love of wisdom) is a form of clear, deep thought; basically putting our thought and language in order. The focus on philosophizing (doing philosophy) as a process guides us to regard philosophy as any language conducted in a certain way. According to Kant, philosophy is a science of the human being, its ability of representing, reasoning and acting, and presents the human being in all of its components in agreement with its natural features, and its relationship to morality and ethics.

Doing philosophy is the first fundamental activity of stating as precisely, clearly, and convincingly as possible what we *believe* and what we *believe in.* Whenever we deal with issues in depth, continuously asking 'why' and 'how' to critically analyze underlying assumptions and move to the foundations of our complex knowledge structures, then it is necessarily philosophy. The first process in philosophizing is *articulation*, i.e., spelling out our ideas linguistically in words and sentences, in a clear, concise and readily understandable way.

Other characteristics of philosophizing are:

- *Argument* (Supporting our ideas and views with reasons from other ideas, principles or observations).
- *Analysis* (Analyzing the ideas and views in order to identify and clarify their various components).
- *Synthesis* (Merging different ideas and views into a single unified perspective).

Western philosophy has its origin in Ancient Greece.

Thales of Miletus *(7th Century B.C.)* is considered the first philosopher who was also concerned with natural philosophy, presently called *science.* Western philosophers include the following [3–7]:

Milesians *(7th Century B.C.)* These are: Thales (universe is made of water), Heraclitus (universe is made from fire, strife is the father of all), and *Anaxagoras* (there is a portion of everything in everything-the earliest theory of infinite divisibility).

Pre-Socratics *(Late 7th to Early 5th Century B.C.)* These philosophers include *Empedocles* (universe is made of air, fire, and earth, the basic stuff), *Parmenides* (the world is a uniform solid, spherical in shape; '*Being is, Non-Being is not';* empty space cannot exist if all things are made of basic stuff), Zeno (paradoxes of space and motion), Euclid and Pythagoras (logic and mathematical theory).

Socrates, Plato, and Aristotle *(Early Fifth to Late Fourth Century).* These philosophers are known as *classical philosophers. Socrates* developed the 'questioning method' (called the 'maieutic' method) by which the weakness in the interrogated are exposed. *Plato* wrote his 'Dialogues', where he presented the views of his teacher Socrates and founded the Academy of Athens. Plato blended Metaphysics, Epistemology, Ethics and Political philosophy into an integrated and systematic philosophy. His political philosophy was presented in his famous "Republic" in which he criticizes democracy, condemns tyranny and an antidemocratic society composed of workers, guardians ruled over wise Philosophers Kings. *Aristotle* (Plato's student) philosophized on a wide range of subjects: biology, ethics, logic, metaphysics, epistemology, etc. His logic system is the deductive (Aristotelian) logic which emphasizes on syllogism which was the dominant form of logic until the nineteenth Century.

Some other ancient Greek philosophical Schools include:

- *Cynicism* (Diogenes, Antisthenes) Cynicists rejected all conventional desires for health, wealth, power and fame, advocating a life free from all possessions and property as a way to achieve a best exemplified life (virtue).
- *Skepticism* (Pyrrho, Timon), In his philosophy Pyrrho argued that because the true inner substance of things can never be known, we must suspend judgment on everything as a way of achieving *inner peace.*
- *Stoicism* (Zeno of Citium) This philosophy, which was further developed by Epictetus and Marcus Aurelius, suggests *self-control* and *fortitude* as a way to overcome destructive emotions, and develop clear judgment and inner calm, and the final goal to be free *from suffering*.
- *Epicureanism* This philosophy was developed by Epicurus and argues that to achieve happiness, and be calm we should lead a simple, moderate life, cultivate friendships, and limit our desires.
- *Neo-Platonism* (Plotinus) This is a religious philosophy that influenced early Christianity (St. Augustine). According to this philosophy there exists an ineffable and transcendent *One*, which is the source of the rest of the Universe as a sequence of *lesser beings.*

Another School is the school of *sophism* (Protagoras, Gorgias) which was actually promoting a *'pseudophilosophy'*. Sophists hold relativistic views based on the position that 'there is no absolute truth', and they were willing to teach anything (whether it was true or not true) to anyone, or argue (using *rhetorical tricks* called *sophisms)* anyone's cause (whether their cause was just or not) for a fee.

Medieval Philosophy *(Late Fifth Century A.D. to Middle Fifteenth Century A.D.)*. During the first period up to 11th Century (known as *Dark Ages*) very little new thinking was developed. A renewed flowering of thought both in Europe (Christianity) and Middle East (Muslim, Jewish) began in the eleventh Century A.D. Most thinkers of this time were attempting to prove the *existence of God* and reconcile Christianity and Islam with the classical philosophy (mainly Aristotelianism). Dominated ideas of this time include questions of universals (with nominalists, e.g., *William of Okham),* rejection of metaphysical concepts of Forms, various 'proofs' of God's existence *(Anselm, Aquinas),* and the belief that reason alone cannot save a human being (for this, faith in God and revelation are required).

Birth of Modern Science *(Late Fifteenth to Late Seventeenth Centuries)* Thinkers of this age include Copernicus (a Polish astronomer), Kepler (a German mathematician and astronomer), Galileo (an Italian physicist), and Francis Bacon (an English thinker). *Copernicus* challenged the Ptolemaic view, saying that the sun is the center of our solar system, and that earth and other planets revolve around it. *Kepler* provided an early mathematical proof of Copernicus views. Galileo, combined mathematics and physics to shape a new scientific world view. *Francis Bacon* promoted to the inductive method of science, particularly to *empiricism,* i.e., pursuit of knowledge by experiment and observation, not through reason alone. This period marked the end of scholasticism, and the advancement of intellectual freedom and curiosity. In overall, the principal focus was on liberation from theology and the other fields of humanities.

Modern Philosophy *(Seventeenth Century to Early Eighteenth Century)* Principal thinkers of this era are Hobbes, Descates, Newton, Spinoza, Leibnitz, Locke, Berkeley, Hume, Rousseau, and Kant. *Thomas Hobbes* (an English philosopher) tried to develop a 'master science' of nature man, and society, departing from Bacon and Gallileo. He moved away from empiricism and formulated principles of human conduct. *Descartes,* who is considered the father of modern philosophy, revisited the ideas of skepticism. The only thing that he couldn't doubt was 'himself thinking', concluding with his famous quote *'cogito ergo sum'*, *'Je pense donc je suis', 'I think, therefore I am'*. He developed the worldwide known and used Cartesian geometry. On the metaphysics side he believed in God and the material world (the so-called Cartesian mind-body dualism). He also developed the 'discourse on method' philosophy. *Spinoza and Leibniz* developed their own metaphysical systems, believing in a rational benevolent God. Spinoza wrote the Ethics denying final causes. He equated God with creation. Leibniz developed the monadology, i.e., the metaphysical units that make up substance. He argued that *monads* are the elements of all things, physical and mental. *Locke* rejected the thesis that people are born with innate knowledge ('human beings are born with '*tabula rasa*', i.e., with empty state). All subsequent knowledge is acquired sensory experience. *Berkeley* (a Bishop) rejected Locke's view of knowledge and proposed an *idealist system (esse est percipi',* i.e., 'to be is to be perceived'). *Hume* attacked Berkeley's views of knowledge and reality, arguing that *reason* cannot provide certain knowledge ('*there is no proof of causality*'). *Rousseau* was concerned more in the fields of *ethics* and *political philosophy,* and less in the field of epistemology. He believed that people are born with morality, but are influenced by societal corruption. Similar to Locke, Rousseau has advanced the *'Social contract theory'. Immanuel Kant*, the most famous German philosopher, argued that the world of 'things-in-themselves' is unknowable, whereas the 'world of appearance', the 'phenomenal world', governed by laws, is knowable. He rejected 'empiricism' (all knowledge is acquired from sensory experience) and accepted that causality, necessity, and unity enable humans to obtain coherent knowledge of the world.

Humanistic Philosophy *(Nineteenth Century)* Thinkers of this period include among others *Auguste Compte* (who developed positive philosophy or positivism), *John Stuart Mill* who defended liberty of expression and fought for women's rights, and *Darwin* who developed the theory of evolution which is based on the doctrine of natural selection ('those best adapted to their environment are most successful in reproduction and hence, the propagation of their kind'). Darwin's theory rejects the 'intelligent design' argument, which implies the existence of God from order, design, and purpose in the world).

American Philosophy *(Nineteenth and Twentieth Centuries)* This philosophy includes the doctrine of *pragmatism (C.S.Pierce)* which sees truth as the effectiveness of an idea, the landmark contributions in psychology of *William James* who argued passionately in flavor of religious faith, the works of *Santayana* who feels repulsed by the pragmatism doctrine, and the contributions of *Dewey* (a pragmatist) to education and his criticism on American philosophy.

Continental Philosophy *(Twentieth Century)* This philosophy includes the *ordinary language philosophy* (W.V.O Quine, Donald Davison), the *analytical philosophy* (G.E. Moore's 'Principia Ethica'), the *phenomenology* (Edmund Husserl), and the *'social context'* philosophy of Martin Heidegger, which was presented in his book 'Being and Time; (Sein und Zeit) [19]. Heidegger argued that existence was inextricably linked with *time,* and that '*being*' is really just an ongoing process of *becoming* (contrary to the Aristotelian idea of a fixed *essence*). He writes: 'Time is not a thing, thus nothing which is, and yet it remains constant in its passing away without being something temporal like the Being in time'. 'We name time when we say: everything has its time. He also writes: 'The meaning of Being is beyond any kind of definite answer, if being is to be conceived in terms of time, and if indeed its various modes and derivatives are to become intelligible in their respective modifications and derivations by taking time into consideration'. He puts the philosophical question 'Why are there beings at all, instead of nothing? One of his conclusions is the world-wide well-known statement: 'we do not say being is, time is, but rather: there is being and there is time' He notes that 'The German language speaks being, while all the others merely speak of being'. Other philosophies developed in this period include *existentialism* (Jean-Paul Sartre) who argued that 'existence is prior to essence', and *structuralism,* i.e. the belief that all human activity and its products (even perception and thought itself) are constructed and not natural, and that everything has *meaning* only through the language system in which we communicate.

2.3 Principal Branches of Philosophy

Traditionally, philosophy has been broken in several branches bounded together in the sense that it is difficult to examine a question in any one of them without the use of concepts and ideas from the other branches. The five principal branches of philosophy are: logic, metaphysics, epistemology, ethics, and aesthetics [1, 2, 8–10] (Figure 2.1).

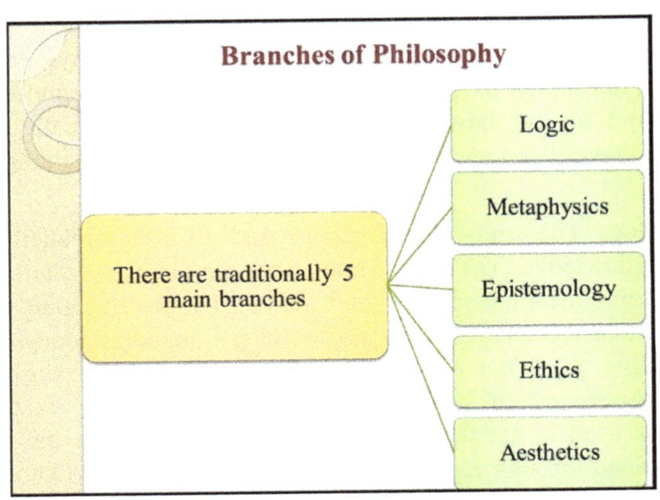

Figure 2.1 Five principal branches of philosophy.

- *Logic* studies the nature of thinking.
- *Metaphysics* explores the nature of reality and existence.
- *Epistemology* investigates knowing and knowledge.
- *Ethics* reveals the nature of moral values and human action.
- *Aesthetics* studies the nature of beauty and art.

In the following we give a brief outline of these philosophical branches:

Logic The word *'logic'* comes from the classical Greek word *'λόγος'* (logos=word/what is spoken, study). It is the study of arguments through correct thinking or reasoning. Logic addresses questions like 'what is rationality?'; which arguments are good ones? In any case the task of logicians is the same: namely: to advance an account of valid and deceptive inference in order to allow one to distinguish.

Metaphysics (ontology) it is the study of existence and the nature of existence, i.e., the study of the reality of all things in an attempt to obtain a comprehensive picture of the world. The word *metaphysics* is derived from the Greek word *'μεταφυσική'* which is composed of the two words *'μετά'* (meta=follow) and *'φυσική'* (physics) which means the *'study that follows physics'*. The usual meaning of metaphysics as *'beyond physics'* is misleading. Metaphysics is concerned with the study of *'first principles'* and *'being'* (ontology), i.e. the study of the most general issues of reality and nature (substance, identity, the nature of the mind, and the free will). The study of *'being'* by ontology tries to clarify what we mean when we say that something *'is'*, and list in order of priority the various kind of entities that make up the universe (physical objects, properties, actions, events, bodies, mind, God, etc.). Metaphysics is the foundation of philosophy. Without an explanation or an interpretation of the world around us, we have not any help to deal with reality. Reality is absolute having a specific nature independent of our thoughts or feelings.

- **Epistemology** The study of knowledge and all issues related to the nature of knowledge. The term comes from the Greek words *'επιστή μη* ('episteme' meaning science) and *'λόγος'* ('logos' meaning study). Questions that are answered by epistemology include, among others, the following: (i) *What can we know?* (ii) *How do we know?* (iii) *What is the nature of knowledge?,* (iv) *What is truth?* Traditional approaches used in epistemology are: *rationalism* (where the knowledge is gained through reasoning), and *empiricism* (where knowledge is acquired through sensory observations and measurements). In many practical cases, both approaches are combined since they complement and correct each other.

Regarding the question *'what is truth?'* There are two theories of truth [10]:

- *Correspondence theory,* according to which 'truth is a relation between two things, namely: *beliefs* (sentences or propositions), and *reality*'. A belief is true if and only if it corresponds to reality (i.e., it 'matches' or precisely characterizes or describes reality).
- *Coherence theory,* according to which a set of propositions (beliefs, sentences) is true if and only if: (i) they are mutually consistent, and (ii) they are supported by or consistent with all evidence, i.e., they 'cohere' with each other and with all evidence.

In other words, epistemology is the explanation of how we think and it is needed to enable us to get knowledge of the real world. The degree to which our epistemology is correct is the degree to which we could understand reality, and the degree to which we could use that knowledge to promote our lives and goals. Without epistemology we would not be able to think, so that we would have no reason to believe our thinking was productive or correct. The key elements of a proper epistemology are:

- *Validity:* Our senses are valid, and the only means to obtain information about the world.
- *Reason:* Our method of gaining knowledge, and acquiring understanding.
- *Logic:* Our way of maintaining consistency within our set of knowledge.
- *Objectivity:* Our means of associating knowledge with reality to determine its validity.
- *Concepts:* Abstracts of specific details of reality, or other abstractions.

Ethics The study (science) of morality is concerned with the ethical behavior (i.e., *'what is good and bad', 'what is right and wrong'),* and with the establishment and defense of rules of morality and good life (moral philosophy). Ethics belongs to '*analytic philosophy*' and is distinguished in [11]:

- Metaethics.
- Normative ethics.
- Applied ethics.

Metaethics studies the nature of morality, in general, and what justifies moral judgments. Basic questions of metaethics are: 'Do moral truths exist? What makes them true? 'Are they absolutely true or always relative to some individual or society culture?

Normative ethics attempts to find what is for an action to be morally acceptable (i.e., rules and procedures for determining what person should do or not do), and involves the theory of 'social justice' (i.e., how society must be structured, and how the social goods of freedom and power should be distributed in a society).

Applied ethics investigates the application of ethics theories in actual life. Information ethics, medical ethics, automation ethics, system engineering ethics, cybernetics ethics, control systems ethics, etc. belong to applied ethics. Key theories of ethics are: virtue theory (*Aristotle*), deontological theory (*Kant*), utilitarian theory (*Mill*), and justice as fairness theory (*Rawls*).

Ethics develops the following practical entities:

- Moral principles.
- Values.
- Rules and regulations.
- Rules of conduct.
- Ethical practices.

Ethics applied to a group of people is the branch of philosophy called *'politics'*. Politics is concerned with the legal rights and government. Politics tells us how a society must be set up and how we must perform within a society. The need for a political system is that the individuals within that system are allowed to fully perform according to their nature. The primary goal of a political system should be the presentation and enabling of the faculty of reason.

Aesthetics This is the philosophical study of the nature of act and the concepts of beauty, pleasure and expression, including the experiences we get when we enjoy the arts or the beauties of nature and science of technology. Aesthetics has its roots in ancient Greece. The word *aesthetics* comes from the Geek word '*αισθητικός*' meaning 'esthetic, sensitive, sentient pertaining to sense perception'. The Greek verb '*αισθάνομάι*' (aisthanomai) means 'I perceive, feel, sense'. Aesthetics is applied not only to art but also to objects of culture. Mathematicians consider mathematical beauty a desirable quality in their work. Comparisons are often made with music and poetry. The most influential of early philosophers in aesthetics, at the end of eighteenth century, *was Immanuel Kant* who is considered as a *formalist* in art theory due to his view that the content of a work of art is not of aesthetic interest. He argued that by design, art may sometimes obtain the cognitive power of free play. It is then 'Fine Art'; but for Kant not all art has this quality.

2.4 Further Branches of Philosophy

Early branches of philosophy besides the five core branches discussed above include the following:

- *Philosophy of Language* The philosophy that is primarily concerned with how our languages affect our thought. It is an ancient branch which gained prominence in the last century under Wittgenstein. Famous works include Plato's Cratylus, Locke's Essay, and Wittgenstein's '*Tractatus Logicophilosophical'* where he argued that the limits of our language mark the limits of our thought.
- *Philosophy of mind* It is concerned with the study of the mind, attempting to explore and exactly state what the mind is, how it interacts with our body, do other minds exist, how the mind works, etc. It is probably the most popular branch of philosophy right now and has been enhanced to include philosophical issues of artificial intelligence.
- *Philosophy of mathematics* It is concerned with the study of questions about the nature of axioms and symbols of mathematics which we use to understand the world. Other questions of philosophy of mathematics include: 'do perfect mathematical forms exist in the real world? 'given that mathematical objects do not have causes or effects, how can we refer to them? 'given that mathematical objects do not have causes or effects, how do we have any knowledge of them?'
- *Philosophy of Law (or Jurisprudence)* It is concerned with the nature of law and questions like 'what are the best laws?' 'How laws came into being in the first place', 'Should we always obey the laws', 'How the different laws influence the development of society? The three main categories into which the topics of philosophy of law fall are: (i) analytic jurisprudence, (ii) normative jurisprudence, and (iii) critical theories of law.
- *Philosophy of end and purpose (teleology)* The term *teleology* comes from the Greek word '*τέλος*' (telos=end) and very broadly refers to end-directedness, i.e., the idea that some things exist, have certain traits, or do certain things for the sake of some end (purpose). Commonly we apply teleological concepts to the parts and features of organisms. For example, the heart is said to have the proper function of pumping the blood, which takes place for the sake of blood circulation and the biological purposes of nutrient distribution and waste removal. In general teleology is concerned with the aims and purposes of 'what we do', and 'why we exist'. According to Aristotle the meaning of teleology

is *'final causality'*. He argued that 'a full explanation of everything must consider its final cause, as well as its efficiency, material, and formal causes'.

- *Philosophy of religion (or philosophical theology)* The branch of philosophy concerned with questions such as, 'how our religion should shape our life', and 'what is the authority of religious works and documents (like the Bible, etc.)'. Philosophy of religion studies the existence of God, his capacities, etc. It is also concerned with the critical and comparative analysis of the various religions developed and followed in different cultures.
- *History of Philosophy* The philosophy's subdomain that looks at what famous philosophers of the past believed, and attempts to interpret their views in the framework of current thinking.
- *Axiology or Theory of Value* The term axiology comes from the Greek words 'άξιος' (axios)=worthy and 'λόγος' (logos)=reason/science'. The term 'theory of value' means the philosophical study of 'goodness or value'. Its significance is that it has provided a unified approach to the study of several questions (economic, aesthetic, moral, logical, etc.) which had usually been investigated in relative isolation. Originally, the term 'value' was in use in economics theory (e.g., Adams theory) to mean the worth of something. A broad extension of the meaning of value to wider areas of philosophical interest occurred in the nineteenth century under the influence of Neo-Kantian and other philosophers. Value has been distinguished in *instrumental* and *intrinsic value,* i.e., what is 'good' is distinguished between 'good as a means' and 'good as an end' (John Dewey). Knowledge and virtue are good in both senses. Other philosophers (e.g., Georg Henrik von Wright, and C.I. Lewis) have extended this distinction, differentiating, for example, between *'instrumental value'* (I.e., good for some purpose) and *'technical value'* (i.e., good at doing something) or between *'contributory value'* (i.e., being good as part of a whole) and *'final value'* (i.e., being good as a whole).
- *Philosophy of Science* It studies science concerned with whether scientific knowledge can be said to be certain, can science really explain everything and questions like, 'does causation really exists?' 'can every process or event in the universe be described in terms of physics?' and so on.
- *Philosophy of technology* Philosophy should not ignore technology. In the past, philosophy of technology has most been concerned with the

impact on society and culture. Philosophy of technology was developed as a true branch of philosophy only recently, concerned with the nature and the practice of designing and creating artifacts in the wide sense (tangible and intangible).

Now, for anything we care to be interested in, we have a philosophy that deals with the investigation of its fundamental assumptions, questions, methods, and goals. That is for any X, there is a philosophy (philosophy of X) which deals with the metaphysical (ontological), epistemological, ethical, aesthetic, and teleological issues of X, where X may be the philosophy itself.

Therefore, we have the following philosophies (many of them are now very mature):

- Philosophy of biology.
- Philosophy of physics.
- Philosophy of computer science.
- Philosophy of design.
- Philosophy of feedback control.
- Philosophy of systems.
- Philosophy of automation.
- Philosophy of cybernetics.
- Philosophy of information.
- Philosophy of Artificial Intelligence.
- Philosophy of cyberspace.
- Philosophy of sociology.
- Philosophy of psychology.
- Philosophy of philosophy (meta-philosophy), etc.

In the following we will present a more detailed discussion of philosophy of science and philosophy of technology which provide the fundamental tools for building the philosophies of the above list. [10,12–18].

2.5 Philosophy of Science

2.5.1 General Issues

The *philosophy of science* (**PS**) attempts to understand what are experimental methods, observations and scientific theories or models, and how they have enabled scientists to reveal a wide range of secrets of natural world. In other words, it is concerned with the principles and processes of scientific explanation including confirmation and discovery issues. It is overlapping with metaphysics ontology and epistemology, particularly when it explores the relationship between science and truth.

Question that are studied by philosophers of science include:

- What is a law of nature?
- What kind of data can we use to distinguish between real causes and accidental regularities?
- Why do scientists continue to rely on models and theories which they know are at least partially inaccurate?

These questions appear to be simple, but actually they are quite difficult to answer satisfactorily, with the answers varying widely.

2.5.2 What is Science?

The starting point of philosophy of science is the ontological question *'what is science?'* Very broadly *science* is any physical system of knowledge that aims to systematically study the physical world and its phenomena, where by *'systematically'* we mean the use of systematic or specific methods, such as experimentation or observation to get a deep understanding of the world. The term *'science'* comes from the Latin word *'scire'* which means *'to know'*. Science philosophers (Aristotle, Descartes, Leibnitz, Francis Bacon, and others) studied the world around them first through *ideas* (philosophy), then as *observation* and *discovery,* and finally as *application* (what is now called '*technology*').

Figure 2.2 shows a map of the major branches of Science, namely physical science, mathematical science, earth science, social science, and life science. Each of these major branches of science involves many sub-branches which can be depicted in a particular detailed map. Thus we have the following subdivisions:

- *Mathematical science* (It includes pure mathematics such as logic, sets, algebra, analysis, geometry, and applied mathematics such as probability, statistics, systems analysis, etc.).
- *Physical sciences* (They deal with non-living things, and the study of matter and energy. They include physics, chemistry, astronomy, astrophysics, etc.).
- *Life sciences* (They deal with living organisms and their organization. They include biology, physiology, botany, zoology, ecology, genetics, agriculture, medicine).
- *Earth science* (It deals with the study of earth and includes rocks, soils, oceans, atmosphere, surface features, etc.).

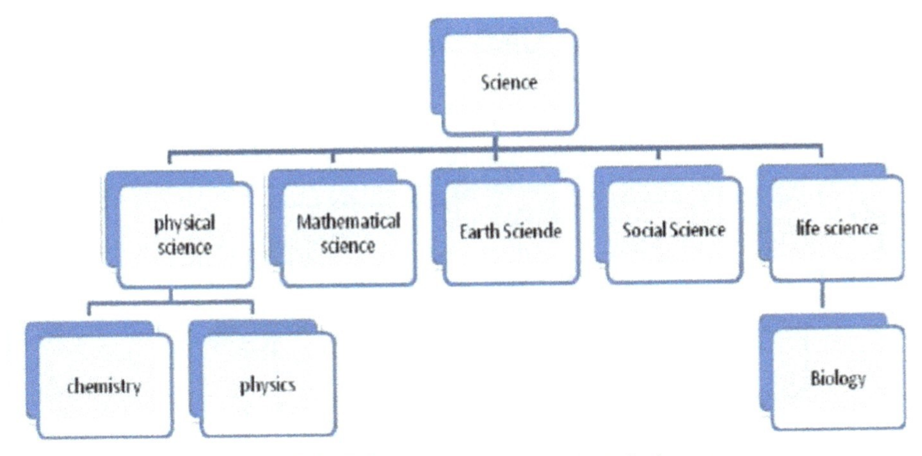

Figure 2.2 Map of the major branches of science.

Source: www.slideshare.act/gdelagdeg/branches-of-science-909005

- Social sciences (They deal with human society and include anthropology, psychology, economics, history, political science, behavioral science).
- *Information sciences* (They deal with all aspects of information production, manipulation and use. They include computer science, artificial intelligence, data science, theoretical computer science, algorithms, programming science, software science, knowledge-based systems, information and library science, etc.).

Mathematics is the most precise and exact of all branches, and started its cultivation in ancient Greece by the greatest philosophers and mathematicians. Plato had become convinced that the road to knowledge lay in exact reasoning, as in mathematics. His pre-eminence of geometry and mathematics was declared by the famous inscription over the entrance of Athens Academy: '*Αγεωμέτρητος μηδείς εισίτω*' (ageometritos mideis eisito), reading: *'let no one who does not know geometry enter here'*. Speaking about mathematics Francis Bacon said: *'Mathematics is the gateway and key to all sciences'*.

The scientific inquiry (method) involves several steps of which the principal ones are: (Figure 2.3):

- Make observations (in the natural world).

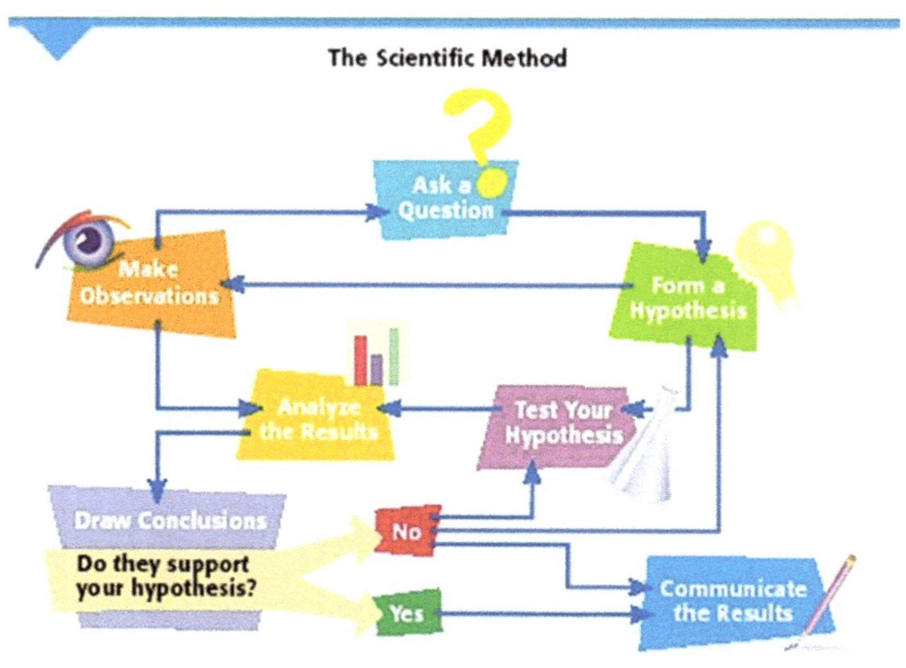

Figure 2.3 The interrelation of the principal steps for implementing the scientific inquiry.

Source: www.super-science-fair-projects.com/scientific-method-for-kids.html#V_dxUMIQRsA

- Ask a question (doing background research).
- Form a hypothesis (based on the research and observations).
- Test the hypothesis (doing proper experiments).
- Analyze the results.
- Draw conclusions.
- Communicate the results (either verifying or rejecting the hypothesis).

For convenience of the reader we provide here a list of big philosophers of science along with a summary of their views.

Aristotle (384–322 BC) He is actually the founder of science and philosophy of science. He studied a wide gamma of topics which are now called logic, physics, chemistry, biology, astronomy, mathematics, and psychology.

Francis Bacon (1561–1626) He developed the scientific method of reasoning based on observations and experiments, making inductive inferences about the processes and patterns in nature.

Rene Descates (1596–1626) He promoted the scientific method which is based on deduction from first principles. His feelings have inspired many thinkers of the scientific evolution (Newton, and others).

Carl Hempel (1905–1997) He developed the theories of explanation and confirmation. According to him 'a phenomenon is explained when we can show that it is the logical consequence of a law of nature.'

Kard Popper (1924–1994) He stated that the falsifiability is both the hallmark of scientific approaches and the proper methods of scientists to employ.

Thomas Kuhn (1922–1996) He argued that empiricism developed by Popper and others does not lead to results resembling the history of science.

Paul Feyerabend (1924–1994) He argued that there is no scientific method, i.e., 'anything goes 'by not based on rational guidelines, scientists do whatever they need to in order to produce new ideas and persuade others to accept them'.

2.6 Philosophy of Technology

2.6.1 Historical Note

The *philosophy of technology* (**PT**) is concerned with the nature of technology and its impact on human life and society. It is a relative young field of philosophy which does not yet exist as a coherent field of research. Because of a lack of consensus about the primary meaning of the term 'technology', the subject covers studies from almost every branch of thinking in philosophy. The philosophical reflection on technology has its origin in ancient Greece [15] with Plato (who argued that technology learns from or imitates nature), Democritus (who argued that house-building and weaving first realized by imitating swallows and spiders building their nests and nets, respectively), and Aristotle's four causes doctrine (material, formal, efficient, and final) which was referred to technical artifacts such as houses and statues. During the Roman empire and the Middle Ages, philosophical reflection on technology was not growing at a similar rate (classical works at these times were concerned more to practical aspects of technology and very little to philosophy). During Renaissance, philosophical reflection on technology and its impact to society showed an increase.

Francis Bacon was the first modern thinker who put forward the philosophical reflection on technology and society that was maintained during the nineteenth century and the first half century of the industrial revolution.

During the last quarter of the nineteenth century and most of twentieth century a criticism of technology was predominated in the philosophical reflection of technology schooled in the humanities or social sciences. The first thinker that coined the term *philosophy of technology* was *Ernst Kapp* in his 1877 book 'Eine Philosophie der Technik'. Most of the authors who wrote critically about technology and its socio-cultural impact during the twentieth century were general philosophers (e.g., Martin Heidegger, Arnold Gehlen, Andrew Freenberg) with a background in social science or humanities. Their views were collectively called 'humanities philosophy of technology', by *Carl Mitcham* (in his 1994 book 'Thinking Though Technology').

An alternative form of philosophy of technology is the so-called *'analytic philosophy of technology'* (emerged in the 1960's) which is concerned both with the relations between technology and society and with technology itself. This form of the philosophy does not book upon technology as a 'black box' but as a field that deserves study and the practice of engineering. It studies philosophically this practice, its concepts, its goals, and its methodologies. A definition of the philosophy of technology which aims to be a general and inclusive one was given by *Kaplan* in 2004 [13]. This is the following:

'Philosophy of technology is a critical reflective examination of the nature of technology as well as the effects and transformation of technologies in human knowledge, activities, societies, and environments. The aim of philosophy of technology is to understand, evaluate, and criticize the ways in which technologies reflect as well as change human life individually, socially, and politically'.

2.6.2 What is Technology?

To define *technology* is not an easy task, and so many attempts were made to provide a definition as more complete and convincing as possible. A first definition can be given through the relation of technology and science. From this point of view *Henryk Skolimowski* stated in 1966 that 'Science concerns itself with *what is* whereas technology concerns itself *what is to be*'. Similarly, *Herbert Simon* stated that 'The scientist is concerned with how things are whereas the engineer is concerned with how things ought to be'. The above definitions view technology as a continuous attempt to bring the world closer to the way one wishes it to be. Three current definitions of technology are [14]:

- Technology as tools and machines.
- Technology as rules.
- Technology as system.

Technology as tools and machines seem to be the most obvious, concrete and easily graspable definition. But this definition cannot cover technology that does not use either tools or machines (hardware), for example the artificial intelligence, knowledge-based systems, and software technologies.

Technology as rules This definition of technology is due to *Jacques Ellul* and treats technology as rules rather than tools. It embraces both hardware and software (algorithms, artificial intelligence, etc.) and considers that hardware and software differ only in their emphasis, as technology involves patterns of means-end relationships.

Technology as system This is the most general definition and is based on the fact that technology is aiming at 'creating artifacts' and 'artifact-based services'. The concept of technology as (a technological system) includes *materials* (hardware), *knowledge* (rules, guidelines, and algorithms), designers, inventors, software engineers, operators, managers, programmers, customers, marketers, economists, etc.). This definition suggests that for an artifact or piece of knowledge to be technology, it needs to be considered together with the people who use, maintain, repair, and operate it.

Table 2.1 shows a non-exhaustive list of modern branches of engineering and technology. Science overlaps considerably with engineering and technology, and engineering overlaps with technology. There is also a small common overlap of all three; science, engineering and technology. Science seeks to understand the natural world, engineering uses scientific discoveries to design products and processes that meet the society needs, and technologies (product and processes) are the result of engineered designs.

2.7 Fundamental Questions in Philosophy of Science and Philosophy of Technology

2.7.1 Philosophy of Science

Philosophy of science attempts to answer both general questions like: 'what is the purpose of science?', 'what counts as science? and 'what is reliability and validity of scientific theories?' and more specific questions which include the following:

- How science is law of nature?
- What kinds and how much evidence are needed in order to accept hypotheses?
- Can science reveal the truth of unobservable things?
- Can scientific reasoning be justified at all?
- Can one scientific discipline be reduced to the terms of another?

Table 2.1 Modern branches of engineering and technology

Aeronautical engineering	Fault identification/tolerance
Aerospace engineering	Geological engineering
Agricultural engineering	Industrial engineering
Applied mechanics	Information technology
Architecture Information technology	Intelligent systems engineering
Audio engineering	Manufacturing
Automation	Materials engineering
Automotive engineering	Mechanical engineering
Bioengineering	Mechatronics
Bioinformatics	Microwaves and antennas
Biomedical engineering	Modeling and simulation
Chemical engineering	Nuclear engineering
Civil engineering	Optical networks/networking
Computational engineering	Parallel computing
Computational intelligence	Pattern recognition
Computer engineering	Power engineering
Control engineering	Radar and optical communications
Digital speech processing	Robotics
Satellite communications	System engineering
Electrical engineering	Signal and image processing
Electronics engineering	Telecommunication engineering
Embedded systems engineering	Wireless networks and communications

- Is science philosophy?
- Could (should) philosophy be scientific?

The problem of reliably, *distinguishing* science from non-science, is known as *'demarcation problem'*. It is generally agreed by many modern philosophers of science that no single and simple criterion exists for demarcating the boundaries of science, while other philosophers see this problem as unsolvable and uninteresting.

Karl Popper called this problem the *'central question'* of philosophy of science, and argued that the core property of science is *'falsiability'*. In his 1934 book 'Scientific Discovery' he advocated that scientific ideas can *only* be tested through falsification, never through a search for supporting evidence. That is, his approach was not to acquire evidence to prove a theory to be true, but rather to find evidence that would prove a theory to be false. One way to distinguish between science and non-science is *logical positivism* according to which science is grounded in observation, whereas non-science is not observational. Logical positivism (also called *verificationism*) was articulated by A.T. Ayer in his book; 'Language, Truth and Logic' (1946).

Referring to the question: *'is science philosophy?'* it is remarked that the empirical or experimental methodology of science is not deductive, but provides highly likely conclusions, and, usually, it is the best we get. It can be said that *science is philosophy,* as long as experiments and empirical methods are considered to be rational and lead to truth [10]. Science cannot be viewed as philosophy if experiments are not considered and do not count as being rational but only as logic. Science is also not philosophy if philosophy is regarded as the search of universal or necessary truths, i.e., things that would be true irrespectively of what results are found by science or what basic assumptions are made.

Regarding the question *'could philosophy by scientific'* (i.e., more scientific or more experimental than it is), it can be said that the area of *'experimental psychology'* can be considered as part of scientific world view. In this area philosophers perform scientific/psychological experiments to find out what people who are not philosophers think about certain philosophical topics (although this can considered as belonging to *cognitive science*) [10]. *Colin Mc Ginn* argues that philosophy is a science, i.e., a systematically organized body of knowledge, calling it *'ontical science'* that philosophy and science belong to the same continuum which means that philosophy is aware, and makes active philosophical use of scientific results. In this way philosophy can help science in its development (not simply restricted to doing philosophy). This kind of philosophy is called *'naturalistic philosophy'*.

Other principal philosophical methods (besides logical positivism) are shown in Figure 2.4 and outlined below.

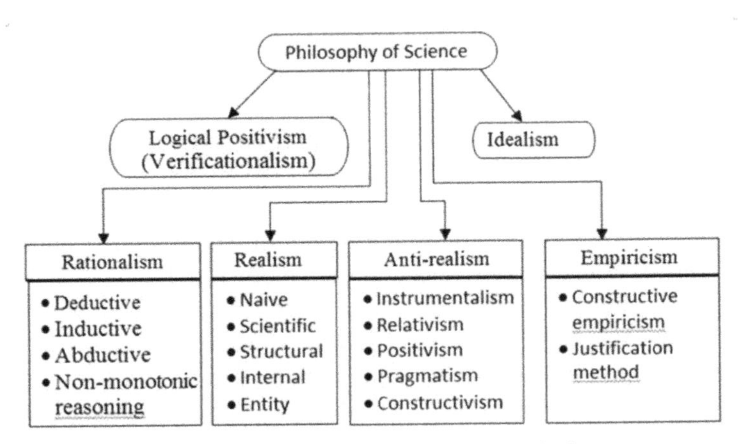

Figure 2.4 Philosophy of science methods.

Rationalism This method was established by *Descrates* (1967) and is based on deduction from first principles. Deduction is the reasoning method in which a conclusion is logically obtained from premises. The works of *Rene Descartes* and *Francis Bacon* (who promoted the alternative scientific method in which scientists collect many facts from observations and experiments, and then make inductive inferences about patterns in nature) have put the foundations of the modern scientific method. The four main forms of logical rationality are:

- *Deductive reasoning* (which represents the absolutely certain rationality)
- *Inductive reasoning* (also called probabilistic reasoning).
- *Abductive reasoning* (which is a probabilistic reasoning, also known as 'inference to the best explanation').
- *Non-monotonic reasoning* (which is more 'psychologically real' than the others and was called by *Herbert Simon 'satisficing reasoning' or 'being satisfied with a reasonable answer' method).*

Realism This philosophical approach was initiated by Pythagoras and was further developed by Russel, Popper, Ladyman, Donnett, Rosenberg and others. It is distinguished in the following main kinds: (i) *nave realism* (The world I see is real. What are you all arguing about?), (ii) *Scientific realism* (science makes real progress in describing real features of the world), (iii) *Structural realism* (Science has identified real patterns, relationships and structures, at least within a regime-in nature).

Empiricism This philosophical school was established by the British philosopher *John Locke* (1632–1704) and advocated that all knowledge comes a *posteriori* from experience. According to Locke, man is born as a blank state *(tabula rasa), and* is subsequently enriched by experiences that in turn becomes knowledge. This was the base of the so-called *justification (explanation) model* of knowledge. Starting with observation one is led to the formulation of a hypothesis which is culminated in a theory about reality and translates into *truth.* The problem of the explanation model is what happens when the thing to be explained cannot be perfectly predicted from what is known. *Wesley Salmon* proposed a model in which a good scientific explanation must be statistically relevant to the outcome to be explained.

Idealism This philosophical method was established by *Immanuel Kant* in his 1978 book 'The Critique of Pure Reason', where he criticized both the rational and empirical traditions and attempted a synthesis of both. Regarding the *justification model,* he argued that the step from observation to theory is impossible and the step from theory to truth is highly questionable.

Instead, Kant proposed that the correct scientific method began with theory in which the observations are made. *Hans Reinchenbach* (1891–1953) and other philosophers argue that a distinction should be made between 'context of discovery' and the 'context of justification'. Once a hypothesis has been proposed, there are rules of logic that determine whether or not it should be accepted *(rules of justification)*. However, no such rules exist to guide someone to formulate the right hypothesis or even hypotheses that are plausible or reasonable.

Paradigm model This philosophical view of science was proposed by *Thomas Kuhn* in his 1962 book 'The Structure of Scientific Revolutions'. He argued that the picture of science developed by logical empiricists, such as Popper, didn't resemble the history of science. He distinguished between *normal science* (where scientists solve problems within a particular framework or paradigm), and *revolutionary science,* when a number of scientific irregularities (abnormalities) occur that do not match the paradigm in consideration. He argued that the processes of observation and evolution take place within a paradigm. Kuhn rejected the view that it may be possible to isolate the hypothesis being tested from the influence of the theory in which the observations are based, and argued that it is impossible to evaluate competing paradigms independently. A scientist that wishes to choose between paradigms should set two or more 'portraits' against the world and decide which is likely more promising. A paradigm shift takes place when a remarkable number of observation irregularities (anomalies) in the old paradigm have rendered the new paradigm more appropriate and useful.

2.7.2 Philosophy of Technology

Similar to the philosophy of science, the questions that are studied in philosophy of technology range from metaphysical/ontological to epistemological, ethical, and aesthetic questions. A non-exhaustive list of questions is the following:

- What is a tool?
- What is engineering?
- What is the purpose of technology?
- Are technological problems self-determined or socially determined?
- Are there problems of philosophy of mind in technology?
- What are the status and the nature of artifacts?
- What is technological knowledge?
- What is technological creativity?

- How does uncertainty inherent in technology (and engineering) differ from scientific uncertainty?
- What is technological (and engineering) design?
- Is there an engineering design method?
- Does technological and engineering design matter? To what? Why? For what purpose?

Let us have a look at the status and character (ontology) of artifacts. Artifacts are man-made objects, and those that are of relevance to technology are made primarily for serving a purpose. The creation of an end artifact is achieved through the creation of intermediate products. Many thinkers of technology argue that an adequate description of an artifact must include both its status as a tangible physical object, and the intentions of the people engaged with it. The concept of functions is very important for the characterization of artifacts, but it is used more widely (e.g., in biology, in cognitive science, etc.).

2.8 Conclusions

Philosophy is about our thoughts, beliefs, and attitudes, about ourselves and the world; essentially putting our thought and language in order. In this chapter we outlined the fundamental issues of Western philosophy which has its origin in the work of Ancient Greek philosophers Socrates, Plato, Aristotle, and others. Greek philosophy is concerned with two principal questions, namely 'τι εστίν' (ti estin=what is), and 'τι το ον' (ti to on=what is being) which refer to the essence of existence. The answers to these questions are derived via an analytic and linguistic understanding process which arises from the recognition that all works in philosophy are expressed and transmitted in some physical form of symbol schemes in a particular framework, and interpreted using a specific language. A typical understanding of philosophy is that of *'questioning search'* and *'knowledge pursuit'*. Since humans are finite and mortal, this search process is endless as Socrates pointed out in his quotation 'Γηράσκω αεί διδασκόμενος' (girasko aei didaskomenos=I grow old learning everyday something new'. In doing philosophy one is usually referring to the view points and arguments of other philosophers, and on their basis develops his/her own new ideas and philosophical results. The chapter has included a separate outline of philosophy of science and philosophy of technology which provide of proper background for doing philosophy in the information-based areas of science and technology. Typically, as it happens in any case of search or exploitation, in doing philosophy we do not know the end of the process. The availability of a kind of *'prior knowledge'* or a kind of *anticipatory feeling'* of what the final outcome might be, would facilitate considerably the searching process itself. Philosophy of science is first concerned with the basic philosophical (metaphysical) questions 'what is science' and 'what is the distinction of science from non-science' and the principles of logic reasoning, rationalism, empiricism and idealism. Philosophy of technology was mostly concerned with the philosophical questions related to the impact of technology on society, development, and culture. Twenty five tutorials on philosophy, covering most topics and philosophies (e.g., the philosophies of Plato, Aristotle, Augustine, Frege, Russel, Wittgenstein, Descates, Kant, Duns Scotus, Ockham, Russel, etc.), are provided in [17].

References

[1] Solomon, R. C., and Higins, K. M. (2010). *The Big Questions: A Short Introduction to Philosophy*. Belmont, CA: Wadsworth Cengage Learning.

[2] Buckingham, W., and Burnham, D. (2011). *The Philosophy: Big Ideas Simply Explained*. London: D.K. Publishing.

[3] A Quick History of Philosophy (The Basics of Philosophy). Available at: www.philosophybasics.com/general/quick_history_of_philosophy

[4] Kenny, D. W. A. (1998/2006). *An Illustated Brief History of Western Philosophy*. London: Blackwell Publishing.

[5] Available at: www.mohamedabeea.com/books/book_9035.pdf

[6] A Short History of Philosophy. Available at: www.questia.com/Online_Library

[7] Available at: www.philosophypages.com/hy

[8] History of Philosophy. Available at: www.philosophicalsociety.com/Archives/

[9] Munitz, M. K. (1981). *Contemporary Analytic Philosophy*. New York, NY: Prentice Hall.

[10] Stroud, B. (2000). *Meaning, Understanding, and Practice: Philosophical Essays*. Oxford: Oxford University Press.

[11] Rapaport, W. J. (2005). Philosophy of computer science: an introductory course. *Sci. Teach. Philos.* 28, 319–341.

[12] Shafer-Landau, R. (1988). *The Fundamentals of Ethics*. Oxford: Oxford University Press.

[13] Durbin, P. (2007). Philosophy of technology: in search of discourse synthesis. Techne 10.
[14] Kaplan, D. M. (ed.) (1972). *Readings in the Philosophy of Technology*. New York, NY: Free Press.
[15] Dusek, V. (2006). *Philosophy of Technology: An Introduction*. Malden, MA: Blackwell Publishing.
[16] Stanford Encyclopedia of Philosophy (2009). *Philosophy of Technology: Stanford Encyclopedia of Philosophy*. Available at: http://plato.stanford.edu/entries/technology
[17] Mitcham, C., and Mackey, R. (eds.) (1972). *Philosophy and Technology*. Lanham, MD: Rowman and Littlefield.
[18] Tutorials: Philosophy (Offered by the University of Oxford to philosophy students). Available at: www.bestmester.com/sso
[19] Available at: http://undsci.berkeley.edu/article/philosophy-of-science
[20] Heidegger, M. (1962). *Being and Time*. (Trans. Sein und Zeit, by John Macquarrie and Edward Robinson). New York, NY: Harper and Row Publishers.

3

Background Concepts: Systems Ontology, History and Taxonomy

We believe that the essence of a system is 'togetherness', the drawing together of various parts and the relationship they form in order to produce a new whole.
J. Boardman and B. Saaser

A system must have an aim. Without an aim there is no system.
W. Edwards Deming

Any system that values profit over human life is a very dangerous indeed.
Suzy Kassem

3.1 Introduction

Over the years scientists, mathematicians and philosophers have been working with the problem of how to understand and make sense of the world. They have put much effort for constructing an exact theory that allows the unification of various branches of the scientific approach. Descriptive theories for understanding the world include the theories of cognition, perception, and thinking. Prescriptive theories follow one of two approaches, namely *reductionism* and *systems theory*. Reductionism advocates the view that the best way to understand natural and social phenomena is to study the properties and operation aspects of their individual parts and elements.

System theory is concerned with the relations between the parts, and instead of reducing an entity into its parts (e.g., the organs, muscles, tissue, bone, cells of the human body) focuses on the relations between parts and their interconnections to understand how they work together as a whole. The properties of a system depend, on the way its parts are organized and how they interact to work as a whole. The behavior of a system does not depend on the individual properties of its parts. Systems theory is a fundamental tool for intelligently engaging change and system complexity. Systems theory is closely related to many other fields, e.g., cybernetics, decision making, management, and operational research. Cybernetics and management systems will be studied later in the book.

The purpose of the present chapter is to provide some background ontological material on the system concept. In particular the chapter:

- Provides a set of definitions of the system concept given by well-known pioneers in the field.
- Presents the basic ontological elements of systems theory (type, environment, boundary, closed/open systems, etc.).
- Gives a brief historical review which includes major representative landmarks and works in the three life periods of systems, namely: precursors, pioneers, and innovations periods.
- Presents a multi-dimensional taxonomy of systems theory, also including a detailed list of systems types studied over the years.

The epistemological aspects of general systems theory (principles, methodology, methods) will be discussed in the next chapter.

3.2 What is a System?

The concept of *system* has its origin in ancient Greece. Plato was using the word *'σύστημα'* (systema) to mean reunion, conjunction or assembly. This concept was reinforced during the seventeenth century, to mean a collection of organized concepts, principally in philosophical sense. *Descartes,* in his work '*Discourse de la Methode*', introduced a coordinated set of rules for achieving coherent certainty, i.e., an epistemic methodology or *systematic study* [1].

The concept of system is difficult (if not impossible) to be defined uniquely and globally. There is actually a great variety of definitions which they both overlap and complement each other.

Some of them given by pioneers in the field are the following:

H. Maturama and F. Varela: System is any definable set of components [2].

L. Bertanlanfly: A system is a set of elements in interaction [3].

R. F. Miles: A system is defined as a set of elements used to satisfy a need or requirement [4].

D. Dori: A system is a value-delivering object [5].

A. Laszlo and S. Krippner: A system connotes a complex of interacting components together with the relationships among them that permit the identification of a boundary-maintaining entity or process [6].

B. Blanchard and F. Fabracky: A system is a bounded region in space-time in which the component parts are associated in functional relationship [7].

INCOSE/D. Hitchins: A system is a construct or collection of different elements that together produce results not obtainable by the elements alone [8, 9].

R. Ackoff: A system is a set of two or more elements that satisfy the following conditions: (i) The behavior of each element has an effect on the behavior of the whole, (ii) The behavior of the elements and their effect on the whole are interdependent, and (iii) however subgroups of elements are formed, each has an effect on the behavior of the whole and none has an independent effect on it [10].

K. MacG Adams: A unified system of propositions made with the aim of achieving some form of understanding that provides an explanatory power and predictive ability [11].

According to *G. M. Weinberg* [13]: 'A system is a way of looking at the world;' This implies that systems don't really exist but they are just a convenient way of looking at phenomena and describing things'.

System theory is a philosophical doctrine which describes systems as abstract concepts independent of substance, type, time, and space. System theories can be viewed both ontologically and epistemologically. The ontological view considers the world to consist of '*systems*' or '*interactive levels*'. The epistemological view implies a '*holistic perspective*' emphasizing the interplay between the systems and their elements in determining their respective functions. According to systems thinkers (e.g., Boulding [14]) 'some of the phenomena were brought to be [...] of almost universal significance for all disciplines.'

With regard to the application in studies of perception, systems theory can model complex interpersonal, intergroup, and human-nature interaction without reduction of perceptual phenomena to the level of individual stimuli. As a field of inquiry which examines the holistic and integrative nature of phenomena and events, system theory pertains to both ontological and epistemological issues, which together can be considered to form what the ancient Greek philosophers called '*γνωσιολογία*' (gnosiology) from the words '*γνώση*' and '*λόγος*' (logos)=study. Gnosiology is generally concerned with the holistic and integrative exploration of phenomena, processes, and events [6].

3.3 Systems' Ontological Elements

Basic ontological elements of the systems theory are:

Environment The environment (*surroundings*) of a system involves all variables that can affect its state. The environment is usually referred to as the context in which the system is found, or as its surroundings. Ackoff points out that the environment of every social system contains three levels of purpose: 'the purpose of the system, of its parts, and the system of which is a part, the *suprasystem*' [15].

Boundary The boundary is the separation between the system and its environment. The actual point, at which the system meets its environment, is called an 'interface'. The boundary of a system specifies those elements and relationships that can be regarded as part of the system, and those which describe the interactions across the boundary between system elements and elements of the environment. In many cases the boundary of a system is not sharply defined. In other cases boundaries are conceptual rather than existing in nature.

Types of Systems In thermodynamics we have three types of systems, namely:

- Closed systems.
- Isolated systems.

A system is said to be *open* when its boundary allows exchange of mass (material) and energy with its environment (surroundings) or system in which it exists. A system is said to be *closed* when only energy is allowed to transfer across the system boundary. There may be flow of energy into or out of the system but no matter flow takes place across the system boundary. A system is said to be *isolated* if it does not interact with the environment, and is not influenced by its surroundings. In this case no mass or energy is allowed to transfer across the system boundary. Figure 3.1 depicts these three types of thermodynamic systems.

A system may or may not exchange information with its environment. Therefore, extending the definitions that refer to a thermodynamic system, we can say that an information system is *informationally open (isolated)* when information is allowed (not allowed) to be transferred across the system boundary.

Inputs and Outputs Mass, energy or information that enters the system from the environment is called an *input*, and mass, energy or information which goes outside to the environment is called an *output* (Figure 3.1). In a system where matter can be exchanged, as it is the case with all living organisms that need oxygen, water and food in order to survive, an exact calculation of future states (with accuracy) is not usually possible. This calculation can be done using their interaction with the environment which has the input and output components.

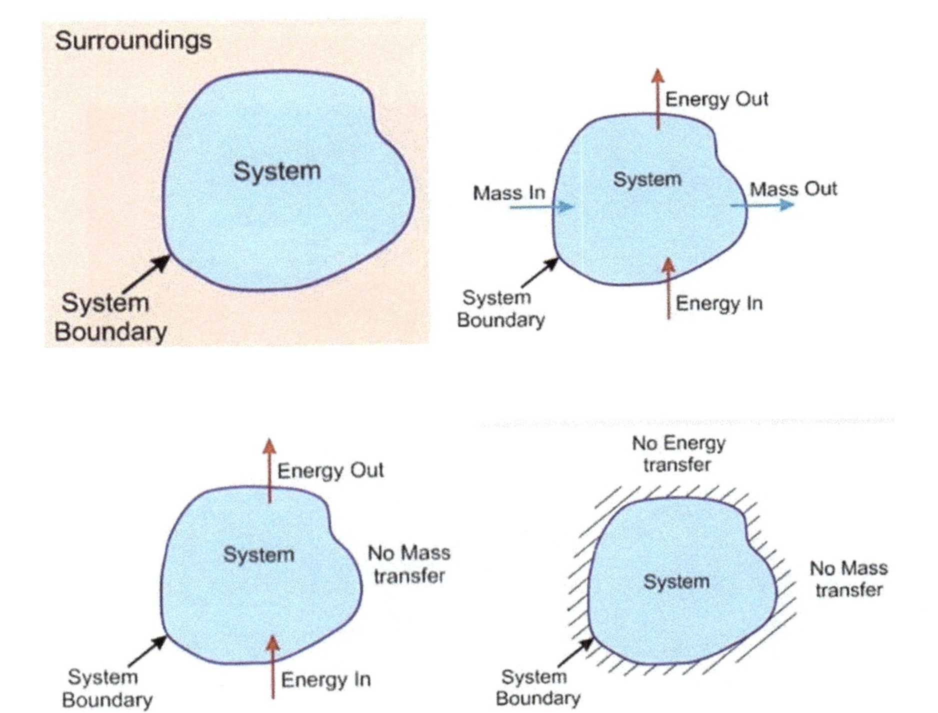

Figure 3.1 Open, closed and isolated thermodynamic systems. *Up left*: A system and its boundary. *Up right*: Open system. *Down left*: Closed system, *Down right*: An isolated system.

Source: www.industrial-items.blogspot.gr

Subsystems A system can be decomposed into parts. Each part is also a system. For each part (subsystem), the remainder of the system is its environment. Subsystems are interrelated and inter-dependent.

Figure 3.2 gives a conceptual block diagram representation of general open systems receiving inputs from the environment, processing and transforming them, providing outputs/products in the environment, and getting feedback from the environment.

A company/organization is an example of open systems. Even if the company has separate departments (subsystems) the employees share data, information, financial resources and technology, and interact with each other on a daily basis during the company's dynamic transformation process. The same happens with other organizations such as suppliers, distributors, and customers residing in the external environment. Figure 3.3 gives a pictorial illustration of the open nature of business/technological organizations which shows examples of inputs received, transformations made, and outputs provided to the environment.

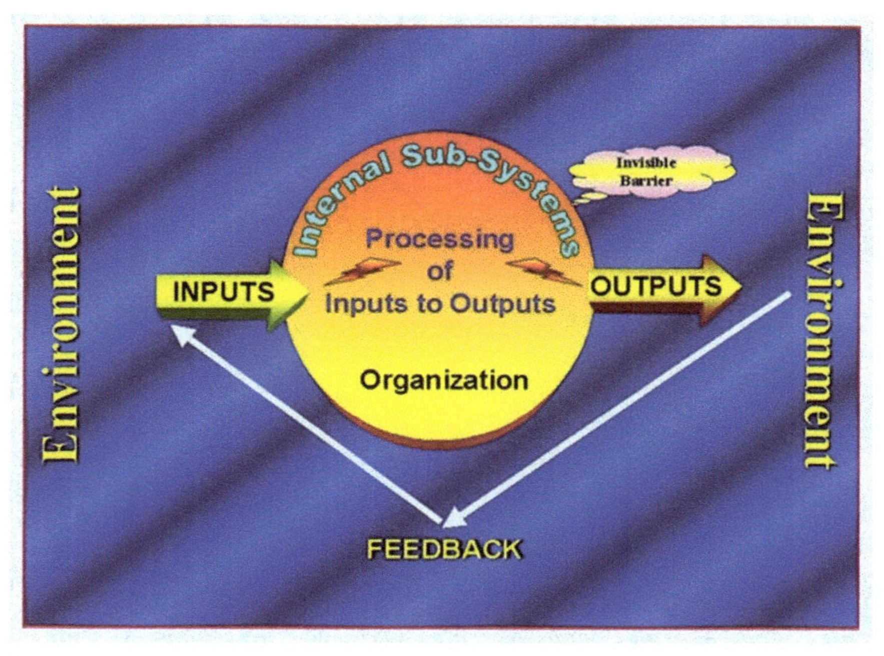

Figure 3.2 Open system/organization with input and feedback from the environment, and output to the environment.

Source: http://brainware.tripod.com (opensys.html).

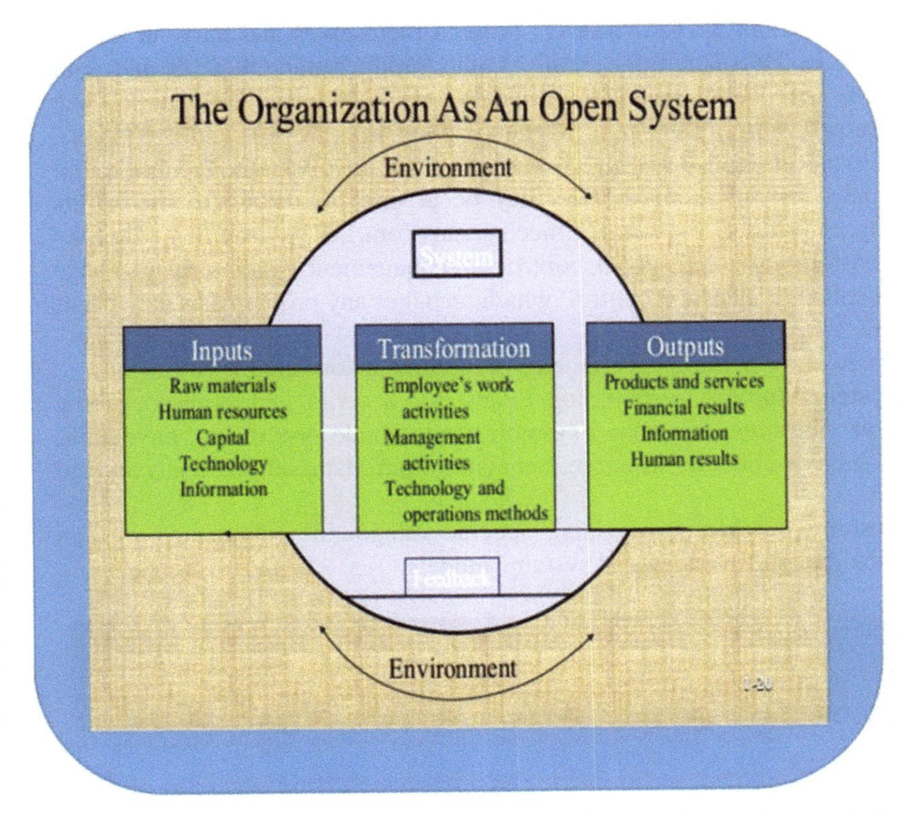

Figure 3.3 An organization is a typical open system with input from and output to the environment.

Source: www.slideshare.net/ZOHANQUR/introduction-to-management-and-organization-p-o-m

Control mechanism This is the mechanism that specifies how the system's behavior is regulated to allow it to endure. The control mechanism involves *feedback* and often is a natural mechanism. Feedback is a circular causal process in which some part of a system's output is fed-back (returned) into the system's input. Systems with feedback (positive or negative) are called *closed-loop* control systems. Feedback enables the system to achieve *homeostasis*, also called '*steady-state or dynamic equilibrium*'. If a system, which has negative feedback is pushed out of its equilibrium state, then it will return to it. Actually, negative feedback has a dampening effect, i.e., the effects are smaller than the causes. Positive feedback is when small system deviations (perturbations) reinforce themselves and have an amplifying effect. Thus small perturbations may produce unpredictable and wild swing in the overall behavior of the system, which can be chaotic. The well-known *butterfly effect* is a phenomenon relating to *chaotic systems*. Open-loop (or feed-forward) control does not involve any feedback, but anticipates any problems before they occur and pre-calculates the inputs directly entering to the control mechanism from the environment. Besides the feed-forward and feedback control, in management systems we also have the so-called 'concurrent control' which manages any problems as they occur (Figure 3.4).

Three other ontological elements of systems theory are the following:

System Model A system model is used to describe and represent all the multiple views that comprise a system. A man-made system may have multi views such as concept, structure, behavior, input, data, output, analysis, and design.

Systems Architecture Using specific integrated architecture we can describe the multiple views of systems models.

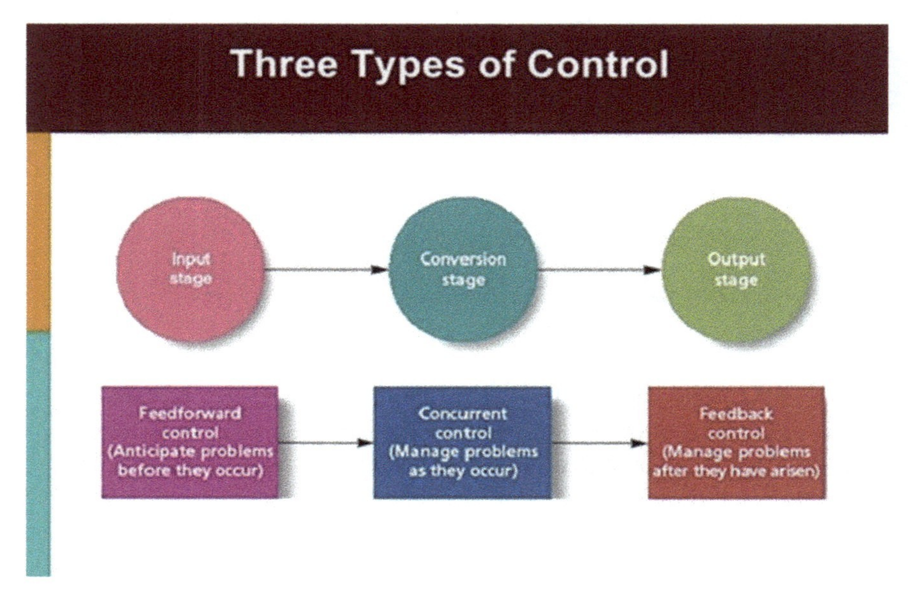

Figure 3.4 The three types of management control: Feed-forward, concurrent, and feedback control.

Source: www.slideplayer.com/slide/7800941

Emergent properties These properties are possessed by a system as a whole but not by any of its parts (agents or subsystems). They are properties that cannot be predicted by studying the parts. The concept of emergent properties leads to the concept of *synergy,* which in everyday language expresses that the system is more than its parts. From a structural point of view, a system is a divisible '*holon*', but functionally is an indivisible unity with emergent properties. Two important issues of emergent properties are [6]:

- They are lost when the system breaks down to its parts (e.g., the property of life does not inhere in organs once they are removed from the body).
- When a component is removed from the whole, the component itself will lose its emergent properties (e.g., a hand severed from the body cannot write, nor can a severed eye see).

Figure 3.5(a) shows an illustrative diagram of how system-wide emergent patterns/behaviors are created out of local agents interactions. Figure 3.5(b) illustrates the emergence process for the case of group behavior in small-world networks.

More details on the emergent properties phenomenon will be *discussed in Chapter 7 (Section 7.7).*

A general diagram of a system that shows many of the above system ontological elements (inputs/outputs, control mechanism, feed forward/feedback control) is given in Figure 3.6.

The relationships between the elements of an open system can be considered as a combination of system structure and behavior. The system structure describes a set of outcomes produced when an instance of the system interacts with its environment. The behavior of a system depends on the effects or outcomes produced by the system-environment interaction. The identification of a system and its boundary is actually the choice of the observer., i.e., any particular identification of a system is a human construct that aims to help make better sense of a set of things and to share that understanding with others if required. Many natural, social and technological systems can be better understood by describing them as open systems that allow the identification and use of shared concepts which apply to them, and give useful insights into them, independently of the type of elements they are made up of [16]. Concepts of this kind are: *structure, behavior, and state,* which form the basis for *Systems Thinking* (see Chapter 4) and the development of theories and approaches in a broad repertory of fields of systems science.

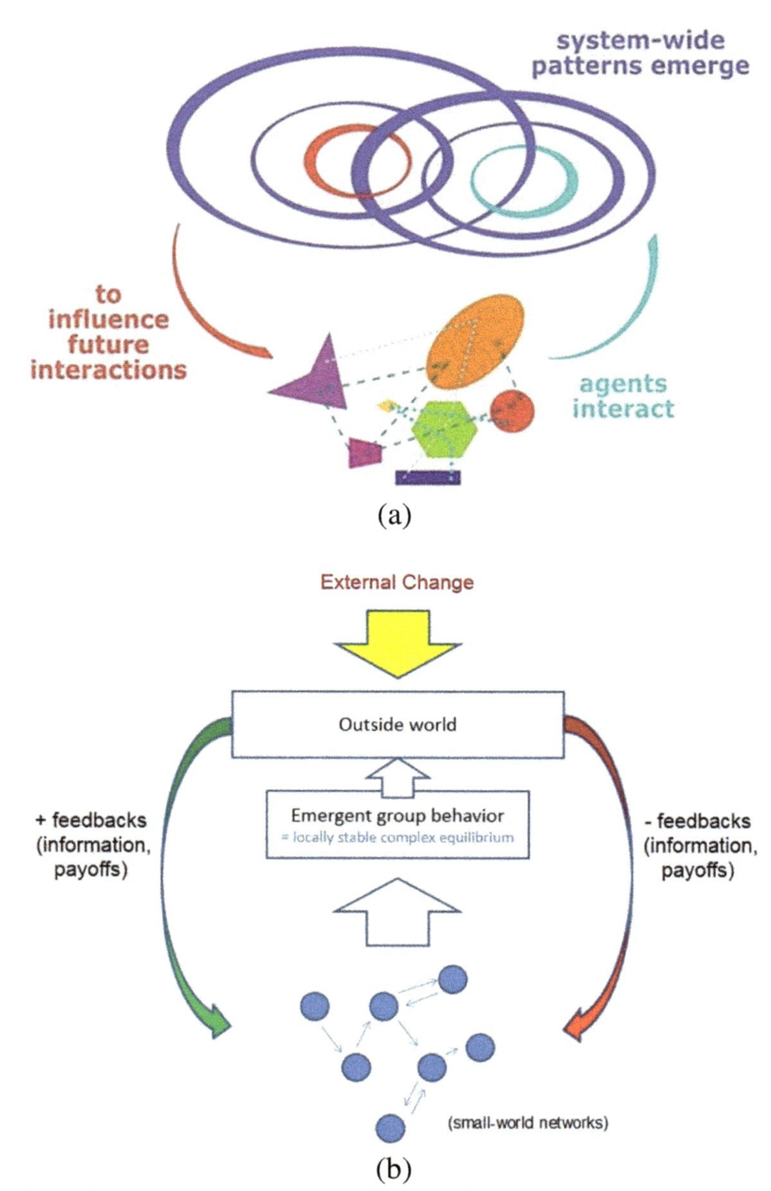

Figure 3.5 Illustration of the creation-mechanism of emergent global patterns and behavior.

Source: www.sldeplayer.com (CAS terminology introduced by members of Santa Fe Institute). https://www.tcd.ie/futurecities/research/energy/adaptations-complex-systems.php www.hsdinstitute.org (resources/complex-adaptive-system.html).

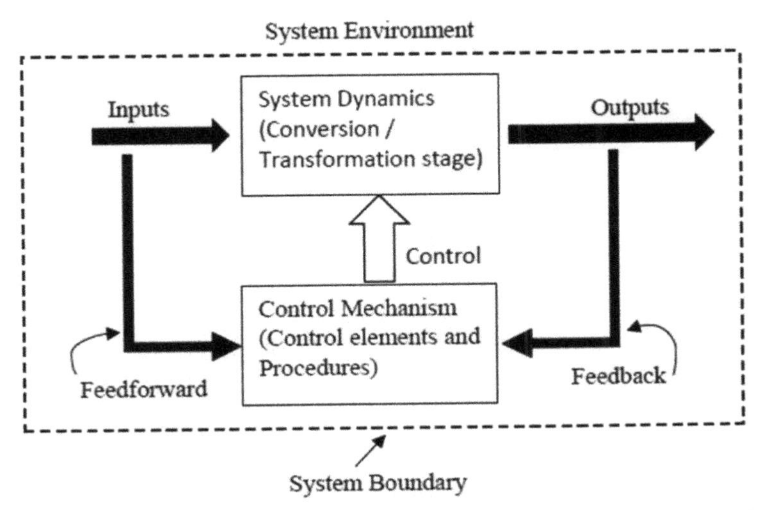

Figure 3.6 Conceptual pictorial representation of a system with feed forward and/or feedback control.

Actually, Figure 3.6 represents an external description of a system, i.e., a black box representation analogous to that of control engineering which includes goals and feedback. Very broadly systems theory can be described in the following ways:

- *Mathematical system theory:* This is based on a set of performance measures that define the state of a system and its qualitative properties (stability, finality, competition, wholeness, and sum, etc.).
- *System technology:* Traditional methods can no more adequately describe and analyze society and modern societal technological processes, because they have become very complex. Systems theory provides the tools for effectively engaging and studying this complexity. Industrial plants, educational systems, urban environments, organizations and companies, ecosystems, etc. all can be analyzed within the general systems framework.
- *System philosophy:* System theory represents a new philosophical strand, which contrasts the classical analytic, mechanistic and causal philosophical view of the natural and man-made entities. System philosophy is to be a new paradigm complete with a new ontology and epistemology that coverts 'real systems, conceptual systems, and abstract systems'.

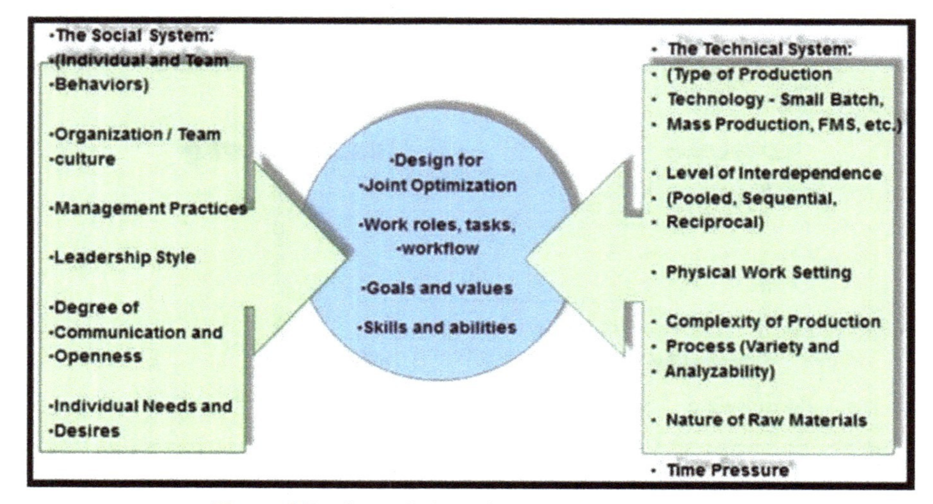

Figure 3.7 General view of socio-technical systems.

Source: IOE421 Midterm Review Part 2 https://www.studyblue.com

All organizations comprise two independent systems: (i) social system, (ii) technical system, i.e. they are *sociotechnical systems*, and to achieve high productivity and employee satisfaction, both subsystems must be controlled and optimized. Changes in one system affect the other system (Figure 3.7).

Socio-technical systems theory aims to achieve joint optimization, with a shared focus on achievement of both excellence in technical performance, and quality in people's work lives. To this end, it takes into account and integrates in the best working way the following aspects:

- Social and political environment.
- Laws, regulations, traditions and practice.
- Organizational values, policies and culture.
- Organizational strategies and goals.

People are working towards '*goals*', follow '*processes*', use '*technology*', operate within a physical '*infrastructure*', and share certain '*cultural norms*'. The six interrelated and interacting elements of sociotechnical systems are represented by the so-called '*socio-technical hexagon*' shown in Figure 3.8(a). The performance of socio-technical systems is specified using several characteristics (requirements). As shown in Figure 3.8(b), these are the following:

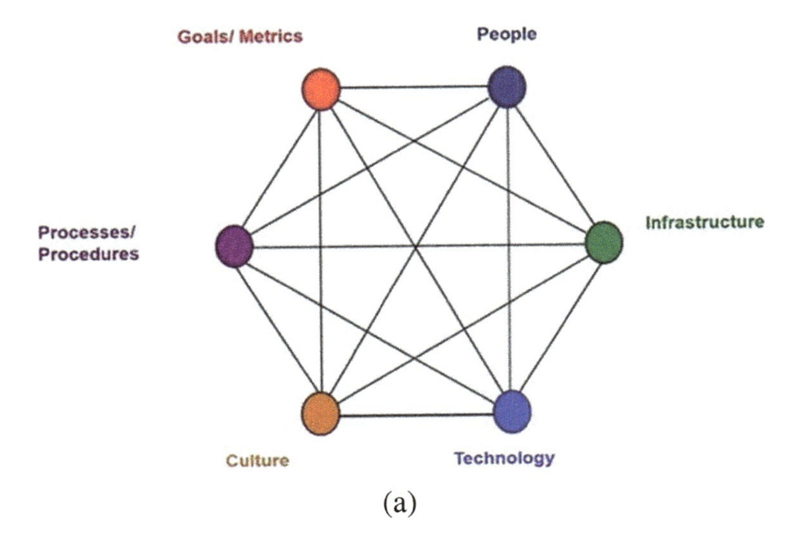

(a)

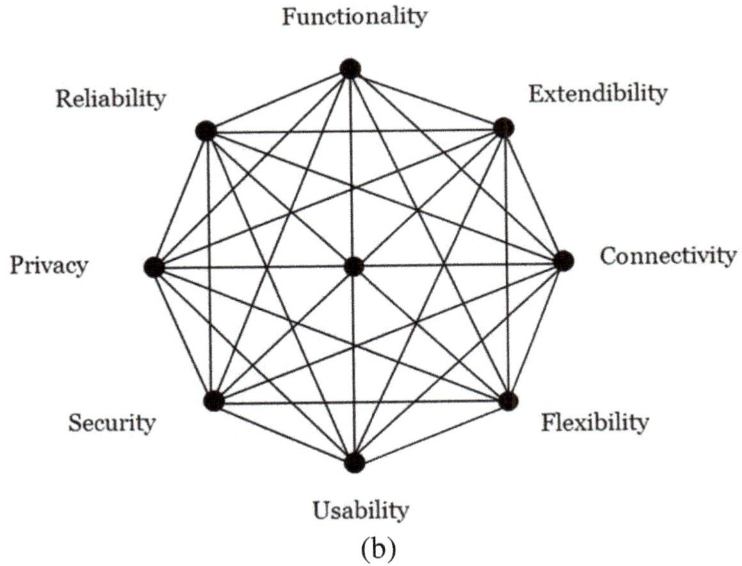

(b)

Figure 3.8 (a) Socio-technical hexagon, (b) Systems general performance characteristics or design requirements space.

Source: (a) https://www.interaction-design.org (socio-technical-system-design). The Encyclopedia of Human Computer Interaction, (b) www.business.leeds.ac.uk

- Functionality.
- Extendibility.
- Connectivity.
- Flexibility.
- Usability.
- Security.
- Privacy.
- Reliability.

In Figure 3.8(b):

- The area represents the overall performance characteristics possessed by the system, i.e., performance in general.
- The shape is the requirements weights, defined by the environment.

The lines are design requirement 'tensions'.

The major characteristics of socio-technical systems are:

- *They are non-deterministic* (i.e., they do not always give the same output when presented with the same input, because the system's behavior is partially dependent on human operations).
- The extent to which the system supports organizational objectives does not just depend on the system itself.
- The overall system possesses emergent properties.

3.4 Brief Historical Review of Systems

Systems theory has started almost synchronously with cybernetics (1948–1950) and has been developed in parallel with it. Systems theory overlaps with cybernetics in many aspects and usually researchers in one field are also working in the other one. The history of cybernetics will be discussed in the respective chapter. Here we will give some principal landmarks of systems (general systems theory). According to Francois [1] general systems theory history can be divided in three main periods:

- Precursors' period (before 1950).
- Pioneers' period (1950–1960).
- Innovators' period (after 1960).

3.4.1 Precursors' Period

The dominating scientific predecessor ('prodrome') of systems theory is Descartes (the 'father' of modern philosophy) who, in an attempt to demarcate between knowledge derived from science and philosophy, described a scientific method which is based on the 'skeptical inquiry principle', which says 'never accept anything for true if you don't clearly know to be such'. Other principles that Descartes established in his *analytic method* are:

- Division of each of the difficulties under examination into as many parts as possible, and as might be necessary for the problem adequate solution.
- Starting with the simplest and easiest known objects proceed step-by-step to knowledge of the more complex assigning in thought a certain order.
- In every case make enumerations so complete, and reviews so general, so it may be assured that nothing was omitted.

1824: The French physicist, *Sadi Carnot*, who studied thermodynamics, initiated the study of '*systems*' in the natural sciences. He thoroughly studied a system which he called the '*working substance*'(consisting of a body of water vapor) in steam engines, exploring fully the system's ability to do work when heat is applied to it (i.e., what is called the '*engine efficiency*'). The working substance could be put in contact with either a *boiler*, a receptable of cold water (called '*cold reservoir*') or a *piston* to which the working body could do work by pushing on it.

1850: The German physicist *Rudolf Clausius* generalized this concept to include the concept of the surroundings and coined the use of the term *working body* for the system. He stated the second law of thermodynamics as: 'The entropy in an isolated system never decreases (i.e., $dS(t)/dt \geq 0$, where S denotes the entropy), and, as a result, the entropy of the Universe tends to a maximum.

1854–58: The French biologist *Bernard* established the existence of the so-called 'internal millieu' in the 'living organisms' arguing that in the 'living being's organism, a harmonic set of phenomena (balanced interrelations) must be taken into account'. Actually, Bernard made clear the difference between what happens 'inside' and what is now called the 'environment'.

1862: *Herbert Spencer* (an English philosopher) introduced the general idea of increasing complexity in systems.

1866: The brothers de Cyon (France) discovered the first example of biological regulator and formulated it (for the first time) in an implicit systemic context.

1892–99: *Poincare, in his work 'Analysis Situs',* introduced a new type of mathematical study which was the start of 'topology'. His results are one of the very first steps towards the establishment of a new type of qualitative mathematics suitable for the study of complex systems.

1931: *Godel* formulated his 'incompleteness theorem' which states that any formal (mathematical) system contains statements that cannot be proved within that formal system. For systems, this theorem implies that systemic models can be designed and used, but never offer an absolute value of truth. Godel's theorem can be seen as the foundation stone for van Gigh's concept of 'meta system' leading to the concept of 'ontological skepticism' [1].

1932: *Cannon* established the concept of 'biological homoestasis' (the biological mechanism of maintaining internal stability in living organisms), which is an important extension of Bernard's stability concept of the *internal millieu.* This was actually the birth (20 years later) of biological cybernetics.

3.4.2 Pioneer's Period

1950: *Von Bentalanfly* published the foundation of his main contribution on 'General Systems Theory' (GST) [17]. He clearly formulated the central concept of systems, which played the catalyst of the systems view. He argued that there exist 'isomorphic laws in science' giving convincing examples. He presented GST 'as an important regulative device in science' which should lead to the '*unity of science*'. He regarded the unity of science as reaction to the reductionist mechanistic worldview, and rejected the crude additive machine theories of organic behavior, which treated wholes as nothing more than linear aggregates of their components. However, he only mentioned *emergence* and hierarchy in passing, adopting almost unchanged the theory of 'biological emergentism' developed in the 1920s. The concept of 'hierarchy' within GST was introduced and studied in later years by other researchers (e.g., H.A. Simon [19]).

1950: *Jay Forrester* created the field of 'system dynamics' (SD), as a result of an MIT project on electrical feedback control systems. The results of this project were applied to develop an aircraft flight simulator, and an education system simulator. System dynamics has been used to study urban dynamics, economic systems dynamics, industrial system dynamics, and world dynamics. System dynamics includes, among others, discrete-event simulators (DES), agent-based modeling (ABM), complex adaptive systems (CAS) modeling, and multi-actor system (MAS) modeling. Forrester, with his system dynamics theory put the foundations of systems thinking which will be discussed in the next chapter.

1950: *Von Neumann* developed automata theory. Automata are in the boundary of collections of unorganized elements and truly complex systems, thus providing the bridge to one of the greater conceptual gaps in systems theory. Automata have been used to model the brain and self-critical systems, and are continually receiving increasing attention.

1956: *Boulding* made particular influential extensions to the view that the hierarchical nature of systems is a logical consequence of the way a system is defined in terms of its constituent parts, since the parts may also meet the definition of systems. Boulding in his 'Skeleton of Science' book [14] provided a hierarchical view of systems, which is analogous to the complexity of the 'people' working in the various empirical fields. He presented a generalized hierarchy with nine levels from physics and chemistry through biology, psychology, sociology, and metaphysics. These nine levels are as follows [20, 21]. The parentheses include in sequence the characteristics of each level, examples, and the respective discipline.

- *Level 1: Structure* (static; crystals; any discipline).
- *Level 2: Clock works* (predetermined motion; machines, solar system, etc. physics/chemistry).
- *Level 3: Control* (feedback control, thermostats, mechanisms in living organisms, etc.; control system/cybernetics).
- *Level 4: Open system* (self-maintaining; biological cells, flames, etc.; biology/metabolism, information theory).
- *Level 5: Lower organisms* (organized whole with reproduction, etc.; plants; botany).
- *Level 6: Animals* (brain/ability to learn; birds *and beasts,* zoology).
- *Level 7: Humans* (self-consciousness, knowledge, etc.; humans; psychology/human biology).
- *Level 8: Socio-cultural systems* (roles, communication, values, etc.; families, societies, organizations, nations; sociology/anthropology).
- *Level 9: Transcendental systems* (inescapable unknowable's; god; metaphysics/theology).

According to Boulding 'general systems theory (and systems science in general) aims to provide a framework or structure on which to hang the flesh and the blood of particular disciplines and particular subject matters in an orderly and coherent corpus of knowledge'.

1960: *Ackoff* [21] pointed out that physical or conceptual systems could be studied either a 'systems' or as 'non-systems'. A system is treated as system, if the analyst represents the behavior of the system as a result of the interaction of its parts. Otherwise a system is treated as non-system, i.e., as a simple object. According to Ackoff, systems are still real, but their analysis and corresponding representation can be systemic or not. About this, Shchedrovitsky, [22], posed in 1975 the question: 'Since we can treat and represent 'systems' as systems, why not assume that reality is neither systemic nor non-systemic, and only our methods of treating real objects and corresponding representations of them are either systemic or not?'

3.4.3 Innovators' Period

1962: *Simon* [19] coined a nested hierarchy, which was used later within the highly abstract generalized systems theory [23]. He used a simple probability argument to support the ubiquity of hierarchies in natural complex systems. This argument says that the time it takes for the evolution of complex form depends critically on the number of potential intermediate stable forms, which actually insulate the process of system assembly against the effects of environmental influences. Because a hierarchy of building blocks can be assembled orders of magnitude faster than a non-hierarchical assembly process, among complex forms, hierarchies are the ones that have the time to evolve. In other words, Simon revealed the criticality as a feature of quasi-systems, making clear that structured organizations in hierarchical levels is a cardinal feature of systems.

1963: *Maryhama* [24] introduced the concept of 'second cybernetics', i.e., the deviation amplifying mutual causal processes' to describe the role of positive feedback, particularly in the structuring of growing and competing/antagonistic systems. Thus he highlighted the presence of antagonism between growth and limiting factors, something which was in line with Lotka and Volterra findings in 1920.

1965: *Miller* started publishing his research results on behavioral systems (i.e., living systems) and provided a descriptive systems' taxonomy covering the whole spectrum of systems from the *cell* to the *man-planet* system (but not including physicochemical and ecological systems). He developed a method for discovering cross-level 'isomorphies' thus making systems theory an important, significant and workable research field. Many other taxonomies were later provided, but it is recognized that none of them is as satisfactory or structured (horizontally and vertically) as Miller's taxonomy, nor by far, as widely embracing (see [25]).

1972: *Laszlo* [26] presented the philosophical aspects of GST, advocating the 'view of thing as a whole', and seeing the world as an 'interconnected interdependent field continuous with itself'. His synthetic approach was aiming at providing a powerful *antidote* to the intellectual fragmentation implied by compartmentalized research and systems analysis based on 'parts' (system pieces or components). He used a version of system theory, to model the natural world, involving two kinds of interacting hierarchies, namely: *micro-hierarchies* and *macro-hierarchies.* His efforts on emphasizing knowledge and goals, brought to the fore aspects of system theory that are alleged to the unite Aristotelian insights with modern theories of complexity.

1983: *Durkheim* in his doctoral thesis and later in his book 'The Division of Labor in Society' (1984) [27] explained that in the highly organized systems, the division of labor contributes to the maintenance of societies. In complex societies, workers perform various roles that while they lead to specialization and segmentation, also create a high degree of mutual interdependence between units. He argued that although the individuals performing these tasks and roles will change, these roles persist over time and maintain a society in the macro sense. Durkheim believed that human beings experience a unique social reality not experienced by other organisms. Therefore he argued that order can only be maintained via the consent of individuals within the group who share the same morals and values.

1990: *Senge* [28] extended Forrester's work and studied thoroughly systems thinking in what he calls 'learning organizations'. In his analysis of corporations as learning organizations, Senge provides several 'archetypes', namely: (i) balancing process with delay, (ii) limits to growth, (iii) shifting the burden, (iv) shifting the burden to the intervener, (v) eroding goals, (vi) escalation, (vii) success to the successful, and (viii) tragedy of the commons. These archetypes will be discussed in the section on system thinking. According to Senge, system thinking can allow the individual, group or organization to have an overview of the structure and the dynamics of the local system and adjust the behavior accordingly.

1994: *Gell-Man* [29] provides an alternative way to consider unity of the sciences. In place of Bertanlaffy's '*isomorphism*' he introduced the concept of '*staircases*' *(or bridges)* between the levels of science.

1997: *Bar-Yam* [30] suggests and studies, universal properties of complex systems independent of composition. Self-organization or as otherwise called '*antichaos*' by Kauffman (1991) due to its stabilizing dynamics is a substitute for '*equifinality*'.

1997: *Mario Bunge* [32] introduced in the social sciences both the *systems approach* and the *mechanism-based explanation:* He has previously worked on the prickly philosophical issue in the social sciences over the primacy of *individualism or holism* as a philosophical constant [33]. He had the belief that '*systemism*'is the alternative to both individualism and holism. He pointed out that there is a strong distinction to be made between system and mechanism where he states that mechanism is a process in a system. He argued that 'systemism' is an approach, a systemic approach, used to explain complex things of any kind, and also that all mechanisms are system-specific, i.e., there is no such thing as a universal or substrate-neutral mechanism [34].

2004: *Mario Bunge* [34] explains the distinction between the term 'systemic approach' and 'system theory' and uses the former term for two reasons. 'The first is that there are nearly as many systems theories as systems theorists. The second is that 'systems theory' that became popular in the 1970s (e.g., [26]) was another name for old holism and got discredited because it stressed stasis at the expense of change and claimed to solve all particular problems without empirical research or serious theorizing. He had the opinion that the constructs proposed by Bertalanfly and Laszlo were inadequate in providing a foundation for systems theory.

2012: *Kevin MacG. Adams* [11] developed an *axiom set* arguing that it is a sufficient construct for systems theory. He showed the multidisciplinary nature of the propositions, derived from the fields of science that support the axioms of systems theory. This set of axioms may remove the ambiguity raised when the term 'systems theory' is used in a variety of disciplines having multiple meanings. He also gave a historical classification of systems theory based on the literature.

3.5 Systems Taxonomy

In general, *taxonomy or classification* is the process of grouping things on the basis of their similarities. The term taxonomy, comes from the Greek word '*ταξινομία*' where '*τάξις*' (taxis)=arrangement, and '*νομία/νόμος*' (nomia/nomos)=law/method. Actually, for systems there are several taxonomies according to general criteria. Klir [35] argued that 'no system classification is complete and perfect for all purposes'.

According to Ralph Stacey [36] there are four general system typologies or ontologies in the system complexity space, namely (Figure 3.9(a, b)):

- *Simple or known systems* where there is one or more clear 'best practices' that always apply for their treatment and use (e.g., in disinfecting and monitoring water reservoir).
- *Complicated or knowable systems* in which cause and effect can be perceived only analyzing the system mechanisms. Such systems are, for example, encountered in business and local ecosystems. For complicated systems there exist only good practices.
- *Complex or unknowable systems* in which cause and effect relationships cannot be perceived in advance but only in retrospect. Complex systems involve only emergent practices.
- *Chaotic systems (or simply chaos)* in which no cause and effect is knowable or there just many perhaps an infinite number of interrelated variables. For chaotic systems we need to develop novel practices.

In Figure 3.9(a) the horizontal axis refers to the *degree of certainty* on how the system works, and the vertical axis to the *level of agreement* on the system behavior (e.g., in group of people or project team, in business, and other negotiations). Figure 3.9(b) explains in an alternative way the four system typologies. Practically, the Stacey matrix provides a tool for selecting appropriate management functions in complex adaptive systems based on the degree of certainty and the level of agreement on the same issue in question.

Here we will provide a multiple systems taxonomy on the basis of several more detailed criteria (dimensions).

This multidimensional taxonomy is as follows (Figure 3.10).

Dimension 1 [16] *(Natural vs. social and technological systems)*

1. *Natural systems* (objects or concepts existing outside human control, e.g., real numbers system, solar system, planetary atmosphere circulation system, etc.).

2. *Social systems* (abstract human types or social entities/groups or concrete individuals).

3. *Technological systems* (man-made systems including computer hardware/software, manufacturing systems, robotic system, etc.).

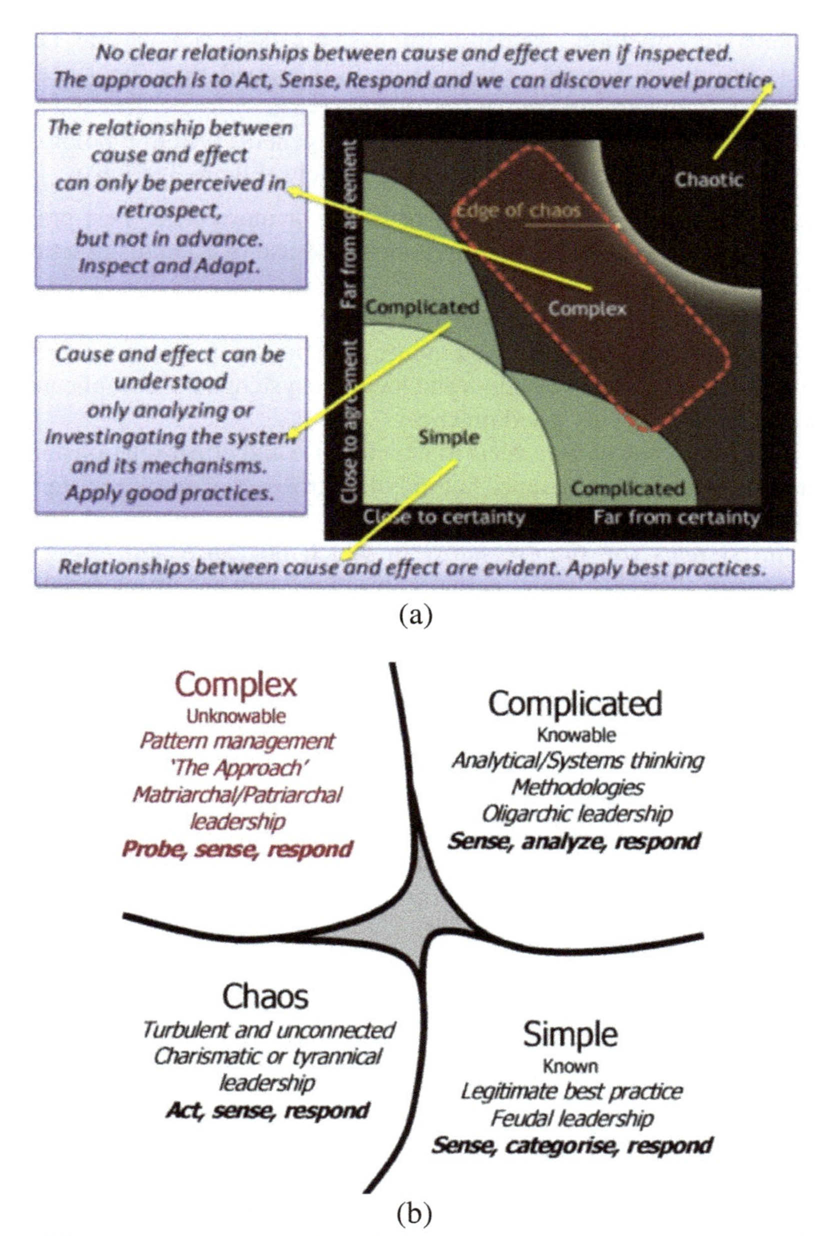

Figure 3.9 Two representations of the 'Stacey Matrix' describing the system typologies in the system complexity space.

Source: http://emilianosoldipmp.info/tag/stacey-matrix www.slideplayer.com (complexity theory in health care).

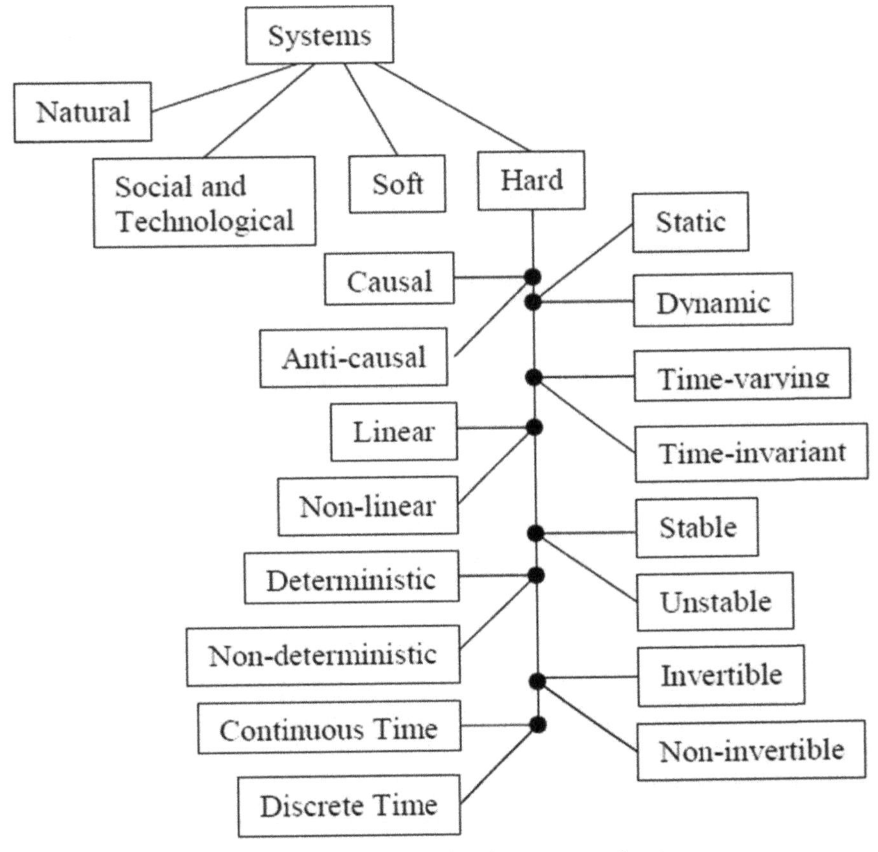

Figure 3.10 Multi-dimensional taxonomy of systems.

It is remarked that actually there are not hard and fast boundaries between these classes of systems (e.g., natural systems may involve social systems or human systems, and typically depend also on technical aspects to fully realize their goals). As already seen, systems that contain technical and either human or natural subsystems are called '*socio-technical-systems*'.

Dimension 2 [37, 38] (Hard, soft, mixed, and ethical systems)

1. *Hard systems* These systems can be easily defined and have clear-cut aims and purposes. They are typically designed and implemented by engineers concerned with real-world problem-solving, machines, process plants, power plants, aircrafts, robots, etc.).

2. *Soft systems* These systems have as their principal elements human beings. They cannot be easily defined and have not clear-cut and agreed purposes or goals. Usually, soft systems involve real-world problems that cannot be formulated as a search for an efficient means of achieving a defined end. At the individual human level there are multiple processes of perception, interpretation, representation and communication that push and pull at our individual and collective behavior. At the multi-person organization level there are often different and conflicting goals that are attempted to be achieved simultaneously. In both individual and multi-person situations, the images and goals of the system, even if agreed upon, may change or modified over time [6].
3. *Hard/soft (mixed) systems* This class of systems includes decision-making processes such as those used in operations research management, economics, etc.
4. *Ethical systems* These systems involve the moral values which are used in the creation and use of systems of any kind.

The taxonomy dimensions that follow are typically applicable to hard systems.

Dimension 3 *(Causal vs. anti-causal systems)*

1. *Causal systems* A system is called causal if its output (response) depends on present and past inputs only and not on future inputs. Since causal systems do not include future input samples, their model can practically be implemented (realized). The condition for causality is: $y(t') = f(u(t);\ t \leq t')$: f is a linear or nonlinear mapping which says that if the input $u(t)$ is applied at $t = t'$, then the output $y(t)$ at $t = t'$ will depend only on the values of $u(t)$ for $t \leq t'$, and the initial value y(0) of y(t).
2. *Anti-causal systems* A system is called anti-causal if its present output depends on future values of the input. Since anti-causal systems involve future input values, they are practically not realizable.

Dimension 4 *(Static vs. dynamic systems)*

1. *Static systems* The output of a static system at any instant of time depends on the input value at the same time, e.g., $y(t) = Au(t)$ or $y(t) = Au^2(t) + Bu(t) + C$ where A, B, and C are constants. Static systems are memoryless, since they don't need to store and use delayed (past) or advanced (future) input values. Static systems are described by algebraic equations (models).

2. *Dynamic systems* The output of a dynamic system at any instant of time depends on the input value at that time as well as at other past times, e.g.: $y(t) = u(t) + 6u(t-1)$ or $y(t) = 4u(t) + 5u(t-2)$. Dynamic systems have memory since their output depends on past input values, and so they need a memory for storing these values. Dynamic systems involve always differential or difference equations in their model (and possibly algebraic ones).

Dimension 5 *(Time invariant vs. time variant systems)*

1. *Time invariant systems* These systems have input/output characteristics that do not change with time. They are also called *shift invariant* because their characteristics do not change with time shifting. Specifically, suppose that an input $u(t)$ applied to a system produces the output $y(t)$, and a shifted input $u(t-\tau)$ produces the output $y(t,\tau)$. If $y(t,\tau) = y(t-\tau)$ then the system is time invariant.
2. *Time variant systems* In these systems the input/output characteristics change with time and with time shifting. In the above system, if $y(t,\tau) \neq y(t-\tau)$ then the system is time varying. For example, if $y(t,\tau) = u(t) + tu(t-\tau)$ then $y(t-\tau) = u(t-\tau) + (t-\tau)u(t-2\tau)$, i.e., $y(t,\tau) \neq y(t-\tau)$. Therefore, the system is time varying (because it involves the time varying coefficient t in $u(t\tau)$.

Dimension 6 *(Linear vs. nonlinear systems)*

1. *Linear systems* These systems obey the 'superposition principle', i.e., if the input $u_1(t)$ produces the output $y_1(t)$ and the input $u_2(t)$ applied to the same system produces the output $y_2(t)$, then the input $au_1(t) + bu_2(t)$ produces the output $ay_1(t) + by_2(t)$. In linear systems zero inputs produces zero outputs.
2. *Nonlinear systems* These systems do not obey the superposition principle, i.e., if $y_1(t) = f(u_1(t))$ and $y_2(t) = f(u_2(t))$ and $f(au_1(t) + bu_2(t)) = ay_1(t) + by_2(t)$ the system is linear, otherwise if $f(au_1(t) + bu_2(t)) \neq ay_1(t) + by_2(t)$ the system is nonlinear.

Dimension 7 *(Stable vs unstable systems)*

1. *Stable systems* A system is bounded input/bounded output (BIBO) stable if and only if any bounded input produces a bounded (non-infinite) output, i.e., an input $u(t)$ with $|u(t)| \leq a < \infty$ produces an output $y\ (t)$ with $|y(t)| \leq b < \infty$, where a and b are given finite valued constants.

2. *Unstable systems* In this case bounded inputs produce unbounded (infinite) outputs. Unstable systems show erroneous and extreme performance, and when they are realized cause overflow.

Dimension 8 *(Deterministic vs nondeterministic systems)*

1. *Deterministic systems* These are systems in which the output can be exactly predicted, i.e., with 100 percent certainty. This means that an initial state of a deterministic system completely determines the future states of the system. If a deterministic system is given some initial inputs, the system will produce the same states every time, i.e., no randomness or other uncertainties are involved in the production of future states of the system. Examples of deterministic systems are the physical laws, and the quantum mechanics systems, where the Schrodinger equation describes the continuous time evolution of the system's wave function. Also the systems of chaos theory are deterministic, despite the fact that they appear to be nondeterministic.
2. *Non deterministic systems* These are systems in which the output cannot be predicted because there are multiple possible outcomes to each input. Non-deterministic systems involve some kind of randomness which determines the future states. If a nondeterministic system is given some initial inputs, the system will produce a different state for each run. Non-deterministic systems include probabilistic systems, stochastic systems, and fuzzy systems.

Dimension 9 *(Invertible vs. non-invertible systems)*

1. *Invertible systems* These are systems where inputs lead to distinct outputs, i.e., systems for which there exist inverse transformations (one-to-one mappings). Let S be the input-output transformation (mapping) of a system. The system is invertible if there exists the inverse transformation S^{-1} that satisfies the relation $S \cdot S^{-1} = S^{-1}S = I$, where I is the unit transformation ($y = Ix = x$). Thus in an inverse system we have $y = Sx,\ y = S \cdot S^{-1}y = y$.
2. *Non-invertible systems* In these systems distinct inputs lead to the same output, and so an invertible (one-to-one) mapping does not exist.

Dimension 10 *(Continuous-time vs. discrete time systems)*

1. *Continuous time systems* In these systems the inputs and outputs are continuous ($u(t)\ and\ y(t),\ t = real\ number$). In state space a dynamic continuous time system is described by a model of the type:

$$\dot{x}(t) = \mathbf{f}(\mathbf{x}, \mathrm{t}, \mathbf{u}),\ \mathbf{y}(\mathrm{t}) = \mathbf{g}(\mathbf{x}, \mathrm{t}, \mathbf{u})$$

where $\mathbf{x}$ is the n-dimensional state vector, $\mathbf{u}$ the m dimensional input vector, $\mathbf{y}(t)$ the p-dimensional output vector, and $\mathbf{f}(.,.,.)$, $\mathbf{g}(.,.,.)$ the input state, and input-output transformations.

2. *Discrete-time systems* In these systems the inputs and outputs are discrete time signals $\mathbf{u}(k) = \mathbf{u}(kT), \mathbf{y}(k) = \mathbf{y}(kT)$, $k = 1, 2, 3, \ldots$, where T is a given sampling period. The model of a dynamic discrete-time system is:

$$\mathbf{x}(k+1) = \mathbf{F}(\mathrm{x}(k),\ k,\ \mathbf{u}(k)),\ \mathbf{y}(k) = \mathbf{G}(\mathbf{x}(k), k, \mathbf{u}(k)$$

3.6 A Comprehensive List of System Categories

Here, we give a more detailed (alphabetic) list of modern systems that fall within the systems science (or general system theory) framework. [39–41].

- Abstract (formal) systems; Adaptive systems; Applied multi-dimensional systems, Automata.
- Behavioral systems; Bioengineering systems.
- Chaotic systems; Complex systems; Computer-aided systems; Control systems; Critical systems; Cultural systems.
- Distributed-parameter systems; Dynamic systems; Discrete-event systems;
- Ecological systems; Economic systems; Electric power systems.
- Family systems, feedback systems, Fuzzy systems.
- General systems.
- Human systems.
- Infinite dimensional systems.
- Large-scale systems; Liberating systems; Linear systems; Living systems.
- Mathematical systems; Modeling systems; Modern control systems; Multidimensional systems.
- Nonlinear systems, nonlinear stochastic systems.
- Open systems.
- Physical systems; Probabilistic systems.
- Random systems.
- Social systems; Sociotechnical systems; Social rule systems; Statistical systems; Systemics.
- Time-delay systems.
- Word systems theory.

Representative references for all these system types can be found in [41].

Here we will have a brief look on the '*world systems theory*' which differs from the other system classes in that it is actually a *sociopolitical economic theory*. This theory was initiated by Immanuel Wallerstein and lies somewhere between the theories of Marx and Engels.

Actually, world systems theory (also known as 'world systems analysis') is a multidisciplinary macro scale approach to world history and social change which gives emphasis on the world system (and not nation states) as a primary, but not exclusive, unit of analysis. For Wallerstein '*a system is a unit with a single division of labor and multiple cultural systems*', and in the world systems theory there have been three kinds of societies, namely *mini-systems* and two types of word systems, viz. *single-state world-empires*, and *multiple-polity world economies*. Wallerstein distinguishes three classes of countries, viz. core countries, semi-periphery countries, and periphery countries, as shown in Figure 3.11. The systematic flow of surplus from

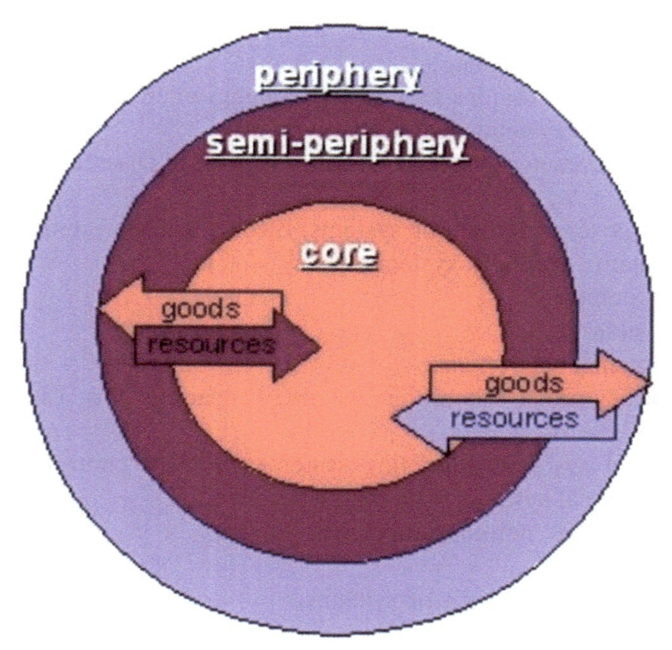

Figure 3.11 Pictorial illustration of Wallerstein's World Systems Theory. Periphery and semi-periphery countries provide to core countries resources (labor, raw materials, etc.), while core countries provide finished high-valued goods and equipment.

Source: www.faculty.rsu.edu/users/f/felwell/www/Theorists/Essays/Wallerstein.htm

the periphery or semi-periphery to the industrialized high technology core countries is what Wallerstein calls 'unequal exchange' that leads to capital accumulation at a global scale. Core countries focus on higher skill, capital-intensive production of goods and technological products, whereas the rest of the world concentrates on low-skill, labor-intensive production and extraction of raw materials. This persistently reinforces the dominance of the core countries. Of course, the system has dynamic features due, for example, to transportation progress. Therefore individual states can lose or gain their core or semi-periphery or periphery status over time. The world-system theory has inspired a large number of research programs, the most well-known of which concerns the study of 'long-term business cycles'. As an interdisciplinary theory, the world systems theory has attracted the attention of sociologists, anthropologists, cultural scientists, economists, historians, and other investigators. World systems theory has strongly questioned the theory of 'globalization' and 'free market'.

3.7 Conclusions

Systems research naturally crosses all fields of classical science and makes the existing boundaries of them (often established on the basis of properties of thin good) completely irrelevant and superficial. It is less concerned to rhetorically unify science as development of quantitative models that have an interdisciplinary application, and has richer experience with real-world systems applications. The approaches that are based on exact mathematical model can be broadly characterized as hard approaches and cannot successfully treat biological, managerial, and social problems. Therefore, the division of systems approaches in hard approaches and soft methods allows a deepest consideration and analysis of natural and societal process.

Soft system methodologies provide a pluralistic view, where a system model is taken to say more about the modelers and their assumptions than the system of interest itself. Proponents of soft system approaches maintain that hard approaches to social systems are problematic and can be dangerous, because they do not take into account the special nature of human self-consciousness and free will. Due to the strong philosophical differences between hard and soft systems research communities, there exists little constructive interaction between them. The boundaries of soft social systems can be partially determined by social norms, values, and customs, and not by mathematical features as in hard systems. A system grows through an exchange of *tangible and intangible* energy between the system and its boundary, a process which can take place only if the boundary possesses '*permeability*'. Tangible energy (resources) include food, money and other things, whereas intangible resources include information as exemplified when a member of the system is educated or has important knowledge for the system.

This chapter has discussed fundamental ontological issues of systems, namely ontological elements, historical landmarks, and a multi-dimensional taxonomy of natural, social, technological, hard, and soft systems. For completeness an extended list of modern system types was also included.

References

[1] Francois, C. (1999). Systemics and cybernetics in a historical perspective. *Syst. Res.* 16, 203–219.

[2] Maturama, H., and Varela, F. (1980). *Autopoiesis and Cognition*. Boston, MA: D. Reidel.

[3] von Bertalanfly, L. (1968). *General Systems Theory: Foundations, Development, Applications*. New York, NY: Braziller.

[4] Miles, R. F. (ed.). (1973). System Concepts, New York, NY: Wiley and Sons Inc.

[5] Dori, D. (2002). *Object-Process-Methodology: A Holistic System Paradigm*. Berlin: Springer.

[6] Laszlo, A., and Kripper, S. (1998). "Systems theories: their origins, foundations, and development," in *Systems Theories and a-Priori Aspects of Perception*, ed. J. S. Jordan (Amsterdam: Elsevier), 47–78.

[7] Blanchard, B., and Fabrucky, F. (1998). *Systems Engineering Analysis*. Upper Saddle River, NJ: Prentice Hall.

[8] INCOSE (2012). *Systems Engineering Handbook: A Guide for System Life Cycle Processes and Activities*. San Diego, CA: International Council on Systems Engineering.

[9] Hitchins, D. (2009). What are the general principles applicable to systems? *INCOSE Insight* 12, 59–63.

[10] Ackoff, R. L. (1971). Towards a system of systems concepts. *Manag. Sci.* 27, 661–671.

[11] MacG, K. (2012). Adams, Systems theory: a formal construct for understanding systems. *Int. J. Syst. Syst. Eng.* 3, 209–224.

[12] Hjørland, B. and Nicolaisen, J. (eds). (2005). "Epistemology and Philosophy of Science for Information Scientists", in *Epistemological*

Lifeboat, Royal School of Library and Information Science (Copenhagen: Royal School of Library and Information Science).

[13] Weinberg, G. M. (ed.). (1981). *An Introduction to General Systems Thinking Silver Anniversary*. New York, NY: John Wiley & Sons.

[14] Boulding, K. E. (1956). General systems theory: the skeleton of science. *Manag. Sci.* 2, 197–208.

[15] Ackoff, R. L. (1981). *Creating the Corporate Future*. New York, NY: John Wiley & Sons.

[16] SEBOK (2016). *What is a System*? Available at: http://sebokwiki/org/w/index.php?title=What_is_a_System%3F&oldid=52182 (accessed March 25, 2016).

[17] von Bertanlaffy, L. (1950). An outline of general systems theory. *Br. J. Philos. Sci.* 1, 134–165.

[18] Matthews, D. (2004). *Rethinking Systems Thinking: Towards a Postmodern Understanding of the Nature of Systemic Inquiry*. Ph.D. thesis, University of Adelaide, Adelaide, SA.

[19] Simon, H. A. (1962). The architecture of complexity. *Proc. Am. Philos. Soc.* 106, 467–482.

[20] Ryan, A. (2008). *What is a Systems Approach*? Ithaca, NY: Cornell University Library.

[21] Ackoff, R. L. (1960). *Systems, Organizations and Interdisciplinary Research. Gen. Syst.* 5, 1–8.

[22] Shchedrovitsky, G. P. (1996). Methodological problems of systems research. *Gen. Syst.* 11.

[23] Allen, T. F. (1996). *A Summary of the Principles of Hierarchy Theory*. Available at: http://www.isss.org/hierarchy.htm

[24] Maruyama, M. (1963). The second cybernetics: deviation-amplifying mutual causal processes. *Am. Sci.* 89, 164–179.

[25] Miller, J. G. (1978). *Living Systems*. New York, NY: McGraw-Hill.

[26] Laszlo, E. (1972). *Introduction to Systems Theory: Toward a New Paradigm of Contemporary Thought*. New York, NY: Torchbooks.

[27] Durkheim, E. (1984). *The Division of Labor in Society*. New York, NY: MacMillan.

[28] Senge, P. (1990). *The Fifth Discipline: The Art and Practice of the Learning Organization*. New York, NY: Doubleday Currency.

[29] Gell-Man, M. (1994). *The Quark of the Jaguar: Adventures in the Simple and the Complex*. New York, NY: Henry Holt and Company.

[30] Bar-Yam, Y. (1997). *Dynamics of Complex Systems*. Boulder, CO: Westview Press.

[31] Bunge, M. (2004). How does it work? The search for explanatory mechanisms. *Philos. Soc. Sci.* 34, 182–210.
[32] Bunge, M. (1997). Mechanism and explanation. *Philos. Soc. Sci.* 27, 410–460.
[33] Bunge, A. (1979). Systems concept of society: beyond individualism and decision. *Theory Decis.* 10, 13–30.
[34] Bunge, M. (1999). *The Sociology-Philosophy Connection.* New Brunswick, NJ: Transaction Publishers.
[35] Klir, G. J. (1969). *An approach to General Systems Theory.* Princeton, NJ: Van Nostrand.
[36] Zimmerman, B. (2001). *Stacey's Agreement and Certainty Matrix.* North York, ON: Schulich School of Business.
[37] Checkland, P. (1981). *Systems Thinking, Systems Practice.* New York, NY: John Wiley.
[38] Checkland, P. and Scholes, J. (1990). *Soft Systems Methodology in Action.* New York, NY: John Wiley.
[39] Mesarovic, M.D. (1989). *Abstract Systems Theory.* Berlin: Springer.
[40] Mulej, M. (2006). Systems, cybernetics and innovations. *Kybernetes*, 35, 939–940.
[41] Wikipedia *List of Types of Systems Theory.*
[42] Wallersten, I. M. (2004). *World System Analysis: An Introduction.* Durham, NC: Duke University Press.
[43] Wallerstein, I. M. (1992). The West capitalism and the modern world system. *Review* 15, 561–619.
[44] Moore, J. N. (2003). The modern World System as an environmental history? Ecology and the rise of capitalism. *Theory Soc.* 32, 307–377.

4

General Systems Theory and System Dynamics

If someone were to analyze current notions and fashionable catchwords, they would find 'systems' high on the list. The concept has prevaded all fields of science and penetrated into popular thinking, jargon and mass media.
Ludwig von Bertalanffy

All science is concerned with the relationship of cause and effect. Each scientific discovery increases man's ability to predict the consequences of his actions and thus the ability to control future events.
Lawrence Peter

Systems theory is an attempt to provide a general explanation for social behavior; certain types of this theory are referred to as grand theory.
Michael Pezart Jackson

4.1 Introduction

The originator of *General Systems Theory* (**GST**) is *Ludwig von Bertalanffy* (1901–1972), and the originator of *System Dynamics* (**SD**) is *Jay Forrester (*1918–2016). The two fields started their development in the 1950s and were advanced in parallel. The systems theory field also includes the *systems analysis* (**SA**) and *systems engineering* (**SE**) areas. As we will see in the chapter, GST is the science (or methodology) of *open systems,* and is linked to the science of *'wholeness'*. Von Bertalanffy, a biologist, advocated the view that an organism is a whole or a system, taking into account the fact that 'the whole is more than the sum of its parts' as first stated by Aristotle. Von Bertalanffy was influenced by a number of sociologists including *E. Durkheim* and *Max Weber* who were early pioneers in the field of sociology. *System dynamics* is a branch of systems theory which is concerned with the study of the dynamic behavior of complex systems. It is a computer-aided approach to policy analysis and development which uses computers to simulate real life systems such as complex decision-making managerial, social, economic, urban, or ecological systems. *Systems analysis* was developed within the framework of operations research which was concerned with the study and analysis of several practical models such as stock/inventory, queuing, decision making, Markov chains, and network optimization models. *Systems engineering* is an engineering discipline aiming at creating and implementing a multidisciplinary process to guarantee that the customer and stakeholder's needs are satisfied in a high quality, cost efficient, and schedule compliant manner during the entire system's life.

According to INCOSE (International Council on Systems Engineering), SE integrates all the disciplines and speciality groups, and considers both the business and the technical aspects of all customers with the goal of providing a high quality product that meets the user needs.

The purpose of the present chapter is to provide the fundamental ontological and epistemological (methodological) issues of GST and SD. Especially, the chapter addresses the ontological questions:

- What is the general systems theory?
- What is isomorphism in GST?
- What is system dynamics?
- What are the systems thinking?

Regarding epistemology the following epistemological/methodological aspects about GST and SD are discussed at a conceptual level without mathematical details:

- Principal propositions (principles) of GST.
- Bertalanffy's view of systems and GST.
- Forrester's causal loop diagrams with particular examples.
- A representative scheme for performing system dynamics modeling and simulation.
- The systems thinking continuum.
- Systems thinking performing rules.
- Axioms of general systems theory.

4.2 What is the General Systems Theory?

According to *Von Bertalanffy, 'general systems theory'* **(GST),** is a general science of *'wholeness',* which in elaborate form would be a logical-mathematical discipline, in itself purely formal but applicable to the various empirical sciences' [1]. In other words, GST is the transdisciplinary study of the abstract properties and organization of natural phenomena independent of their substance and kind or temporal and spatial scale of existence. It examines both the principles that govern all complex entities, and the models (usually mathematical) that can be employed to describe them.

Von Bertalanffy emphasized the fact that real systems are *open* to and *interact* with their environments, and that they can elicit qualitatively new properties through *emergence* which results in a continual evolution. He argued that 'it is necessary to study not only parts and processes in isolation, but also to solve the decisive problems found in organization, and order unifying them. These problems result from dynamic interaction of parts, and make the behavior of the parts different when studied in isolation or within the whole'. According to *Rudolf Stichweh,* 'System theory is a science which has the comparative study of systems as its object'.

Von Bertalanffy explained that the main propositions (principles) of general systems theory include:

- *The open systems principle* (Systems require the flow of matter, energy, and information between the system and its environment).
- *The isomorphism principle* (There are isomorphisms between the mathematical structures in different scientific disciplines that could integrate and unify the sciences).
- *The equifinality principle* (In open systems the same final state may be reached from different initial conditions and via different paths).
- *The teleology principle* (Teleological performance directed towards a final system state or goal is a legitimate phenomenon for systemic scientific inquiry).
- *The scientific theory principle* (A scientific theory of organization is needed, to include the issues of wholeness, growth, differentiation, hierarchical order, competition, dominance, and control).

Von Bertalanffy's goal was to provide an alternative foundation for unifying science, which contrasts the reductionist mechanistic worldview.

Other fundamental principles of GST are:

- Systems are an idealization.
- The whole is greater than the sum of its parts.
- A change in any part of the system affects all parts.
- All parts of the system are interconnected.
- The systems must be understood as a whole.
- All systems are made up of subsystems.
- Though each subsystem is a self-contained unit, it is part of a wider and higher order.

- All systems have boundaries.
- All systems seek stability and balance.
- Systems are enduring.
- Systems and their environments affect each other via feedback (input and output).
- Systems are organized and exhibit emergent properties.
- Systems offer a way of looking at the phenomena, but they are not real objects.
- The portion of the world studied (system) must exhibit some predictability.
- Systems have non trivial behavior.
- The boundary of a system can be redrawn at will by the system expert.
- A system includes a set of goals and their relationships.
- To be viable, a system must be strongly goal-directed, ruled by feedback, and capable to adapt to changing situations.
- The function of any system is to convert or process materials, energy, or information for use within the system or its environment or both. In fact, for survival a system must save some of the outcome or product to maintain the system.

Non-systemic approaches to studying natural phenomena are based on the following basic assumptions:

- The system is closed.
- Superpositionality is valid.
- Local structure is smooth.
- Different levels are independent.
- Causation is linear.
- Space can be ignored.
- Control is centralized.
- Averaging over time and across individuals can be applied.

The assumption of *closed system* guarantees that one can design reproducible experiments with predictable results from the same initial conditions. In closed systems the entropy is always non-decreasing, and equilibrium states dominate.

The *superposition* assumption implies that the whole is the sum of the parts. The interactions within the system can be studied by calculating the sum of pairwise interactions between parts. This simplifies considerably the modeling process.

The *centralized control* assumption enables the improvement of modeling efficiency, and linear, sequential, and synchronous control.

An open system requires more information than the initial conditions of the system to predict the future behavior, for example, what kind of feedback (positive, negative, or mixed) the system has.

Bertalanffy provided a full study of GST. For the convenience of the reader we give here some central excerpts from his book [1].

- "Modern science is characterized by its ever-increasing specialization, necessitated by the enormous amount of data, the complexity of techniques and of theoretical structures within every field. Thus science is split into innumerable disciplines continually generating new sub disciplines. In consequence, the physicist, the biologist, the psychologist and the social scientist are, so to speak, encapsulated in their private universes, and it is difficult to get word from one cocoon to the other [. . .].
- Entities of an essentially new sort are entering the sphere of scientific thought. Classical science in its diverse disciplines, be it chemistry, biology, psychology, or the social sciences, tried to isolated the elements of the observed universe – chemical compounds and enzymes, cells, elementary sensations, freely competing individuals, what not – expecting that, by putting them together again, conceptually or experimentally, the whole or system-cell, mind, society-would result and be intelligible. Now we have learned that for an understanding not only the elements but their interrelations as well are required [. . .].
- In our considerations we started with general definition of 'systems' defined as a set of elements standing in interrelations. No special hypothesis or statement was made about the nature of the system, of its elements or the relations between them. Nevertheless from this purely formal definition of 'system' many properties follow which in part are expressed in laws well-known in various fields of science, and in part concern concepts previously regarded as anthropomorphic, realistic or metaphysical. The parallelism of general conceptions or even special laws in different fields, therefore, is a consequence of the fact that those are concerned with 'systems' and that certain general principles apply to systems irrespective of their nature.
- There appear to exist general system laws which apply to any system of a particular type, irrespective of the particular properties of the systems and the elements involved. Compared to the analytical procedure of

classical science with resolution into component elements and one-way of linear causality as basic category, the investigation of organized wholes of many variables requires new categories of interaction and organization. General systems theory should be a means of investigating the transfer of principles from one field to another, so that it would no longer be necessary to duplicate the discovery of the same principles in different fields [...].

- These considerations lead to the postulate of a new scientific discipline which we call *general system theory*. Its subject matter is a formulation of principles that are valid for 'systems' in general, whatever the nature of the component elements and the relations or 'forces between them'.

The above ideas and concepts give a representative picture of Bertalanffy's view of systems and *general systems theory*. Figure 4.1 summarizes seven

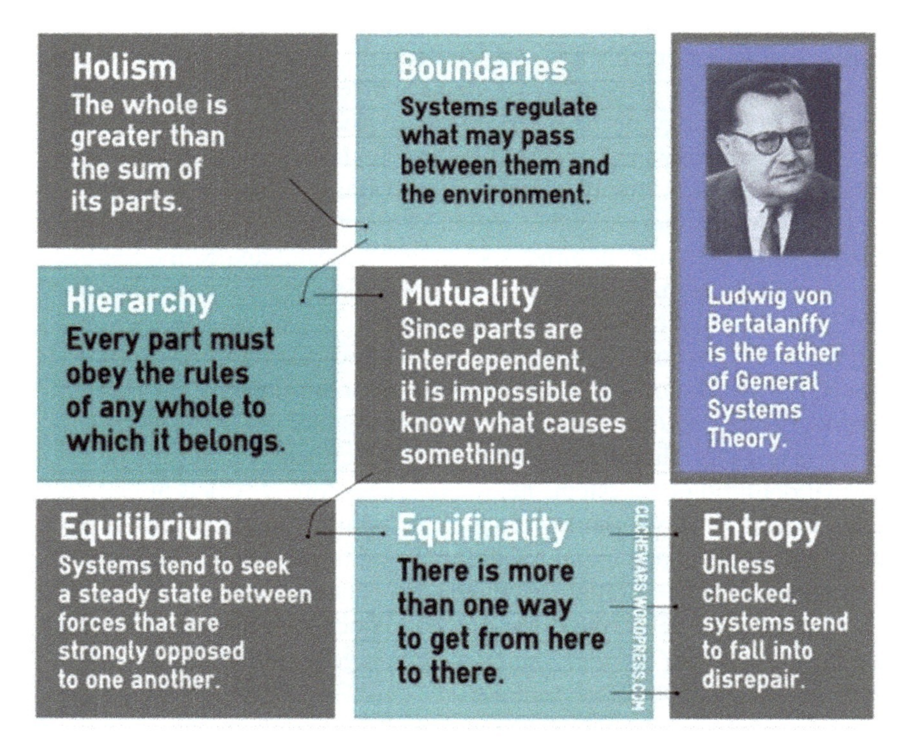

Figure 4.1 Summary of seven fundamental concepts of GST.

Source: Digital Fabrication. Cybernetics: Systems, Purpose, and Control. http://dfabclasss.com/uncategorized/cybernetics-systems-purpose-and-control

fundamental concepts of GST, namely: holism, boundaries, hierarchy, mutuality, equilibrium, equifinality, and entropy.

4.3 What is Isomorphism in GST?

Isomorphism is the formal mapping between complex structures where the two structures involve equal parts. The term *isomorphism* comes from the Greek composite word *'ισόμορφος'* (isomorphos) where the word *'ίσος'* (isos) means *'equal'* and the word *'μορφή'* (morphe) means *'shape'*. In mathematics the term *'isomorphism'* or *'isomorphic'* express a fundamental concept. When used in science it helps in identifying isomorphic structures which enable us to gain deeper knowledge of complex entities. In biology it helps to map complex cell structures to sub-graphs which are regarded as equally related objects.

Von Bertalanffy recognized the existence of isomorphisms between different fields of science which comes from the existing *'analogies'*, and has introduced it as a principal concept in general systems theory which turns out to be an important regulative tool in science. The existence of similar laws and structures in different fields enables us to use models which are simpler or more accurate for more complex and less manageable phenomena.

Formulating exact criteria, GST will protect the system's analyst to use superficial analogies that are useless in science and harmful in their practical consequences. Isomorphisms can be used both in logical-mathematical strict form and in ordinary language form whenever they cannot be expressed in mathematical ways.

The usefulness of establishing isomorphisms between differing systems is obvious, since the GST experts try to find concepts, principles and patterns between such systems which can be readily employed and transformed from one system to another. Systems can be mathematically modeled and so the level of isomorphism can be determined. As a result, event graphs and data flow graphs can be constructed to represent the system's behavior. Identical vertices and edges within the graphs are sought to identify equal structure between systems. Identifying such an isomorphism between modeled systems enables the discovery/establishment of shared abstract patterns and principles which are applied to both systems. In other words, isomorphism is a powerful element of GST which propagates knowledge and understanding between different groups of system expands. The amount of knowledge acquired for each system is increased. This helps decision makers and leaders to make correct choices about the systems which they study. Clearly, as future performance of a system is better understood, optimal decisions can be made concerning the potential balance and so the operation of the system is improved.

Isomorphisms play a central role in research aiming at comparing an artificial (synthetic) model of a natural system and the real existence of that system in nature. System theories work to develop and construct such isomorphic features between the synthetic models they create and the real world phenomena. The synthetic model's principles must directly parallel the natural world. These models can be used to potentially solve important or critical problems occurring in business, engineering and other societal fields, and to obtain valid representations of the real world.
However, as argued in Dubrovsky [2]:

- The identification of isomorphism common to systems studied in different sciences.
- The formal construction of theoretical models, and
- Systems research.

cannot (in principle) produce *'general system principles'* applicable to all systems. It seems, that the claims that the derivation of isomorphism can give *'general system principles'* are not valid since there may be a confusion of what principles can be produced.

In opposition to the method of identification of isomorphisms, *Ashby* [3] suggested a *'non-empirical'* method, which starts *'at the other end'*. Instead of studying first one system, then a second, then a third, and so on, he goes to the other extreme. He considers the set of all conceivable systems and then reduces this set to a more reasonable size. By the term *'conceivable systems'*, Ashby meant an arbitrary set of variables such as temperature and humidity in a room, or speed and distance traveled by a car, etc. Ashby argues that scientists always select their systems by discarding a large number of possible combinations of variables, mostly on the basis of intuition.

Ackoff [4] claimed that the method of identification of *isomorphism* is *ineffective,* due to the GST's implicit assumption that the structure of reality is isomorphic to the structure of science, so reality can be divided into physical, biological, sociological entities, and so on. He proposed an alternative *'systems research'* approach that studies real systems *'as they are'* and that produces knowledge which cannot be separated according to existing division of science. According to Ackoff *unification of science,* or *interdisciplinary* can be achieved by developing *integral theoretical models* in the process of systems research conducted by interdisciplinary teams of experts. These theoretical models go beyond generalization of existing theories. Ackoff' alternative approach at his time was represented by *operational research* **(OR)** and systems engineering (e.g., inventory systems, queuing systems, network systems, etc.).

Responding to *Ackoff's* criticism to GST approach to the *unification of science, Rapoport* [4] remarked that system phenomena cannot be studied in a *'conceptual vacuum'*, and therefore, 'some conceptual theoretical assumption should be made to serve as the basis for formal models of systems research'. In fact, Ackoff has studied inventory management, systems, and has suggested that a metabolic process in a living organism, water supply in a certain geographical area, and any other object can be studied as an inventory control system as long as its variables behave *'inventory-like'* [2–4].

All GST theorists have the *'realistic view'* of systems, i.e., the view that the *ontological* picture of their discipline depicts the reality as it is, independent of their biases, research method, etc. Some of them hold that the world is filled with real physical, chemical, biological, and sociological systems, and that the systems' isomorphisms are real features of these systems. Others hold that open, and closed systems, machines, inventory control, etc. are the models of real systems as they are. However, all 'systems scientists' believe that the reality itself is a hierarchy of systems with the universe at the top and the elementary particles at the bottom. This realist view determines the ontological picture of GST [2].

4.4 What is System Dynamics?

System dynamics **(SD)** was founded by *Jay Forrester* in the 50s. He was able to apply his engineering view of electrical systems and digital computers to the field of human systems. He used computer simulation to study social systems and predict the consequences in a series of formal models. His first book (which is still classic) is entitled *'Industrial Dynamics'* (1961) [5], and was followed by the books *'Urban Dynamics'* (1969) [6] and *'World Dynamics'* (1971) [7].

System dynamics provides a positive answer to the fundamental question: 'Can complex systems such as companies, organizations, and societies be influenced?' His positive answer was based on the fact that systems dynamics studies and manages complex feedback systems as found in companies and other social systems.

System dynamics is an aspect of *system theory* which provides a methodology to understand the dynamic behavior of complex systems using *stocks, flows, internal feedback loops, and time delays* [5]. SD is presently being used throughout the public and private sector for decision making, analysis, and design problems. SD methodology enables the development of models that can solve the problem of *simultaneity* (mutual causation) by updating all variables in small time increments with positive and negative feedback, and time delays structuring the interactions and control.

The system dynamics methodology involves the following concepts [8]:

- *System dynamics* The problems are defined dynamically and their modeling is based on maps. Modeling should be followed by steps for building confidence in the model and its policy implications.
- *Simulation* The basic structure of a formal system is expressed mathematically and the simulation model involves a set of first-order differential equations.
- *Feedback thinking* Diagrams of information feedback loops and 'circular causality' are useful tools for the conceptualization of the structure of a complex system and the communication model insights.
- *Loop dominance* The feedback loop concept is not sufficient for the system analysis. The concepts of 'active structure' and 'loop dominance' enhance the explanatory power and insight-fullness of 'feedback understandings'.
- *The endogenous point of view* The concept of endogenous change is fundamental to the system dynamics approach. The causes are contained within the system. Exogenous disturbances are considered to be triggers of system behavior.

Forrester has captured the above ideas and provided the following hierarchical organizing framework for system structure:

- *Closed boundary* around the system
    - *Feedback loops*
        - Levels
            - Goal
            - Observed condition
            - Discrepancy
            - Desired action

The closed boundary signifies Forrester's endogenous point of view, and

- It comes before feedback loops, stocks and flows, graphs over time, and all the rest of what we do.
- It has top billing.
- It is the deep foundation of systems thinking.

According to Forrester, 'The image of the world around us, which we carry in our head, is just a model. Nobody, in his head imagines all the world, government or country. He has only selected concepts and relationships between them, and uses those to represent the real system'.

Stocks (levels) and *flows* (rates) that affect them are essential components of system structure. Stocks (accumulations and state variables) represent the memory of a dynamic system and are the sources of its *disequilibrium* and dynamic performance. The system performance is a consequence of its structure. The systems dynamics approach provides a continuous-time view, which looks beyond events to see the dynamic patterns underlying them. Also, the continuous view does not focus on discrete decisions but on the policy nature underlying decisions. Events and decisions are regarded as surface phenomena that ride on an underlying tide of structure and behavior [9].

The structure of the system is represented by a *causal loop diagram* **(CLD)** which is a simple map of the system with all its constituent components and their interactions. By capturing the interactions and thus the feedback loops, a causal loop diagram reveals the structure of a system. Moreover, understanding the structure of a system enables us to ascertain the behavior of this system over a certain time period.

The alternative way of representing systems in the SD approach is the so-called *stock and flow diagram* **(SFD)** which shows the existing feedback from stock to inflow (s) and from outfloor (s) to stock.

Pure positive feedback which is symbolized by **R** (from *reinforcement)* or by '+' *produces growth.* Pure negative feedback which is symbolized by **B** (from *balancing*) or by '–' limits *growth.* The combination of negative and positive feedback is the source of complex (nontrivial) behavior. Figure 4.2 shows pictorially three examples of the effect of positive, negative, and combined positive/negative (multiple) feedback with reference to a production system.

Figure 4.3 shows the *'stock and flow diagram'* **(SFD)** of a simple general system which involves a stock, an inflow, and an outflow.

In Figure 4.4, the SFD of Figure 4.3 is specialized for the case of a *'population*' system, in which the *'population'* is the *'stock' variable, the 'inflow'* is the *births'* variable, the *'outflow'* is the *'deaths'* variable, and *'birth rate'* and *'death rate'* are two *'auxiliary (external) variables.'*

The causal loop diagram corresponding to Figure 4.4 is shown in Figure 4.5. Three other possible system behaviors, besides exponential growth, goal seeking, and S-shaped growth, are S-shaped growth with overshoot, oscillation, and overshoot and collapse, as shown in Figure 4.6.

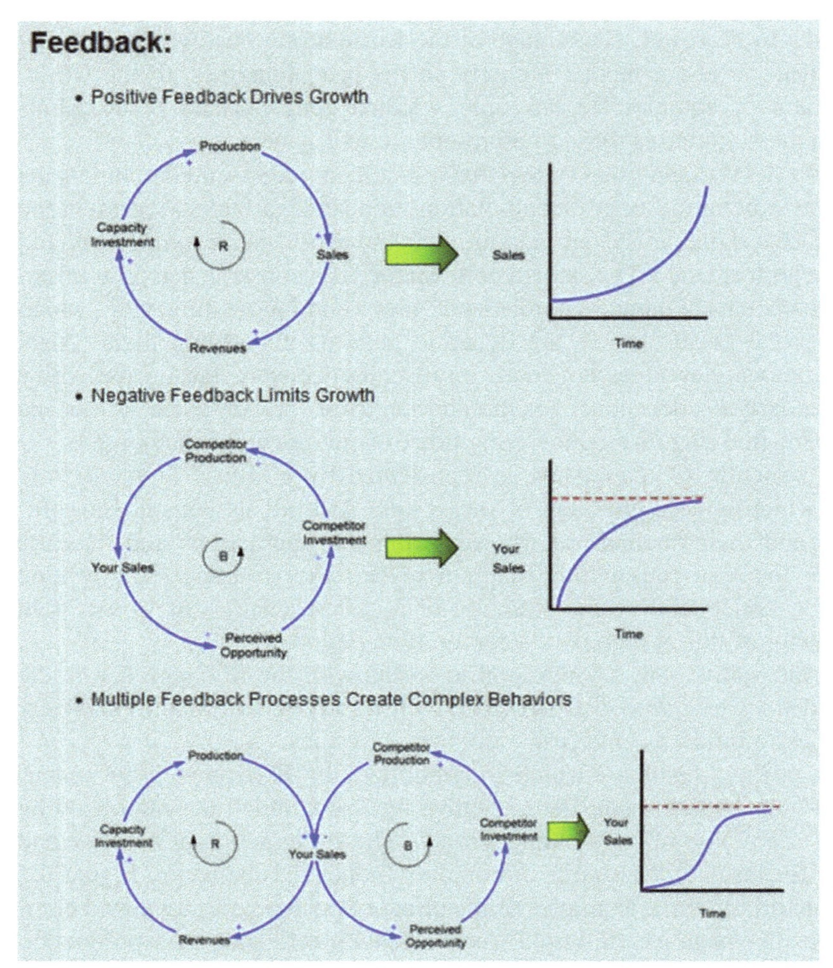

Figure 4.2 Causal loop diagrams of a system with one positive feedback loop (R) one negative feedback loop (B), and two loops. Positive feedback leads to growth, the negative feedback limits growth, and the combination of positive and negative feedback leads to complex behavior.

Source: www.dynamic.forecasting.com (DynSysDyn.html).

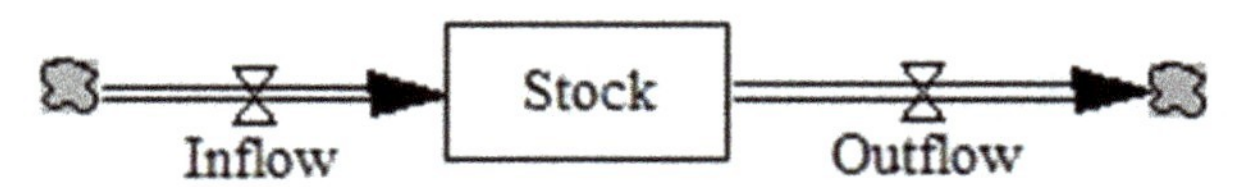

Figure 4.3 A general stock and flow diagram, with one stock, one inflow, and one outflow.

Source: www.file.script.org/Html/4-1480067_40744.htm

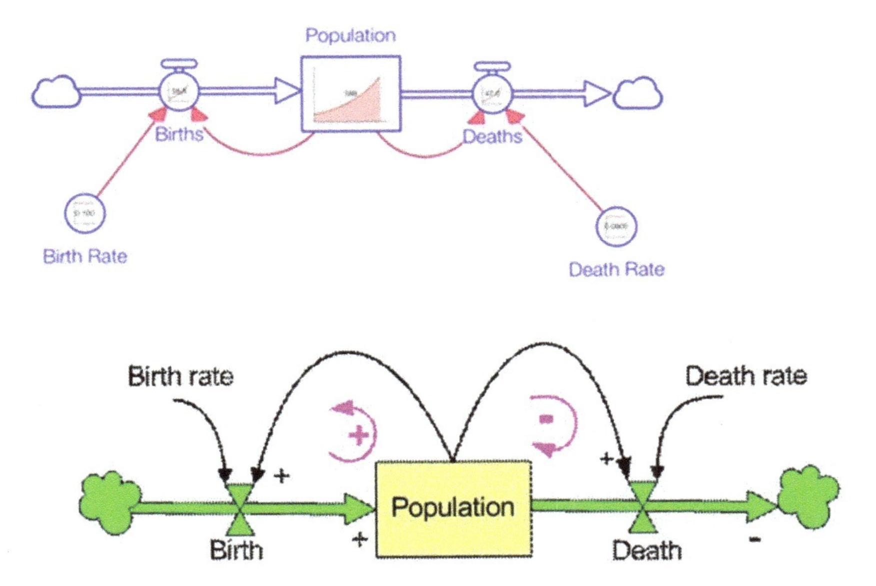

Figure 4.4 Stock and flow diagram of population system (two equivalent representations).

Source: http://people.revoledu.com/kardi/tutorial/SystemDynamic

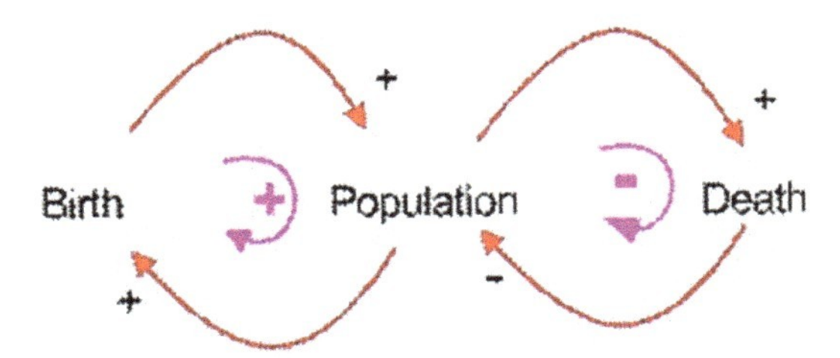

Figure 4.5 Causal loop diagram of the two – feedback – loop population system of Figure 4.4.

Source: http://people.revoledu.com (kardi/tutorial/System/Dynamics).

Figure 4.7(a) shows the **CLD** for a population control system with two positive and two negative feedback loops for the case of a resource constrained society. The aim of the system is to achieve a desired family size as the population increases the resources per person decrease and the life expectancy falls resulting in increase of the rate of deaths. Figure 4.7(b) shows the **SFD** model of a new product adoption process. The model has

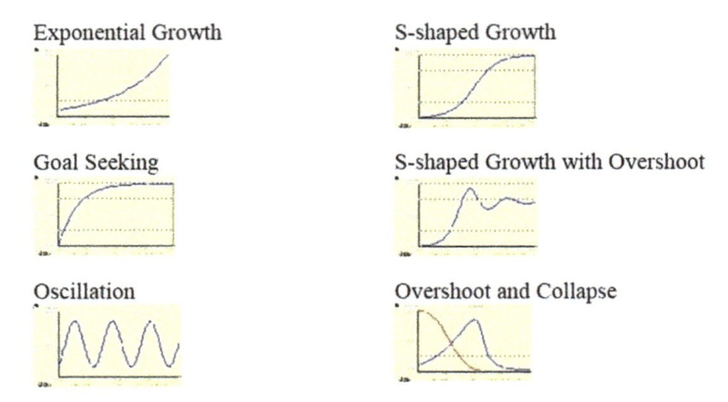

Figure 4.6 Basic system behaviors. These behaviors are produced by combinations of positive and negative feedback.

two stocks (potential adopters P and adopters A) and one flow (the adoption rate AR). There are three feedback loops namely the word of mouth positive loop (R) and two negative feedback loops (B) for the market saturation to adoption from advertising and advertising effectiveness. There are three external variables, namely total population N, adoption fraction I, and contact rate c. Figure 4.7(c) shows the **SFD** of an inventory control system with inflow variable the *production rate,* outflow variable the *shipment rate*, stock variable the *commodity inventory*, and external variable the *order rate* [30].

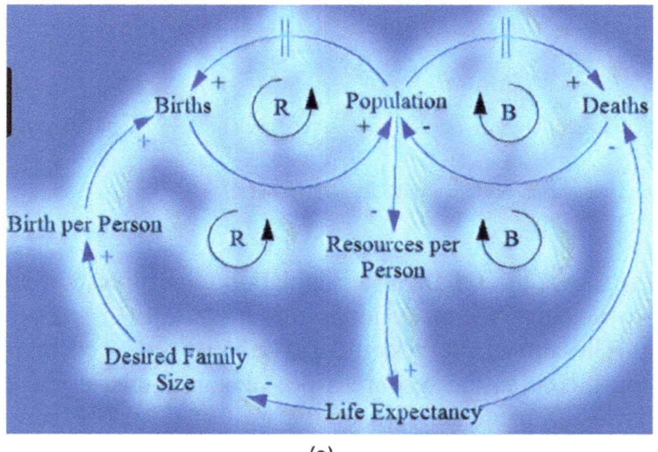

(a)

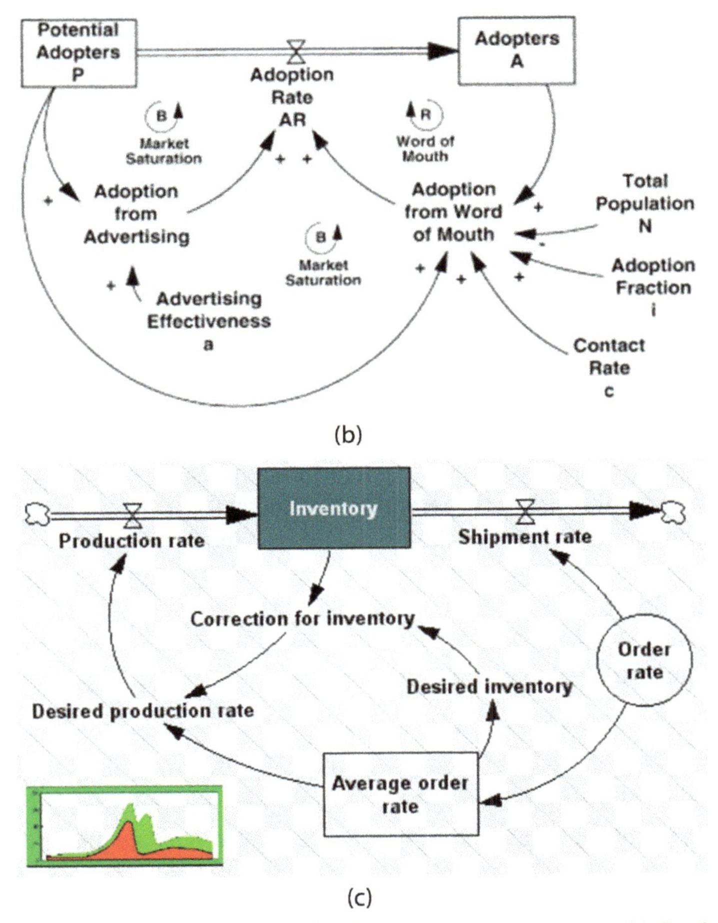

Figure 4.7 (a) Causal loop diagram of a family-size control system, (b) Stock and flow diagram of 'new product adoption' system, (c) SFD model of an inventory control system.

Source: (a) https://systemsandus.com (2012/08/15/learn-to-read-clds).
(b) https://www.sg.ethz.ch/teaching/systems-dynamics-and-complexity
(c) www.dinamica-de-sistemas.com (wdse.html).

System dynamics sees a system as a business system, and the dynamic modeling can be used in numerous management situations in which adaptability during change process is promoted and are tracked down and eliminated. This results in organizational effectiveness. Systems dynamics employs *push-strategies* rather than *pull-strategies.* The customer service is focused on understanding customer demands and/or customer complaints, and does not assume that customers are willing to wait until companies change their processes.

A company receives optimal feedback from its customers, and needs to understand it (e.g., what customers think is important).

The systems dynamic modeling and simulation is performed in six steps as follows:

Step 1 Specify the problem (system) boundary.
Step 2 Find the most important stocks, and the flows that alter these stock levels.
Step 3 Identify sources of information which affect the flows.
Step 4 Identify the dominating feedback loops.
Step 5 Draw a causal loop diagram for the stocks, flows and information sources, and their interconnections.
Step 6 Write down the equations which determine the flows and estimate the initial conditions, and the parameters of the system.
Step 7 Perform the simulation of the model and evaluate the results.

System equations

A formal system is described by a state-space model (1st-order vector differential equation):

$$\dot{\boldsymbol{x}}(t) = \mathbf{f}(\boldsymbol{x}, \boldsymbol{a}), \quad \dot{\boldsymbol{x}}(t) = \mathrm{d}\boldsymbol{x}/\mathrm{dt}$$

where $\boldsymbol{x}$ is a vector of levels (stocks or state variables), $\mathbf{f}$ is a linear or nonlinear vector valued function, and "a" is a vector of parameters. A linear function $\mathbf{f}(\boldsymbol{x}, \boldsymbol{a})$ has the form $\mathbf{f}(\boldsymbol{x}, \boldsymbol{a}) = \mathbf{A}(\mathrm{a})\,\mathbf{x}$, where $\mathbf{A}$ is a square matrix. For example the money $\mathbf{x}$ of a person in the bank is given by $\dot{x} = r \cdot x$, where r is the interest rate, and x_o is the initial amount of money (initial condition). This system's model is shown in Figure 4.8.

The solution of a state-space model is obtained numerically by computer simulation [9]. Any efficient and accurate numerical method can be used for performing the simulation. In the literature and the market there are available simple or sophisticated software programs (tools). One of them freely available can be down loaded from [10], which is accompanied by a tutorial [11].

The discrete-time dynamic equation for the population system of Figure 4.8 is the following:

$$\begin{aligned} x(t+1) &= x(t) + B - D \\ &= x(t) + bx(t) - dx(t) \\ &= (1 + b - d)\,x(t). \end{aligned}$$

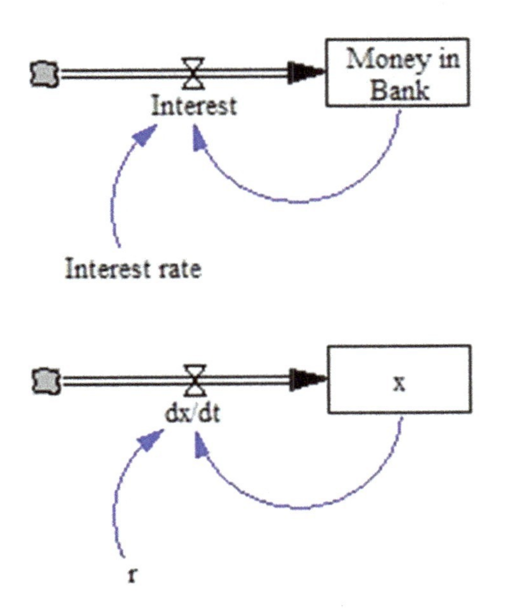

Figure 4.8 Bank deposit system dynamics model.

Here, $x(t)$ is the population level, the time increment (sampling period) is 1 year, *B* is the birth level given by $B = bx(t)$ (b is the birth rate), *D* is the death level given by $D = dx(t)$ (d is the death rate). Assuming that the birth rate is $b = 0.08$ and the death rate is $d = 0.02$, then, we get the simulation results shown in Table 4.1 and Figure 4.9. These results were produced by using Microsoft-Excel which can be downloaded from [32].

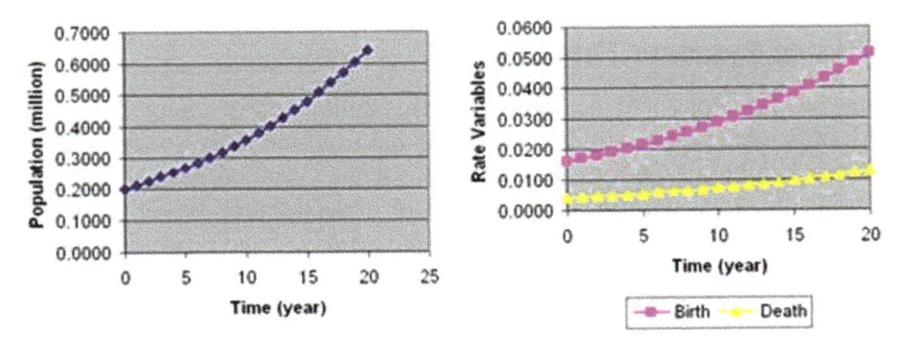

Figure 4.9 Plots of population level $x(t)$ (*left*), and birth and death levels $B(t)$ and $D(t)$ (*right*).

Source: http://people.revoledu.com/kardi/tutorial/SystemDynamic

Table 4.1 Time evolution of population level, and birth and death levels

	Constant1	Birth rate	0.08
	Constant2	Death rate	0.02
	Level	Rate1	Rate2
Time	Population	Birth	Death
0	0.2000	0.0160	0.0040
1	0.2120	0.0170	0.0042
2	0.2247	0.0180	0.0045
3	0.2382	0.0191	0.0048
4	0.2525	0.0202	0.0050
5	0.2676	0.0214	0.0054
6	0.2837	0.0227	0.0057
7	0.3007	0.0241	0.0060
8	0.3188	0.0255	0.0064
9	0.3379	0.0270	0.0068
10	0.3582	0.0287	0.0072
11	0.3797	0.0304	0.0076
12	0.4024	0.0322	0.0080
13	0.4266	0.0341	0.0085
14	0.4522	0.0362	0.0090
15	0.4793	0.0383	0.0096
16	0.5081	0.0406	0.0102
17	0.5386	0.0431	0.0108
18	0.5709	0.0457	0.0114
19	0.6051	0.0484	0.0121
20	0.6414	0.0513	0.0128

We see that with the birth and death rates given above the population will reach 500 thousand after 16 years and 640 thousand after 20 years.

4.5 What is Systems Thinking?

The concepts, principles and methods of systems thinking have emerged from the work of general systems theory scientists, and from the practitioners who applied the insights of GST to real-world problems. Very broadly systems thinking is concerned with the development of understanding based on the integration of *analysis* and *synthesis* (called *anasynthesis)* and represents the study of the whole, the parts, and their interactions. This means that we are doing systems thinking if we consider something on its totality together with its features and its interactions with the environment, as well as the interactions between parts. Systems thinking is sometimes referred to as *'thinking systematically'*. Systems thinking enables us to understand better the world around us.

Some definitions of *'systems thinking'* are the following [12]:

- Systems thinking is a way of seeing and talking about reality that helps us understand better and work with systems to influence the quality of our lives (*Kim*).
- Systems thinking is a 'new way of thinking' to understand and manage complex systems (*Bosch/Cabrera*).
- System's thinking is a discipline for seeing wholes. It is a framework for seeing interrelations rather than things, for seeing patterns of change rather than static snapshots (*Peter Senge,* www.quotesnsayings.net/quotes(7475) [13]).
- Systems thinking is recognizing systems, subsystems, components of systems, and system interrelationships in a situation (*James O'Brien/George Marakas,* http://snipr.com/u3ec).
- System thinking happens when one considers the entirety of the object of interest and mutual interactions with others (I.e., see the forest and the trees). As result, the thinker should better understand why things are the way they are, and how to design interventions which are more likely to succeed, and improve the situation (*Cecilia Haskens*).
- Systems thinking is the process of predicting on the basis of anything at all, how something influences another thing. It has been viewed as an approach to problem solving by 'viewing problems' as part of the overall system (*Wikipedia*).
- Systems thinking is a way of looking at, learning about, and understanding complex situations (*Wilson*).
- Systems thinking is a method of formal analysis in which the object of study is viewed as comprising distinct analytical sub-units (*China.org,* http://snipr.com/u3e65).
- Systems thinking is a state of mind. Once you have this state of mind you realize how each part of your business flow into the other (*Lisa A. Mininn*).

Figure 4.10 gives an overall comparison of a number of systems thinking definitions on the basis of the features they contain from the following set of features [31]:

- Wholes rather than parts.
- Dynamic behavior.
- System as the cause of its behavior.

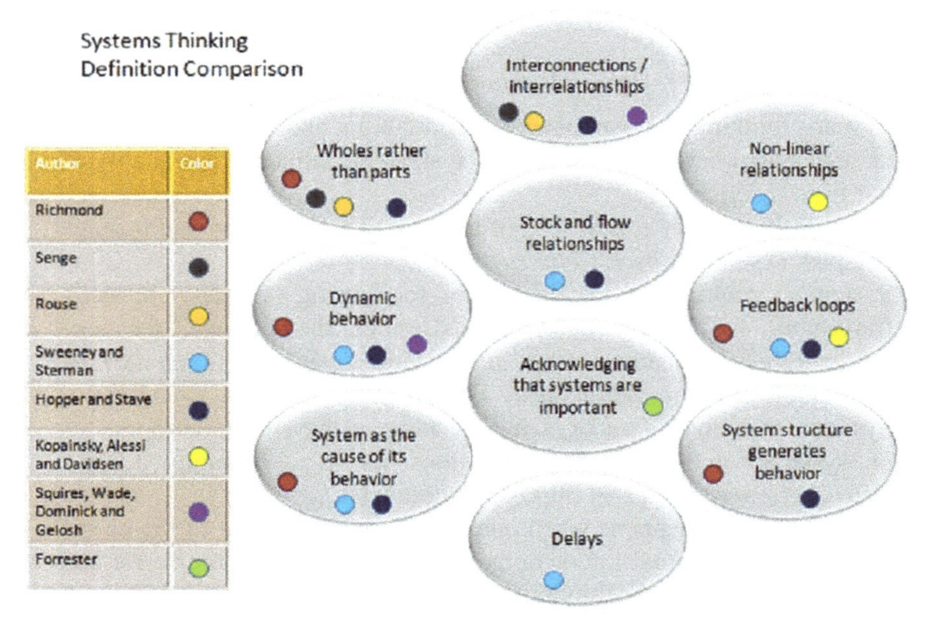

Figure 4.10 Comparison of systems thinking definitions in terms of the features they contain.

Source: www.researchgate.net

- Interconnections/interrelationships.
- Non-linear relationships.
- Stock and flow relationships.
- Feedback loops.
- Delays.
- System structure generating behavior.
- Acknowledgement that systems are important.

According to *Klir* [16] 'systems thinking focuses on those properties of systems and associated problems that emanate from the general notion of 'systemhood', while the divisions of the classical science have been done largely, on properties of 'thinghood'. All the above definitions say in one way or the other that the principal goal of systems thinking is to consider and study objects and systems as a whole including their parts and the associated interrelations, i.e., systems thinking follows a *holistic* approach taking into account observed phenomena such as emergence (The term 'holistic' comes from the Greek term 'όλον' (holon) which means 'entire/integral/total').

Systems thinking requires the consciousness of the fact that we deal with models of our reality and not with the reality itself [14]. Systems thinking is distinguished from systems theory and systematic perspectives [12]. Specifically:

- *Systems thinking* is the broad paradigm.
- *Systems theory* is the science (laws, principles, theorems).
- *System perspectives* are the methods (soft systems, methodology, viable systems models, etc.).

A ten-step model for performing 'systems thinking' in practice is the following:

Step 1: Vision/purpose: Describe/refine your ideal state.
Step 2: Share and test key mental models in the system.
Step 3: Sketch key trends and name the essential variables driving the trends.
Step 4: Make the system visible (via causal maps).
Step 5: Look for leverage.
Step 6: Share, test, and revise maps.
Step 7: Use leverage to identify stakeholders and prioritize actions.
Step 8: Engage stakeholders and collaborate across boundaries.
Step 9: Take actions and get results.
Step 10: Share results, get feedback, learn.

Schematically this model for systems thinking in action is shown in Figure 4.11.

A diagram that shows the evolution of systems thinking over the years (with names of principal thinkers) is given in Figure 4.12(a) [15]. The systems thinking continuum is as shown in Figure 4.12(b) (*Richmond* [21]):

An explanation of the items involved in the systems continuum of Figure 4.12(b) is as follows:

- *Cause and effect (or event thinking)* Here the cause and action are identified, without any reference to the underlying structure of relationships or behavior patterns.
- *Patterns of behavior* Here we look beyond cause and effect to recognize how things have changed over time and take this into consideration before proceeding to any action.
- *Systems Perspective* Going sufficiently back far (in both space and time) the systems thinker tries to see the web of ongoing reciprocal (cycling) relationships that produce the system's patterns of behavior.

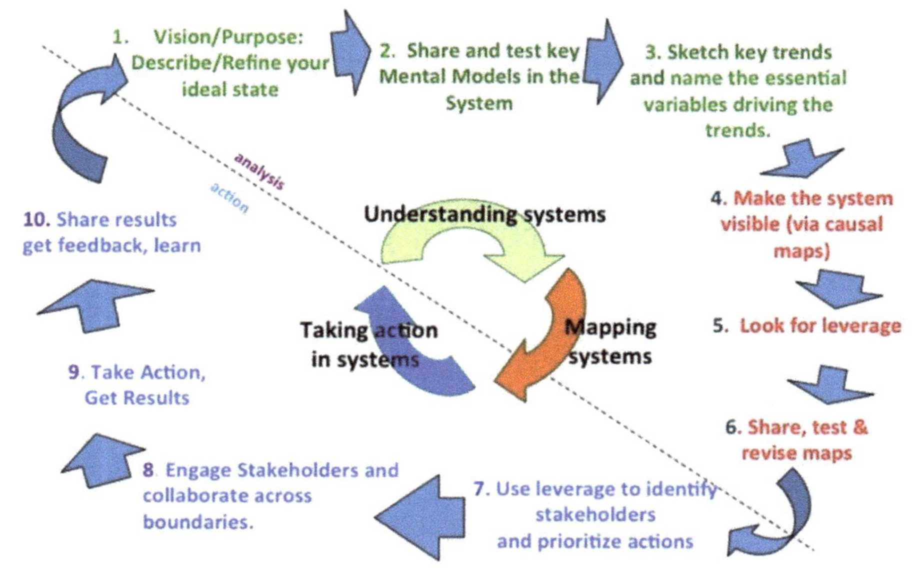

Figure 4.11 Steps for systems thinking in action.

*Source:*http://seedsystems.net/systems-thinking-course.html

- *Influence diagram* This is a simple map of the reciprocal relationships that the thinker believes to be principally responsible for the exhibited behavior patterns of the system.

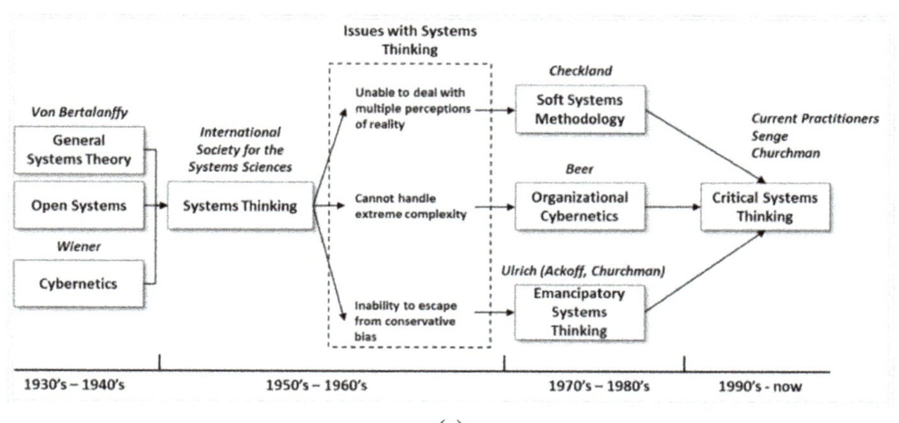

(a)

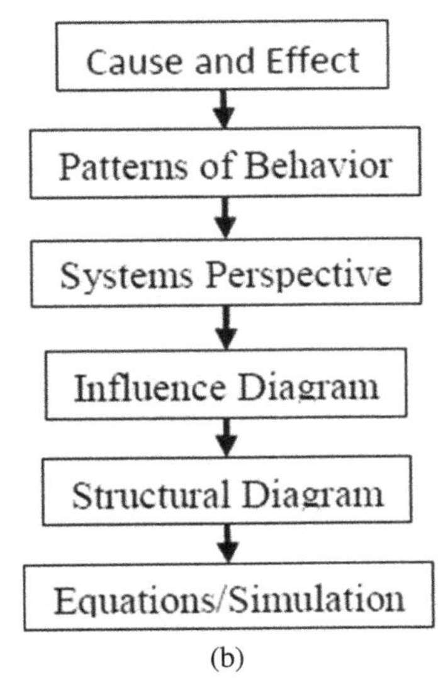

(b)

Figure 4.12 (a) Evolution of systems thinking and (b) Systems thinking continuum.

- *Structural diagram* This is a more disciplined map which shows what really makes a system tick (Here the systems thinker is laying out the mechanisms he/she thinks are used by the system to control itself).
- *Equations/Simulation* Here the structural diagram is expressed by a set of equations, that characterize the nature of the relationships put down in the structural diagram (including the numerical values that define the strength and direction of these relationships). This step is completed by simulating on a computer these relationships, and analysis/evaluation of the results, and modifying one or more of the previous steps as it is deemed to be required.

A causal loop diagram which expresses the above systems thinking procedure was given in SEBOK [12]. This is shown in Figure 4.13.

A question about *systems thinking* is: 'How one can take a holistic approach while still being able to focus on changing or creating the system?' According to [17] this can be done through '*separation of concerns*' which provides a balance between considering parts of a system problem or solution while not losing sight of the whole'. [12]. Abstraction is the process of moving away characteristics from an entity/object so as to reduce it to a set of fundamental characteristics. When trying to understand complex problems, it is easier to concentrate our attention to problems, whose solutions still remain unknown to the greater problem [18].

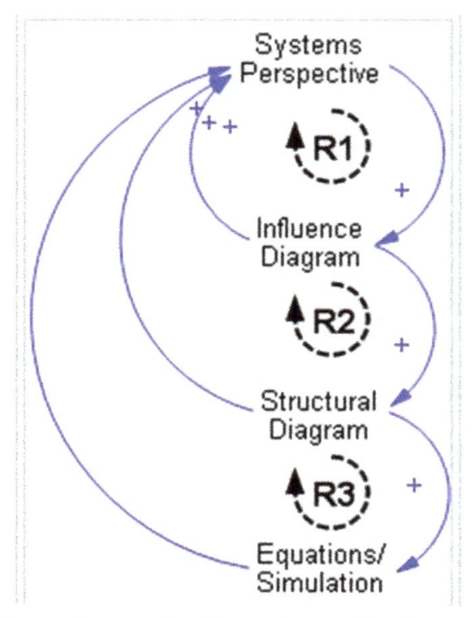

Figure 4.13 Causal loop diagram for the systems thinking procedure (from systems perspective to equations/simulation).

Source: ThinkingDefinitions-SystemsWiki.htm

Systems thinking is performed on the basis of fundamental principles (general rules) which (besides holism, abstraction, and boundary identification) include the following:

- *Dualism* (Recognition of dualities and use of them in the context of a larger whole).
- *Equifinality* (In open systems different initial conditions may lead to the same final state).
- *Interaction* (The system behavior, capabilities and features are the result of interaction between parts).
- *Hierarchy* (The hierarchical structure of complex systems facilitates their study).
- *Modularity* (Related system parts should be joined together, and unrevealed parts should be separated).
- *Parsimony* (The simplest explanation of a phenomenon requiring the fewest assumptions should be used).
- *Regularity* (Use system regularities to enhance the understanding and application of the system in practice, e.g., feedback).

Other principles are *'encapsulation'* (hide the internal parts/interactions from the external environment), *'network'* (which implements togetherness, and dynamic interactions), *'separation of concern'* (decomposition of a large system into a set of smaller problems makes more effective its solution), *'stability'* (use entities and concepts in the stable region of the spectrum), and *'synthesis'* (choose the right parts to synthesize a system with the right interactions).

A full study of formal system models which includes features, dependencies and tradeoffs between the above characteristics was given by *Lipson* [19]. It is however useful to remark that not all principles can be employed to every system or engineering decision. Guidance on how these principles should be applied relating many of them in a concept called the *'conceptagon'* was provided by *Edsoon* [20]. Modularity is a central principle of systems thinking and studies the degree to which the components of a system may be separated and recombined. It can be applied to systems in natural, social, and engineering areas.

Systems thinking is particularly useful (and needed) when dealing with complexity of any type, and increasing or complex system/problem dynamics which may include aspects such as:

- Increasing conflicts.
- Technological developments.
- Global information exchange.
- Social and political developments.
- Social responsibility.
- Climate and environmental changes.
- Differentiation of customer needs.
- Disruption of the value chain.

These aspects are interrelated, and contribute to complexity and dynamics as shown in Figure 4.14.

Senge [13] describes how to use *'system archetypes'* as thinking tools for identifying causes and possible solutions to typical problems. Senge's book does not include any simulation and so it is a good example which shows that scientists in the field of systems thinking are not focusing in deriving exact quantitative results or predictions with their models, but that their main goals are learning and policy formulation. The general critical thinking process can be applied not only to systems thinking but to all problem solving situations, and involves a number of steps which are shown in Figure 4.15.

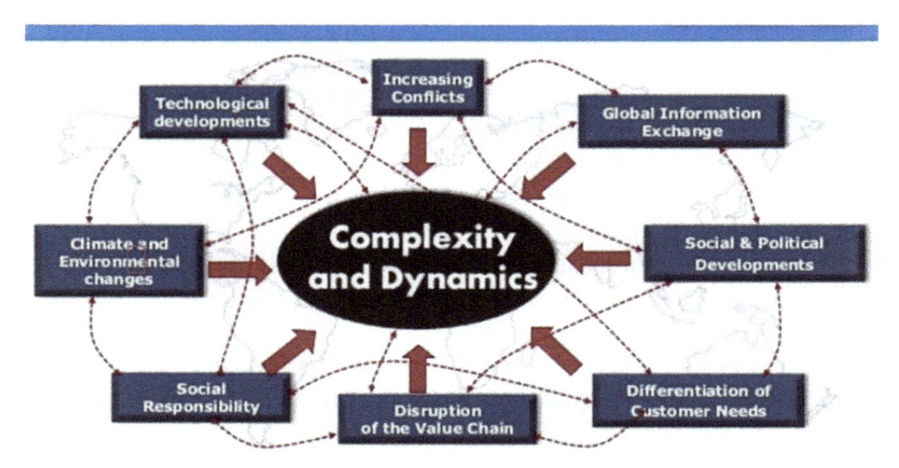

Figure 4.14 Systems thinking can deal with complexity and copy with increasing and complex dynamics.

Source: www.slideshare.net (Roberts, Ross and Smith: Systems Thinking).

Figure 4.15 The general critical thinking process steps.

Source: www.prismabr.com.br

4.6 Axioms of General Systems Theory

Kevin MacG Adams [22–24] has provided a set of axioms which is a sufficient construct for systems theory and *system of systems engineering* **(SoSE).** He argues that systems theory is a unified group of specific propositions which help in understanding systems by both GST scientists and practitioners. This set of propositions hold for all 42 fields of science included in the *Organization for Economic Co-operation and Development* **(OECD)** classification scheme [2007]. In [24] Adams describes his construct for systems theory, the axioms, and propositions of systems theory, also explaining the multidisciplinary nature of the propositions. This set of seven axioms was produced using the axiomatic method [25], by recognizing the constituent propositions involved in the 42 OECD fields of science. These axioms, are pictorially shown in Figure 4.16, and were called the *'theorems of systems theory' by Honderich* [26]. There are presumed to be true by systems theory, from which all other propositions in systems theory may be derived.

As argued by *Adams* neither the propositions nor their associated axioms are independent of one another. This interdependence of the axioms is illustrated by the links of Figure 4.16. It is exactly this group of propositions

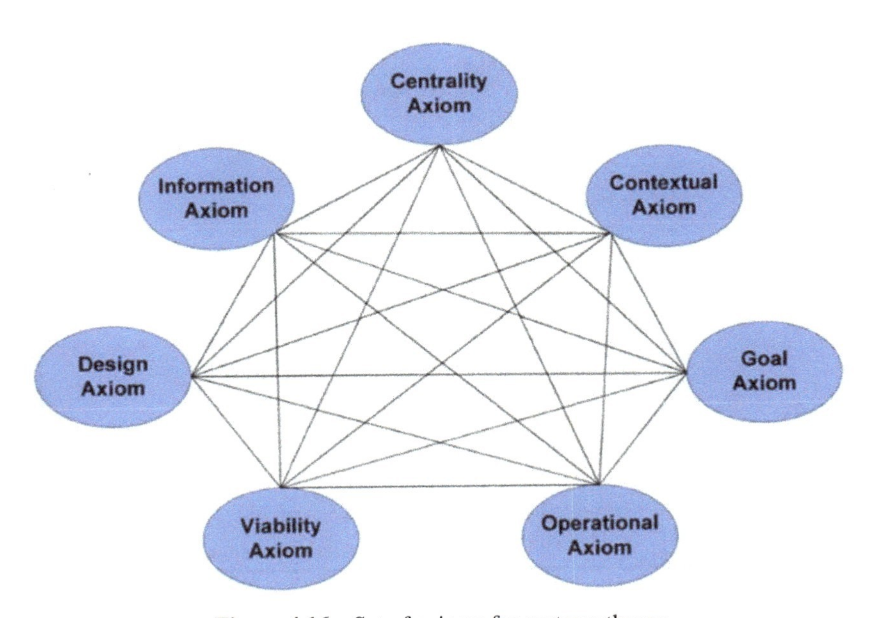

Figure 4.16 Set of axioms for systems theory.

Source: [23, 24].

which enables thinking, decision making, action, and interpretation with respect to systems.

Figure 4.17 shows how the propositions of systems theory are superimposed into the six major scientific sectors (humanities, natural sciences, social sciences, agricultural sciences, medical and health sciences, and engineering) and the 42 fields of science.

From Figure 4.17 one can easily verify that systems theory and its foundation are inherently multidisciplinary. This multidisciplinary construct guarantees wide applicability of systems theory in both its theoretical foundations and applications. Detailed descriptions of the concepts involved in Figure 4.17 are provided in [23].

Here we provide a brief description of these axioms [24]:

Axiom 1 *Centrality* (Central to all systems are two pairs of propositions; 'emergence and hierarchy', and 'communication and control').

Axiom 2 *Information* (Information affects systems, which create, process, transfer, and modify information).

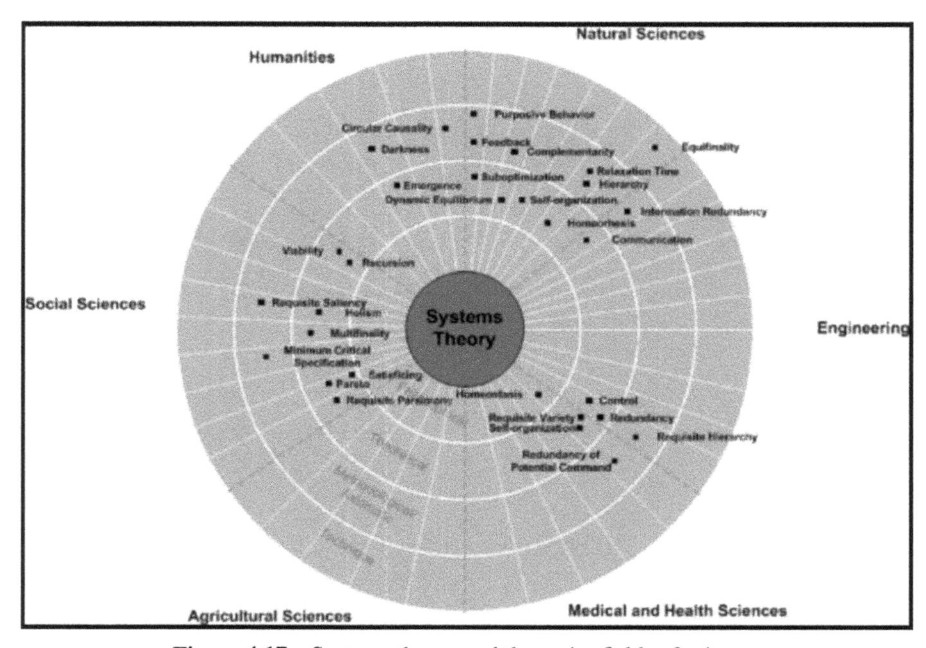

Figure 4.17 Systems theory and the major fields of science.

Source: [23, 24].

Axiom 3 *Design* (This axiom is concerned with the purposive design and provides guides how to achieve it).

Axiom 4 *Viability* (This axiom helps the system scientist to design a system such as changes in the operational environment may be detected and altered to guarantee continual existence).

Axiom 5 *Contextual* (The propositions of this axiom give meaning/context to the system through guidance that assists the system analyst to understand the external factors and circumstances that enable or constraint a particular system).

Axiom 6 *Goal* (The propositions of this axiom are concerned with the pathways and means for implementing systems that can achieve a specific goal).

Axiom 7 *Operational* (The propositions of this axiom provide guidelines to those that must study in situ, where the system is operating to produce the desired behavior).

The propositions of these axioms along with their primary proponents are the following:

Centrality: Communication *(Shannon),* Emergence *(Aristotle)*, Hierarchy *(Pattee).*

Information: Redundancy *(McCulloch), Shannon, Weaver).*

Design: Minimal critical specification *(Cherns),* Requisite parsimony *(Miller),* Requisite saliency *(Bounding).*

Viability: Circular causality *(Korzybski),* Feedback *(Wiener),* Recursion *(Beer),* Requisite variety *(Ashby).*

Contextual: Complementarily *(Bohr),* Darkness *(Cilliers),* Holism *(Smuts).*

Goal: Equifinality *(Bertalanffy),* Multifinability *(Buckley),* Purposive behavior *(Rosenblueth).*

Operational: Dynamic equilibrium (Dalembert), Homeorthesis (Waddington), Redundancy *(Pahl),* Relaxation time *(Holling), Self-organization* (Ashby), Suboptimization *(Hitch).*

Representative references of the above scientists who provided the corresponding propositions can be found in [24].

Another set of axioms for GST was provided by *C.W. Churchman* [27], who argues that 'the purpose of his axioms is to state two things:

- That systems are complexes which can be designed and evaluated.
- That the adjective 'general' in the term 'general systems theory' modifies systems as well as theory.

The nine axioms of Churchman are the following:

Axiom 1: System are designed and developed.
Axiom 2: Systems are component designed.
Axiom 3: The components of systems are also systems.
Axiom 4: A system is closed if its evaluation does not depend on the design of its environment within a specific class of environments.
Axiom 5: A general system is a system that is closed and remains closed for all possible environments.
Axiom 6: There exists one and only one general system *(monism).*
Axiom 7: The general system is optimal *(optimism).*
Axiom 8: General system theory is the methodology of searching for the general system *(methodology).*
Axiom 9: The search becomes more difficult with time and is never computed *(realism).*

Some comments on the above axioms are as follows [27].

Axiom 1 says that 'systems are man-made and designed, which seems seriously restrict and one would say, unduly so, the possible use of systemic models, as applied to various kinds of non-designed concrete'.

Axiom 2 implies that 'the designer conceives a design as consisting of a set of subproblems, the solution to each subproblem leading to a component of the larger system (ibid)'.

Axiom 3 Churchman says that 'this axiom could be called the 'matrioshka postulate'. However, just as the 'matrioshka' (or factals), it should end somewhere and somehow'.

Axiom 4 Being this a tall order, Churchman adds: 'Usually the designer does not attempt to account for all environmental changes. If he does then Axiom 5 applies.

Axiom 5 Here we obviously pass from the model-based on a concrete situation to a totally abstract one.

Axioms 6 through 9 (not discussed here) have a philosophical nature and may be subject to debate [26, 27].

4.7 Conclusions

This chapter has presented basic ontological and epistemological/methodological issues of general systems theory. system dynamics, and systems thinking. Von Bertalanffy has clearly stated and studied the central concepts of systems, and insisted on the existence of 'isomorphism' in science providing convincing examples. He discussed a set of specific aspects as competition between parts, finality and equifinality, and closed and open systems. Forrester's system dynamics approach is applied to any dynamic system which involves interdependence and mutual interaction between components and subsystems, feedback and circular causality [28]. Conceptually, the feedback process is in the core of the system dynamics approach, which involves: defining problems dynamically, seeking an endogenous view of system dynamics, considering all concepts as continuous quantities interconnected in information feedback loops, identifying independent stocks with their corresponding inflows and outflows, formulating behavioral models that can reproduce the system's dynamic performance, gaining insights about the system operational characteristics from these models by simulating them on a computer, and implementing modifications suggested from these insights and understandings. This chapter was indented to provide a clear view of GST and SD approaches at a tutorial level. The interested reader should consult the references that follow [1–29], and other sources to get the mathematical and practical details of general systems theory and system dynamics methodology developed over the years after the original works of Von Bertalanffy and Forrester.

References

[1] von Bertalanffy, L. (1968/1998). *General System Theory: Foundations, Development, Applications*. New York, NY: George Braziller.

[2] Dubrovsky, V. (2004). Toward system principles: general system theory and the alternative approach. *Syst. Res. Behav. Sci.* 21, 109–122.

[3] Ashby, W. R. (1958). General systems theory as a new discipline. *Gen. Syst.* 3, 1–8.

[4] Rapoport, A. (1964). "Remarks on general systems theory," in *Proceedings of 2nd Systems Symposium, Case Institute of Technology*, Cleveland, OH.

[5] Forrester, J. (1961). *Industrial Dynamics*. Cambridge, MA: The MIT Press.

[6] Forrester, J. (1969/1973). *Urban Dynamics*. Cambridge, MA: The MIT Press.

[7] Forrester, J. (1971). *World Dynamics*, Cambridge, MA: Wright-Allen Press.

[8] SDS: Systems Dynamics Society, What is System Dynamics?

[9] Richardson, G. P. (1999/2011). *System Dynamics, Encyclopedia of Operations Research and Management Science*. S. Gass and C. Harris, eds, Boston, MA: Kluwer.
[10] Net logo system dynamics tool Available at: https://ccl.northeastern.edu/netlogo/download.shtml
[11] Netlogo system dynamics tutorial Available at: https://ccl.northeastern.edu/netlogo/docs/systemdynamic
[12] Wiki/SEBOK (2017). Systems thinking definitions-Systems Wiki/SEBOK: Principles of Systems thinking.
[13] Senge, P. M. (2006). *The Fifth Discipline: The Art and Practice of the Learning Organization*. New York, NY: Doubleday Currency.
[14] Ossimitz, G. (1997). *The Development of Systems Thinking Skills using System Dynamics Modeling Tools*. Klagenfurt: University of Klagenfurt, Austria.
[15] Insightmaker (1971). Available at: http://www.insightmaker.com/insight/
[16] Klir, G. (2001). *Facets of Systems Science*. New York, NY: Kluwer/Plenum.
[17] Greer, D. (2016). *The Art of Separation of Concerns*. Available at: http://aspiringcraftsman.com/tag/separation-of-concerns
[18] Earl, T. (2016). *SoA Principles: An Introduction to Service Orientation Paradigm* Available at: http://www.soaprinciples.com/p3.php
[19] Lipson, (2007). Principles of modularity, regularity, and hierarchy for scalable systems. *Biol. Physics Chem.* 7, 125–128.
[20] Edson, R. (2008). *Systems Thinking: Applied a Primer, Applied Systems Thinking (AsysT)*. Arlington, VA: Institute, Analytic Services, Inc.
[21] Isee Systems (2000). Available at: http://www.iseesystems.com/resourses/Articles/ST%204%20Key%20Questions.pdf
[22] Adams, K. M. (2011). Systems principles: foundation for SoSE methodology. *Int. J. Syst. Eng*. 2, 120–155.
[23] Adams, K. M., Hester, P. T., Bradley, J. M., Meyers, J. J., and Keating, C. B. (2014). Systems theory: the foundation for understanding systems. *Syst. Eng*. 3, 225–232.
[24] Adams, K. M. (2012). Systems theory: a formal construct for understanding systems. *Int. J. Syst. Eng*. 3, 209–221.
[25] Audi, R. (ed.) (1999). *Cambridge Dictionary of Philosophy*. New York, NY: Cambridge University Press.
[26] Honderich, T. (2005). *The Oxford Companion to Philosophy* 2nd edn, Oxford: Oxford University Press.

[27] Churchman, C. W. (1963). "General systems theory: some axioms," in *International Encyclopedia on Systems and Cybernetics*, 2nd Edn. Munich: K.G. Saur, Verlag, 1407–1408.

[28] Forrester, J. W. (1995). The beginning of system dynamics. *McKinsey Q.* 4, 1–13.

[29] Teknamo, K. (2016). *System Dynamics Tutorial, Revoledu.com/ Symmetric partnership LLP*. Available at: http://people.revoledu.com/ kardi/tutorial/SystemDynamics

[30] Martin Garcia, J. (2014). *Theory and Practical Exercises of System Dynamics, System Dynamics-Innova.com*. Irvine, CA: Innova Electronics.

[31] Arnold, R. and Wade, J. P. (2015). "A definition of systems thinking: A systems approach," in *Proceedings of 2015 Conference on Systems Engineering Research*, Berlin, 669–674.

5

Cybernetics

You are never too old to set another goal or to dream a new dream.
Aristotle

Try not to become a man of success but a man of value.
Albert Einstein

If cybernetics is the science of control, management is the profession of control.
Anthony Stafford Beer

5.1 Introduction

Cybernetics is a remarkably wide transdisciplinary scientific field which overlaps considerably with general systems theory, and includes information, communication, and control theories with all their manifestations in nature, biology, technology, and society. Cybernetics and systems theory are essentially concerned with the study of the same problem, namely the organization problem, independent of the substrate in which it is embodied. Looking more closely one may think that the two fields seem to be district because systems theory is focused on the *structure* of systems and their models, whereas cybernetics is focused mainly on the *function* of systems, i.e., on how they are controlled and how they communicate with other systems. However, structure and function of a system cannot be understood in separation. Therefore, systems theory and cybernetics represent two facets of the same field. According to *Ashby,* cybernetics is not concerned with *things* but with *'ways of behaving'*, i.e., it does not ask *'what is this thing?'* but *'what does it do?* or *'what can it do?* Because most systems in the natural living, technological and social world can be understood in this way, cybernetics is a multidisciplinary field that crosses many traditionally disciplinary boundaries. Cybernetics is about to have a purpose (goal) and take action for achieving that goal. To know whether you have achieved your goal (or at least getting closer to it) you need to measure your status and employ feedback. Cybernetics emerged from Shannon's information theory which was developed in order to optimize the transmission of information through communication channels, and from the feedback concept used in control engineering.

Our purpose in this chapter is to discuss the fundamental ontological and epistemological/methodological aspects of cybernetics. In particular:

- We review a number of answers to the question 'what is cybernetics?' given by key researchers in the field.
- We give a brief historical review of cybernetics including relevant epistemological issues.
- We discuss first-order and second-order cybernetics both ontologically and epistemologically highlighting the difference between them.
- We outline the basic characteristics of social systems and sociocybernetics including a brief discussion of the inherent processes of cybernetics (self-organization, autopoiesis, and autocatalysis).

5.2 What is Cybernetics?

Due to its generality there are several definitions of *cybernetics* depending on the point of view considered over the years by the scientists who have pursued research in it. As a concept in society cybernetics has been around since Plato to refer to government [1]. In modern times the term *cybernetics* was coined by Norbert *Wiener* coming from the Greek words '*κυβερνήτης*' (kybernetes) meaning 'steersman', and '*κυβερνητική*' (kybernetiki) meaning 'steering'.

The original definition of Cybernetics as a discipline was given by **Wiener** in the title of his book *'Cybernetics: or Control and Communication in the Animal and the machine'* (1948) [2]. Wiener claimed to have been the first to unify control theory with communication theory, but actually this has been done in practice by engineers several years before, i.e. during World War II.

In the following we provide a list of definitions (non exhaustive) given over the years by *cyberneticists* (cyberneticians):

A.N. Kolmogorov (i) Cybernetics is a science of the methods of processing and use of information in machines, living organisms and their combination, (ii) Cybernetics is a science concerned with the study of systems of any nature which are capable of receiving, storing, and processing information so as to use it for control [3].

A.M. Ampere Cybernetics is the art of governing or the science of government [20].

W. Ross Ashby (i) Cybernetics is the art of steering, (ii) Cybernetics is the study of systems that are open to energy but closed to information and control systems that are information tight, (iii) Cybernetics deals with all forms of behavior in so far as they are regular, or determinate, or reproducible, (iv) Cybernetics offers a method for the scientific treatment of the system in which complexity is outstanding and too important to be ignored [4].

Gregory Bateson (i) Cybernetics is a branch of mathematics dealing with problems of control, recursiveness, and information (ii) Cybernetics is the study of form and pattern [5].

Heinz von Foerster One should name one central concept, a first principle, of cybernetics, it would be circularity. Cybernetics introduces for the first time – and not only by saying it, but methodologically – the notion of circularity, circular causal systems [6].

Huberto Maturana Cybernetics is the science and art of understanding [7].

E. Griffin Cybernetics is the study of information processing, feedback, and control and communication systems, where information is the reduction of uncertainty; the less predictable the message is, the more information it carries as described in the equation 'channel capacity = information + noise'. Noise is the *enemy of information* because it interrupts the information carrying capability of the channel between the transmitter and receiver [8].

Henryk Greniewsku Cybernetics is the general science of communication. But to refer to communication is consciously or otherwise to refer to distinguishable states of information input and outputs and/or to information been processed within the boundary some relatively isolated system [9].

Richard F. Ericson The essence of cybernetic organizations is that they are self-controlling, self-maintaining, self-realizing. Indeed cybernetics has been characterized as the 'science of effective organization' [10].

Ervin Laszlo Now *cybernetics* is the term coined by Wiener to denote *'steersmanship'* or the science of control.
Although current engineering usage restricts it to the study of flows in closed systems, it can be/taken in a wider context, as the study of processes interrelating systems with inputs and the outputs, and their dynamic structure.

It is in this wider sense that *cybernetics* will be used here, to wit, as system-cybernetics, understanding by 'system' an ordered whole in relation to its relevant environment (hence one actually or potentially open) [11].

Louis Kauffman Cybernetics is the study of systems and processes that interact with themselves and produce themselves from themselves [12].

Gordon Pask Cybernetics is the science or the art of manipulating defensible metaphors, showing how they may be constructed and what can be informed as a result of their existence [13].

F. H. George Cybernetics is concerned primarily with the construction of theories and models in science, without making a hard and fast distinction between the physical and the biological sciences [14].

T. C. Helvey The meaning of the term 'cybernetics' is today somewhat different from that used by Wiener. McCulloh, Rosenblueth and others used the Greek word 'Kybernetes' or helmsmen, to describe an automatic computer [...] the definition, which I first gave in 1966: 'Cybernetics describes an intelligent activity or event which can be expressed in algorithms [15].

Ludwig von Bertalanffy So a great variety of systems in technology and in living nature follow the feedback scheme, and it is well-known thus a new discipline, called cybernetics, was introduced by Norbert Wiener to deal with these phenomena [16].

Louis Couffignal Cybernetics is the art of ensuring the *efficiency* of action [17].

Felix von Cube Cybernetics in a narrow sense is 'the science and techniques of information transforming machines,' and in the broad sense 'cybernetics is a mathematical and constructive treatment of general structural relations, functions and systems, common to various domains of reality [18].

Stafford Beer Cybernetics is the science of effective organization [19].

Herbert Brun Cybernetics is to cure all temporary truth of eternal triteness, because of its uncanny ability to describe complex systems with simplicity, without being simplistic.

From the above definitions we see that cybernetics was characterized as:

- A science.
- An art.
- A branch of mathematics.
- The study of information.
- The general science of communication.
- The science of effective organization.

- The study of interacting systems (internally and externally).
- The methodology of constructing theories and models.
- An intelligent algorithmic activity or event.

It is, therefore, obvious that there is not a consensus of what actually cybernetics is. All scientists considered cybernetics as focused to their own particular fields of concern. A definition which could cover all the spectrum of systems, approaches and applications of cybernetics is an enhancement of Wiener's definition which also includes society's aspects and applications, namely: *'Cybernetics: or Control, and Communication in the Animal, the Machine, and the Society'*. More definitions and interpretations of cybernetics with comments and remarks can be found in [21, 22]. Regarding the societal/behavioral part of cybernetics a useful source of information is a paper written by *Lawrence S. Bale, which* was focused on Gregory Bateson's work, also providing a good overview of the context and themes surrounding the rise of cybernetics and system theory. Lawrence S. Bale noted that: 'Following Bateson, it is my conviction that the patterns of organization and symmetry embodied in living systems are indicative of mental processes; and, that the cybernetic paradigm – with its focus on communication and information as the key elements of the self-regulation and self-organization – best exemplifies these hierarchical patterns of epistemic organization' [64]. A fundamental law of cybernetics says: 'The unit within the system with the most behavioral responses available in it controls the system philosophically'. A classical book on Cybernetics and management is [19].

5.3 Brief Historical Review of Cybernetics

Cybernetics is a very broad transdisciplinary subject which studies human-machine interaction guided by the principle that numerous types of natural, living, technological, and societal systems involve feedback control and communications. As demonstrated in the previous section scientists advocated and promoted very particular and somehow specialized concepts of what is cybernetics. The acceptance and advancement of cybernetics in the United States and Europe were quite different from that of the Soviet Union before the *'glosmot'* and the *perestroika'* [23].

As a field of scientific activity in the United States, cybernetics began in the years after World War II. Between 1946 and 1953 *'Maccy Foundation'* sponsored a series of conferences on the subject of 'circular causal and feedback mechanisms in biological and social systems', which after the publication of Norbert Wiener's book in 1948 [2] has been renamed as *'Macy Conferences' on 'Cybernetics: Circular, Causal and Feedback Mechanisms in Biological and Social Systems'*. In subsequent years cybernetics influenced many academic fields (computer science, electrical engineering, artificial intelligence, robotics, biology, management, sociology, etc.), and inspired the foundation of relevant cybernetics companies and research institutes unfortunately with short-life span, typically up to the retirement or death of their founders. Previous scientists that influenced cybernetics include:

- *A. M. Ampere* In his essay on philosophy of science (1845) introduced the term 'cybernetique' with the same meaning as Plato, i.e., 'art of governing'.
- *S. Trentowski* In his book on management proposed the term 'kibernetiki' as a new Polish word.
- *C. Bernard* In his introduction to experimental medicine, he revealed the role of 'regulation' in the equilibrium of the body (milieu interieur).
- *W. B. Cannon* He introduced the term 'homeostasis' for Bernard's 'milieu interieur'.

In the initial period, dominant interpretations of cybernetics were provided by Norbert Wiener, Warren McCulloch, George Bateson, and Margaret Mead.

- *Norbert Wiener* considered cybernetics as part of electrical engineering where automatic control systems from thermostats to automated. manufacturing systems were playing dominant role (with relevant societies such as IEEE 'Control Systems Society' and later 'Systems, Man and Cybernetics Society'.
- *Warren McCulloh* attempted to understand the operation of the functioning of the brain and the mind. His work served as the foundation of artificial neural networks.
- *Gregory Bateson* and *Margaret Mead* carried out research in the social sciences (anthropology, psychology, family therapy, etc.), thereby promoting cybernetics of social systems (also promoted later by the American Society of Cybernetics: ASC).

Besides them, other research groups on control systems, bioengineering (biofeedback, neurofeedback, etc.), on medicine, and psychology and on self-organizing systems and cellular automata, were tied their work with cybernetics.

The following is a brief outline of historical landmarks from1940s to the present [24].

1940s

- Norbert Wiener, Arthur Rosenblueth, and Julian Bigelow published '*Behavior, Purpose, and Teleology*', where observed behavior was considered as 'purposeful'.
- Warren McCulloch and Walter Pitts published a paper in which they explained how a network of neurons functions.
- By 1949 three central books were published (i) Theory of Games and Economic Behavior (von Neumann and Morgestern), (ii) Cybernetics: Or Control and Communication in the Animal and the Machine (Wiener), and (iii) Mathematical Theory of Communication (Shannon and Weaver).
- Macy Conferences on Cybernetics were organized in New York City (key participants: Norbert Wiener, Julian Bigelow, John von Neumann, Margaret Mead, Gregory Bateson, Rosh Ashby, Grey Walter, and Heinz von Foerster).

1950s

- Many Cybernetics Laboratories and Research Institutes were founded (e.g., Biological Computer Laboratory: BCL, at the University of Illinois, 1958; Mental Health Research Institute at the University of Michigan: Founders McCulloch and Foerster).
- The Proceedings of many Macy Conferences, edited by von Foerstier were published, inspiring further research on Cybernetics.

1960s

- Several Conferences on Cybernetics and Self-organizing systems were organized.
- The American Society for Cybernetics (ASC) was founded (1964).

1970s

- *Heinz von Foerster* initiated the study of second-order cybernetics and coined the term. Second-order cybernetics (or constructivist epistemology) was used profitably on family therapy systems [25].

1980s

- Joint conferences of the American Council of Learned Societies and the Soviet Academy of Sciences were initiated.
- The second Soviet American Conference was held in Tallin, Estonia (1988) with the original topics of epistemology, management, and methodology been expanded (after 'perestroika') to involve 'large scale experiments'.
- Interest on second-order cybernetics in Europe was strengthened and ASC decided to focus their attention and advance second-order cybernetics.
- Many seminars and tutorials on first-order and second-order cybernetics were delivered by members of ASC.

1990s

- The work on second-order cybernetics was changing. Engineering cybernetics was in the air from mid-1940s to mid-1970s. Biological cybernetics lasted from mid 1970s to mid-1990s. The field of social cybernetics started in the mid-1990s.

Engineering cybernetics was based on the realist view of epistemology that distinguishes reality vs. scientific theories, and advocates that knowledge can be used to change natural process at will for the benefit of people.

Biological cybernetics was based on the biological view of epistemology that distinguishes realism vs. constructivism and advocates that ideas about knowledge should be rooted in neurophysiology. They also argued that if people accept constructivism they will be more tolerant.

Social cybernetics was based on the pragmatic view of epistemology (knowledge is constructed to achieve human goals) that distinguishes cognition vs. the observer as a social participant. It was also based on the view that transforming conceptual systems (via persuasion, not coercion) we can change society.

- Symposia on the transitions of the former Soviet Union continued to be held, in Vienna, within the framework of European Meetings on Cybernetics and System Research, to bring together scientists from East and West.
- A Socio-Cybernetics Working Group was founded within the International Sociological Association.

2000s

- Large annual conferences on informatics and cybernetics were organized in Orlando, Fl., which has motivated work on Cybernetics in Latin America.
- Annual conferences on reflexive control began to be held in Moscow and inspired the founding of the Russian Association of Cybernetics.

In the *Soviet Union,* cybernetics has followed a course that is almost diametrically opposite from the one that was followed in the West [23, 26]. In the West, Wiener's ideas were greeted with enthusiasm, and are now viewed with mixed emotions. In the Soviet Union Wiener's ideas were originally greeted coldly, but now they have been accepted (with some exceptions) enthusiastically in both science and government. The term 'cybernetics' was used by the Soviets with the widest possible scope, embracing fields from control theory and mathematical economics to computer science (artificial intelligence, information/communication theory), and military command and control.

Although, Wiener's book on cybernetics was published in 1948, due to the Stalin's dogma the history of Soviet cybernetics began in 1953, after Stalin death. In 1955 the complexion of Soviet cybernetics changed. Three events took place in this year:

- A paper entitled 'Basic Traits of Cybernetics' was published by S. L. Sobolev, A. A. Lyapunov and A. I. Kitor (professors at Moscow University) in the journal *Questions of Philosophy* (No.4, 1955, 136–147).
- A series of seminars on cybernetics began at Moscow University: Attending the seminar were many people who subsequently became prominent in Soviet cybernetics. This seminar was primarily devoted to the review of Western work on cybernetics.
- A 23-page survey of recommended literature on cybernetics and its applications was published in the *Lenin Library* corresponding to the *Library of Congress* in the US.

The development of cybernetics in the Soviet Union after 1955 is as follows.

1957

In 1957 Soviet delegates attended the first UNESCO-sponsored conference on Cybernetics, and a conference on Cybernetics was organized by the Laboratory of Electromodeling of the Soviet Academy of Science (500 participants from 90 different organizations).

1958

The serious Soviet scientific literature on cybernetics made its first appearance in this year, with volume 1 of *Problems of Cybernetics* (Problemy Kibernetiki) a hard-bound, irregularly published collection of papers on all aspects of kybernetics.

1959

On April 10, 1959 the Academy of Science founded the 'Scientific Council for Complex Problem of Cybernetics'.

1960

- One of the most prominent scientific journals in the Soviet Union (Dklady Academic nauk; Transactions of the Academy of Sciences) introduced a section on 'Cybernetics and Control Systems'.
- A series of hard-cover translations of Western works in cybernetics (Cybernetics Collection) made its appearance.

1961

- A collection of articles entitled 'Cybernetics in Communism's Service' edited by A. I. Berg was issued.
- A collection of articles entitled 'Philosophical Problems of Cybernetics' was also issued.
- The Program of the Communist Party of the Soviet Union adopted by the 22nd Party Congress pointed out that for the next 20 years cybernetics, control systems, and electronic computers will be widely applied in all sectors of the country (industry, production processes, building, accounting, statistics, and management).

1962

A section on Cybernetics was introduced in the Reference Journal-Mathematics (Referativni Zhurnal-Matematika); the Soviet equivalent of Mathematics Reviews.

1963

The 'Academy of Sciences: Technical Section' launched a new journal entitled: Engineering Cybernetics (Teknicheskaya Kibernetika) covering all aspects of automatic control, information theory, adaptive systems, computer design, and reliability.

Some further developments of Soviet Cybernetics (after the 1970s) were included in our discussion above on the Western cybernetics. Marxism in Soviet Union was seen as the *'scientific'* approach to *'organizing society'*, and cybernetics promised to bring the physical, engineering and social sciences all under a unifying framework umbrella.

In Soviet Union there are still departments and professorships in cybernetics. However, in the West as well as in Soviet Union there was never agreement on cybernetics fundamentals a fact that prevented unity.

As *Eric Dent* argues [27], after World War II the systems science dramatically expanded the scientific enterprise in eight directions:

- Causality.
- Determinism.
- Relationships.
- Holism.
- Environment.
- Self-organization.
- Reflexivity.
- Observation.

Each subfield developed its own language, theories, methods, traditions and results, and as so, issues that are important in one field are less important or are not present in other fields. This implies that the various systems subfields did not select unique aspects and dimensions. As stated in [24] 'the above scientific dimensions have both *united* and *divided* the systems sciences. The dimensions unite the systems sciences because each of the subfields of systems science uses at least one of the new assumptions, whereas classical science uses none. The dimensions divide the systems sciences because each subfield emphasizes a different dimension or set of dimensions' [24]. It is hoped that 'in the twenty-first century the progress made in developing the field of cybernetics in many disciplines will be successfully integrated [28].

Additional aspects of cybernetics history can be found in [29–31].

5.4 First- and Second-Order Cybernetics

5.4.1 First-Order Cybernetics

The field of cybernetics, as initiated by Wiener, and discussed ontologically and historically in Sections 5.2 and 5.3, is the so-called *'classical cybernetics'*. After the introduction by von Foerster of the concept *'cybernetics of cybernetics'* or *'second-order cybernetics'* in his 1975 book: 'Cybernetics of Cybernetics', classical cybernetics has been named *'first-order cybernetics'* or *'old cybernetics'*. The term cybernetics of cybernetics was the title of Margaret Mead's inaugural keynote address at the founding meeting of the American Society of Cybernetics (1968) but it was von Foerster who gave her the title and the briefing for her keynote address. Therefore, in the literature, the term 'cybernetics of cybernetics' is linked with Heinz von Foester's name who also was using the term 'cybernetics of observing systems'.

As *Ranulph Glanville* notes [32] the relationship of first-order Cybernetics to second-order Cybernetics is like the relationship between the Newtonian view of the universe, and the Einsteinian view. Just as Newton's description remains totally appropriate and usable in many instances (e.g., flights to the moon), so first-order Cybernetics also maintains its value and very often provides us with all we need (e.g., in many control problems). Also, just as the Newtonian view is now understood to be a special (simplified) restricted and slow form of Einstein's view, in the same way first-order Cybernetics is a special, simplified, restricted and linear version of second-order Cybernetics'.

First-order Cybernetics had an engineering nature and provided a powerful approach for solving various practical problems, particularly of a technological nature although it was also successfully applied to social systems. First-order cybernetics introduced the following concepts [33]:

- Boundaries.
- Sub- and supra-systems.
- Circular causality.
- Feedback (positive and negative).
- Simulation.

Boundaries As we saw in Chapter 3, clearly defining the boundaries of the system under study is very important. This is inevitably always performed in time-dependent, observer-dependent, or problem-dependent ways which the system designer needs to know in advance.

Sub- and Supra-Systems The cyberneticist needs to know what are the supra-systems in which the system under study belongs, and what are the subsystems (parts, components) which should particularly be considered.

Circular causality Cybernetics is always interested in the circularity in which the observer (agent) observes what is happening in the system and acts on that system (of course, the relation between observing and acting should be explored). We recall here that the Macy conferences were on 'Circular, Causal, and Feedback Mechanisms in Biological and Social Systems'. In first-order cybernetics the observer was seen as acting on the observed, but the observed was not understood to act on the observer. There was something called *feedback* by which the results of the observer's actions were returned back to him. However, the circularity of this got lost behind a perceived need to give precedence to a view dominated by Newtonian thermodynamics, in which the very small amount of energy which was to be fed-back to the observer was taken to be negligible. Therefore, the actions of the observed were seen as acted on and amplified by the observer, but not via versa. It was later realized that a world of information is not one dominated by energetic, and that the essential circularity is not tied to energetic. Therefore the logic of information-based circularity began to be recognized. Cybernetics exhibits circularity as its core insight, and, through that the related concepts of circular causality (**CC**), interaction, betweeness, and so on. In cases we cannot see the mechanism we would like to explain, we use the *black box* to overcome this difficulty, because, by definition, we cannot see inside a *black box.*

Positive and negative feedback An example which was studied in first-order Cybernetics is *'thermostat'*. A thermostatic system involves two principal components: a heater and a sensor/switch. The heater provides heat, and the sensor/switch controls the heater turning *on* and *off* according to whether or not the sensor attached to the switch has exceeded a desired (goal) temperature. If the environment is so warm that the desired temperature is exceeded, the switch turns the heater off. When the temperature of the environment drops, below the desired temperature, the switch turns the heater on. Examples of positive feedback include error/deviation-amplifying and morphogenetics feedback loops which can either occur spontaneously or be engineered. First-order cybernetics was primarily concerned with negative feedback loops, because its purpose was generally to steer technological, industrial, and managerial systems by keeping them on a steady course, fluctuating within specified margins around equilibrium.

Figure 5.1 shows the typical sub-processes (or implementation steps) of the cybernetic control process, and exhibits its circularity [66].

From Figure 5.1 it follows that (like all cases of feedback control) the basic steps of cybernetic control are the following [66]:

Step 1: Set the standards, goals, or targets of the system.
Step 2: Measure the actual performance.

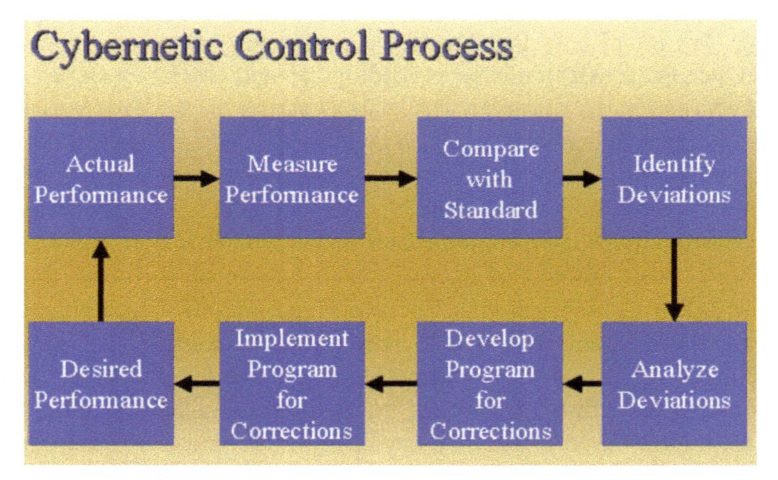

Figure 5.1 Structure of the typical cybernetic control process.

Source: www.swlearning.com (/management/Williams/effective1e/powerpoint/ch06/sld006.htm).

Step 3: Compare the actual performance with the goals and performance standards.

Step 4: Evaluate the result and proceed to proper feedback corrective actions, if required.

The questions that rise here are:

- What should be controlled?
- Is cybernetic control possible? To what extent? (Clearly, if a step cannot be implemented, control may not be possible).
- Is more control necessary?
- Is more control possible?
- What should be done if more control is necessary but not possible?
- What is the degree of dependence? (i.e., how much a company needs particular resources to accomplish its product/service?).

Referring to organizations, Williams [66] indicates the following issues that must be controlled:

- *Balanced scorecard*: To this end, managers set specific goals and apply proper financial measures.
- *Economic value added*: This is the amount by which profits exceed cost of capital cost.

- *Quality*: Three typical ways of measuring quality are excellence, value and conformance to expectations.
- *Customer defections*: To control the rate by which customers are leaving the company, managers don't rely solely to customer satisfaction surveys. Of course it is more difficult to get new customers than retaining the existing ones.
- *Waste pollution*: To control waste pollution managers select good housekeeping methods, perform proper substitutions of materials and products, and suitable process modification.

Simulation This is a technique or methodological tool developed for the computer-based study of man-made or physical systems, which later has also been used for social and behavior-based cybernetics systems.

In first-order cybernetics, the observer is outside the system being observed. The observer treats the observed system as an *artifact,* where neither the artifact nor the observer is changed by the act of observation: all that happens is that a record of the behaviors (states) observed is recorded, supposedly *'as is'*. The goal is always external to the system under study and is actually a state that it is desirable and must be achieved by the system. This means that there is a separation between the system and the goal.

5.4.2 Second-Order Cybernetics

Second-order (new) cybernetics aims to move away from 'old cybernetics' which is tied to the image of the machine and physics. Second-order cybernetics closely resembles organisms and biology and attempts to overcome entropy by using 'noise' as positive feedback. Heinz von Foerster defined first-order cybernetics as the *'cybernetics of observed systems'*, and second-order cybernetics as the *'cybernetics of observing systems'*. The main difference is that the essentially biological paradigm of second-order cybernetics includes the observer(s) in the system under study, which moreover are generally living systems, rather than inanimate man-made systems and artifacts (e.g., robots).

Heinz von Foerster attributed the origin of second-order cybernetics to the efforts of classical cyberneticists to build a model of the mind. He observed that 'researchers realized that [...] a brain is needed to write a theory of a brain [...] the writer of this theory has to account for her or him. Translating this into the domain of cybernetics, it means that cyberneticist, by entering his own domain, has to account for his/her own activity. Cybernetics then becomes *cybernetics of cybernetics* or *second-order cybernetics*.

Humberto Maturana [34] has stated the theorem that 'Anything said is said by an observer'. Heinz von Foerster added a corollary to Maturana's theorem which says: 'Anything said is said to an observer'. Von Foerster points out that with these two propositions 'a nontrivial connection between three concepts has been established. First, that of an *observer* who is characterized by being able to make descriptions. This is because of Maturana Theorem. Of course what an observer says is a description. The second concept is that of *language*. Maturana's theorem and its corollary connect two observers through *language*. But in turn, by this connection a third concept is established namely that of *society:* the two observers constitute the elementary nucleus for a society' [35].

Actually, second-order cybernetics has enhanced classical (old) cybernetics to include the interface with the observer, the 'subjective' feature of cognitive methodology (a theory of the observer based on functional constructivism). That which is observed cannot be neatly abstracted and separated from the observer's own biological, nervous and cerebral structuration. We will now discuss the issues of control and communication within the approach of the Cybernetics of Cybernetics (i.e., when circularity is taken into account) [36].

Control To understand control in the framework of second-order cybernetics we will consider the *thermostat* example included in most classical cybernetics books. In studying the thermostat classical apprentices usually refer to a *switch* that measures the temperature in the room and turns *on* or *off* the boiler that produces and distributes heat in the room. The switch controls the boiler (and hence the temperature of the room). But this is not a complete description, because what it makes the switch turn *on* and *off* is the boiler which supplies heat in the room. This means that the switch controls the boiler (and hence the heat supplied to the room) turning it *on* and *off,* and, similarly, the boiler by supplying heat to the room, turns the switch *on* and *off.* The above process possesses *circularity,* and we can say that in cybernetic systems which involve feedback, control is always essentially circular. Therefore, the question is that: 'if control is circular where is it?' Also, in the case where there is control and a controller, which is which?

Glanville [36] argues 'that control can be neither in the controlled nor in the controller, but lies between them, i.e., it is *shared.*' Furthermore, he says *'there is no control and controller.* These are at best just roles. Each is

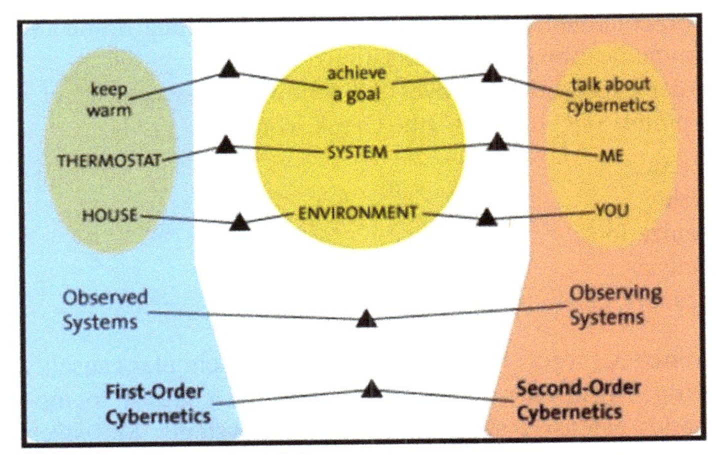

Figure 5.2 Relation of first-order and second-order cybernetics with reference to the *'thermostat'* cybernetic system. First-order cybernetics is concerned with observed systems', whereas second-order cybernetics deals with 'observing systems'.

Source: www.pangaro.com/designsummit/

controller to the other's controlled. Neither is controlled, neither is controller: control is in either (or) both but shared between.'

A pictorial illustration of the relation of first-order cybernetics and second-order cybernetics that refers to the 'thermostat' example discussed above is shown in Figure 5.2.

Communication To exercise control, cybernetic systems need communication. Without intent of communication feedback cannot exist and be implemented. The control intent has also to be communicated. Thus, communications are necessary to the exercise of control, and so to cybernetic systems. As argued in *Glanville* [36] the communication is more than that, i.e., it is a cybernetic system-and a second-order cybernetic system at that. Communication is a second-order (openly reflexive) process because it can communicate about communication. Communication occurs between beings that build understandings (meanings) out of their interpretations of what (they sense that) their conversational partner give to them. This understanding is fed back to their partner(s) in new offerings that the partner(s) in turn interpret compare to their original interaction.

This circular process is a conversation; it is a complex model that operates both as communication and as communication about communication, simultaneously. In the same way as linear control is a particular limited version of circular control, communication (coding) is also a particular limited version of circular communication (i.e., conversation).

Second-order cybernetics involves the following concepts all of which start with the prefix *'self'* or *'auto'* (the Greek word 'self') [33]:

- Self-reference.
- Self-steering.
- Self-organization.
- Auto-catalysis.
- Autopoiesis.

Self-reference All forms of self-reference involve circular causality.

Self-steering This is the bottom-up steering in contrast to top-down steering and finds extensive application in social systems. Self-steering is clearly linked to circular causality.

Self-organization This is the capability of biological, natural and society systems to change their structure by themselves during their interaction process with the environment. Natural systems have a tendency to become more orderly, without external interaction. Again, self-organization is based on circular causality.

Auto-catalysis In molecular chemistry one distinguishes autocatalytic cycles, whereby the product of a reaction catalyzes its own synthesis, and cross-catalytic cycles where two different (group of) products catalyze each other's synthesis. In this case we say that we have a *'collectively autocatalytic'* set of chemical reactions, if a number of those reactions produce as reaction products, catalysts for enough of the other reactions such that the whole set of chemical reactions is self-sustaining given an input of energy and input molecules.

Autopoiesis This concept was introduced by Maturana and Varela (1970s) [7] to mean self-production and differentiates the living from the non-living. It is useful to mention that Niklas Luhmanm defined social systems as consisting of communications constituting autopoietic networks, rather than of individuals, or roles, or actions.

According to *Pask* [37] the differences between first-order (old) cybernetics and second-order (new) cybernetics are the following:

- From information to coupling.
- From the reproduction 'of order-from-order' (Schrondiger) to the generation of 'order from noise' (von Foerster [25]).
- From transmission of data to conversation.

- From external to participant observation (a process that could be assimilated to autopoiesis).

Umpleby [24] collected the following definitions that distinguish first-order and second-order cybernetics:

- *Von Foerster* (First-order cybernetics is the cybernetics of observed systems, second-order cybernetics is the cybernetics of observing systems).
- *Pask* (First-order cybernetics is the purpose of a model, Second-order cybernetics is the purpose of modeler).
- *Varela* (First-order cybernetics is the controlled system, second-order cybernetics is autonomous system).
- *Umpleby* (First-order cybernetics is the interaction between variables or the theories of social systems, second-order cybernetics is the interaction between observer and observed or the theories of the interaction between ideas and society).

5.5 Social Systems and Sociocybernetics

In the following we will briefly present the fundamental issues of *social systems* and *sociocybernetics* [33, 38, 43]. We start with an outline of the contributions of three eminent thinkers on social systems/cybernetics; namely: Talcott Parsons, Walter F. Buckley, and Niklas Luhmann. Then, we discuss the three classes of social systems (macro, meso, micro systems), and we illustrate the application of first- and second-order cybernetics to sociology/social systems.

5.5.1 Social Systems

All social systems receive input from the environment, engage in processes, and produce outputs to serve particular functions. Three principal thinkers that contributed to the foundation of social systems theory are:

- Talcott Parsons (1902–1979).
- Walter F. Buckley (1922–2006).
- Niklas Luhmann (1927–1998).

Talcott Parsons

Parsons' structural–functional theory of society was based on concepts derived from cybernetics. He argued that all systems and their subsystems are decomposable. He developed the so-called **AGIL** method in which system functions were defined through four component functions, namely [41]:

- **A**daptation.
- **G**oal attainment.
- **I**ntegration.
- **L**atency.

Each system's function possesses several subsystems with specific roles. He connected the system functions and subsystems roles to the structure of the human body's organs and functions. In his work he was influenced by the cybernetics concepts presented in the Macy Conferences where he participated. He therefore extended himself beyond traditional sociological topics. Parsons believed that the conditions of AGIL paradigm were static, i.e., not influenced by cultural and societal changes. Because of this, and the fact that his models had a hierarchical structure (whereas later new less hierarchical models were developed) Parsons' theory drew considerable criticism and was characterized as not appropriate (too complicated, too rigid, not willing to consider cultural difference).

Walter F. Buckley

Buckley was among the first sociologists who recognized concepts from GST. Actually, his book 'Sociology and Modern Systems Theory' [38] was the first to apply systems theory to sociology. He successfully combined concepts of information, communication, cybernetic feedback, and control, and applied them to sociological systems also employing the related concepts of self-regulation, adaptation, self-organization, and complex adaptive systems. His feeling was that systems theory provides a general framework and set of tools applicable to any scientific domain.

Niklas Luhmann

Luhmann, a student of Parsons, has developed a theory of social systems adapting the view that social systems are communication networks, and a particular system selects what kind of information it will accept. This creates and maintains the identity of the system. Luhmann's position was that 'every social contact is understood as a system, up to and including society as the inclusion of all possible contacts' [39]. His proposal to describe social phenomena like interactions, organizations, or societies as systems has broken the traditional views in sociology of his colleagues. Although little

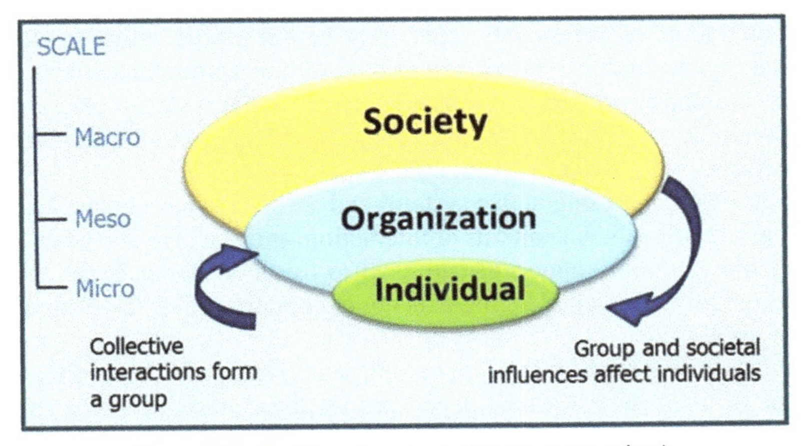

Figure 5.3 Social systems types (macro, meso, micro).
Source: www.taylorpeterson822.blogspot.gr (Talcott Parsons Lecture).

known in US, Niklas Luhmann was recognized as the most important social theorist of the 20th century.

An essential social system is the *family* which has the function of socializing and caring for its members. The theory of family systems examines the dynamic process of a family and intervenes to correct or adjust any malpractice structures or malpractice processes [40]. One essential process for a family and other social systems is communication and information which as we have already seen constitutes an input into a system and an output in the interaction with other systems, while communication regulates and either stabilizes or disrupts a system. A social system belongs to one of the following particular class of social systems (Figure 5.3):

- Macrosystems.
- Mesosystems.
- Microsystems.

- *Macrosystems* focus on large systems (communities, scientific societies, etc.).
- *Mesosystems* focus on intermediate-size systems (organizations, groups, support networks, and extended families).
- *Microsystems* are understood to refer to small-size social systems such as individuals and couples.

This differentiation of systems by size may be somewhat arbitrary. For example, an organization can be viewed as a *meso* unit within the context of its broader community or from a *macro* perspective. Social systems categories include governments, business/manufacturing systems, economic systems, social networks, social webs, political systems, etc.

Families are complex interactive systems and are seen to have goals. They strive to reach their goals/via patterns of interaction, and negative and positive feedback loops. A family may organize itself to foster academic excellence of the children providing challenging educational opportunities that push the children to excel [42].

Although social systems may seem to represent a straightforward concept, they can be difficult to understand depending on the context in which they are considered. In general, social systems theory provides a way of breaking down a larger group (e.g., an entire society), and classifying subgroups in order to understand how their interactions are combined to produce a functional whole.

Using cybernetic methods, sociology attempts to find the reasons of such spontaneous phenomena as smart mobs and riots, and also how communities create rules (e.g., etiquetes) by consensus without formal discussion. McCleland and Fararo [67] review a number of models, including:

- Affect control theory.
- Memetics.
- Sociocybernetics.

Affect control theory was introduced by Osgood [68]. It explains roles behavior, emotions, and labeling theory in terms of homeostatic maintenance of sentiments associated to social categories. It advocates that people maintain affective meaning through their actions and interpretations of events. In affect control theory sentiments are the enduring affective meanings prevailing in society that permit individuals to orient quickly and automatically in different situations. Actually, the activity of social bodies takes place through cultural sentiments [69].

Memetics is a theory of mental model initiated by Dawkins [70] which is analogous to evolution theory. *'Meme'* was defined by Dawkins as a unit of human cultural transmission analogous to the *gene*, advocating that replication also happens in culture but in a different form. Proponents of memetics describe it as an approach to evolutionary models of cultural information transfer, and argue that it provides a framework for handling the most troubling social and military problems at the most causal level.

5.5.2 Sociocybernetics

Sociocybernetics is related to social systems theory in the same way as cybernetics is related to general systems theory. Actually, sociocybernetics can be viewed as *systems science in sociology and other social sciences* or in other words sociocybernetics is dealing with society and social systems, having the same goals as the social system theory. The definition given by Bernard Scott is: "Sociocybernetics is concerned with applying theories and methods from cybernetics and system science to the social sciences by offering concepts and tools for addressing problems holistically and globally". Particularly one of the tasks of sociocybernetics is to map, measure, and find ways of intervening in the network of social forces which drive human behavior. Sociocyberneticists have to understand the control processes that govern the functioning of society (macro, meso, micro systems) in practice, such that to be able to design better ways of more effective and acceptable operation or modify it appropriately. To this end, and to deal with a variety of complicated sociological problems and processes, sociocybernetics employs systems science, systems thinking, and first- and second-order cybernetics methodologies.

In practice, sociocybernetics is mainly based on second-order cybernetics because first-order cybernetics has limited power to deal with social processes where the researcher himself forms part of the subject under study, in contrast with the natural sciences. Sociocybernetics is not restricted to theory but includes application, empirical research, methodology, axiology (i.e., ethics and value theory), and epistemology. In general, research interest in sociocybernetics is focused on studying social phenomena with reference to their complexity and dynamics using the systemic approach, where an observer attempts to trace the variety of interactions in reality instead of analytically isolating individual causal interactions.

Sociology begins when two agents (individuals, groups) are together and start to communicate and interact. Traditional social entities like families, groups, organizations, enterprises, communities, government institutions, etc. can be easily considered as *union of agents* (actor systems) that can themselves be or/not be higher-level non-individual agents. Sociocybernetics is interested on both *upward* (bottom-up) and *downward* (top down) interactions and communication, and not only on a bottom-up 'determination' as in reductionist approaches.

To study social systems using sociocybernetics, all concepts and methods of first- and second-order cybernetics should be employed, e.g., those mentioned in Section 5.4.1 (boundaries, sub- and supra-systems, circular causality, positive and negative feedback, simulation). In addition, we have to include in the system under study the interface with the observer, taking into account the existing circularity in both the communication and control processes.

Sociocybernetics advocates that all living/societal systems go through the following six levels (phases) of social interactions, known as **ABCDEF** social contracts:

- ***A****ggression* (either survive or die).
- ***B****ureaucracy* (apply the rules and norms).
- ***C****ompetition* (your loss is my gain).
- ***D****ecision* (revealing intentions and individual feelings).
- ***E****mpathy* (collaboration to achieve a unified goal).
- ***F****ree will* (govern your own existence without external interference, discrimination or control).

These phases shape the framework for the study of any evolutionary system using sociocybernetics (Wikipedia).

Important for the study of social systems with the aid of second-order cybernetics is to understand and make worthy the inherent processes of self-steering, self-organization, auto- and cross-catalysis, and autopoiesis.

Self-organization is inherent in nature, life, and society. However, only after the 1950s the scientific study of self-organization has taken a concrete shape. Primary thinkers who have worked on self-organization are: W. Ross Ashby, Francis Heylinghen, Chris Lucas, Scott Camazine, A.N. Whitehead, and M.B.L Dempster.

Ashby developed the *'Law of Requisite Variety'* which says [44]:

'Any quantity K of appropriate selection demands the transmission or processing of quantity K of information. There is no getting of selection for nothing'. An alternative expression for the law of *Requisite Variety,* which is called the *first law* of cybernetics is the following: 'The unit within the system with the most behavioral responses available to it controls the system'. Ashby's requisite variety law was inspired by *Shannon's Tenth Theorem* which says: 'If an error correction channel has capacity C, then equivocation of amount C can be removed, but no more'. Both Shannon's and Ashby's theorems are applicable to biological systems by saying that 'the amount of regulatory or selective action that the brain can achieve is absolutely bounded by its capacity as a channel'. Cybernetics 1st Law is particularly relevant to the modern development of management and plays a significant role within systems of all kinds. Specifically, it has enormous significance in understanding the development of stronger personal self-determination, tolerance, and wider variety of responses to situations and people around us.

Heylinghen defined self-organization as the spontaneous emergence of global structure out of local interactions [45]. He developed on the consequences of 'spontaneous' performance (which emerges without any internal or external agent participating in the control of the process).

Lucas defined self-organization as the evolution of a system into an organized form in the absence of external constraints, i.e., as a move from a large region of state space to a persistent smaller one, under the control of the system itself [46].

Camazine defined the self-organization in biological systems as a process in which pattern at the global level of a system emerges solely from numerous interactions among the lower-level components of the system [47].

Whitehead states that self-organization of society depends on commonly diffused symbols, evoking commonly diffused ideas, and at the same time indicating commonly understood action [48].

Dempster worked on the distinction between *autopoietic* (self-producing) and *sympoietic* (collectively-producing) systems. These two contrasting lenses offer alternatives views of the world, forcing recognition of system properties frequently neglected. He believed that it is difficult, probably impossible, to find a precise definition of a self-organizing system and stated that 'On the intuitive level, self-organization refers to exactly what is suggested: Systems that appear to organize themselves without external direction, manipulation, or control' [49].

The fundamental mechanisms of nature which enable self-organization to be achieved are the following [50]:

- Synergetics.
- Export of entropy.
- Positive/negative feedback interplay.
- Selective retention.

The achievement of *synergetic state* is in general a '*trial-and-error*' or '*mutual adaptation*' process. The *export of entropy* self-organization mechanism was revealed by Prigozine and Nicolis [61]. Prigozine states that 'order creation' at the macro level is a way of dissipating (exporting) entropy caused by energy flux at the micro level. The 'interplay' between positive and negative feedback self-organization mechanism works as described in Chapter 4. Finally, the *selective retention* mechanism of self-organization ensures that the outcome of the interactions of the system components is not arbitrary, but shows a 'preference' for certain situations over others [45].

Autopoiesis means 'self-creation' and was initially proposed as a system description which was claimed to define and explain the nature of living systems. A basic example of an autopoietic system is the biological cell. An *autopoietic* system is different than an *allopoietic (*externally creating*)* system such as car factory which makes the cars (organized structures) using raw materials (components) which are something *other* than itself (the factory). However, if the components, workers, suppliers, etc. are included in the factory system, then the car making system could be considered to be autopoietic.

Auto-catalysis Chemical reactions that exhibit autocatalysis start slowly and as more of the reactants are converted to the products there is an increase in the rate of the reaction up to a maximum rate. Then, there is a fall in the rate as the concentration of the reactants decreases and the reaction finally comes to an end. A second-order autocatalytic reaction: $A + B \rightarrow 2B$ proceeds at a rate given by $r = k\,(A)\,(B)$ where (*A*) and (*B*) are the concentrations of A and B. The graph of the way an autocatalytic reaction proceeds has a sigmoid form (Figure 5.4).

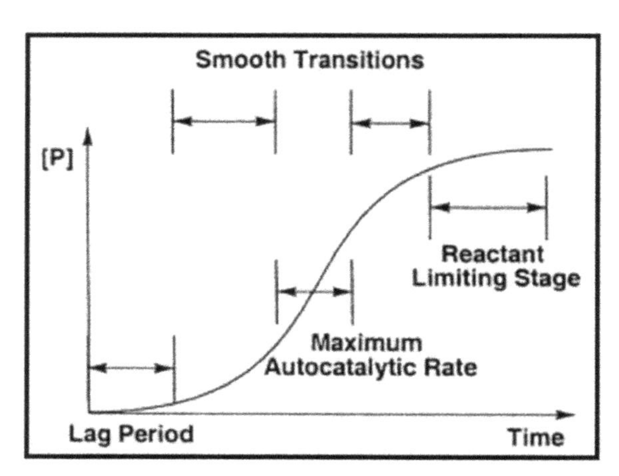

Figure 5.4 Sigmoid evolution over time of the concentration of the product P of an autocatalytic reaction.

The autocatalysis phenomenon occurs also in societies and plays a dominant role, as a central science and technology, in guaranteeing sustainable human life. In a society a catalyst may be: (i) something that causes activity between two or more persons or forces without itself being affected, (ii) a person or thing that precipitates an event or change, (iii) a person whose talk, enthusiasm, or energy causes others to be more friendly, enthusiastic, or energetic (Dictionary.com). Examples of social catalysts are:

- A teacher (for good social performance).
- A financial planner (for social investment).
- Personal values (for social change).
- The Internet (for paradigm shift).
- Sports events (for social development).
- Social media (for social relations and development).
- Social entrepreneurs (for the creation of social enterprises that combine commercial and charitable goals).

5.5.3 Organizational Cybernetics

A modern cybernetics methodology developed for creating viable organizations through the study of structures, roles, and communication and information systems is called '*organizational cybernetics* **(OC)**'. This methodology is based on the theoretical foundations developed by Stafford Beer [19, 62, 63]. As it is always the case with cybernetics systems, the methodology uses communication and feedback tools for getting the error and performing the proper corrective actions in order to achieve the desired goal through dynamic adjustment. Figure 5.5 gives an illustrative picture of organizational cybernetics.

5.6 Conclusions

Cybernetics is the field of inter-causal system processes and navigational networks which constitute the basis for communication, control, and organizational processes in dynamic systems. Since its beginning, cybernetics embraced many approaches, one of which is *circular causality* exhibited by both natural and man-made systems. The principal aim of cybernetics was the exploration of epistemological issues (e.g., how we come to know), and understanding how people and their environment have co-evolved, as well as the phenomena of self-reference, autonomy, purpose, and identity.

Figure 5.5 Pictorial representation of organizational cybernetics.
Source: b67ghttps://www.kbmanage.com/concept/organizational-cybernetics8

Many cyberneticists seek to contribute towards the creation of a more human-minding world, while others attempt to understand the dialogue that takes place between models (or theories) and social systems. From a technological point of view cybernetics is characterized by a tendency to universalize the concept of feedback, seeing it as the underlying principle of the technological and societal world.

Cybernetics has had a significant conceptual, intellectual, and practical impact on a wide spectrum of disciplines and applications world-wide. For example, an important application area of cybernetics is the 'cyborg/implants' area which may help impaired people to get back artificially vital functions they have lost for various reasons (e.g., eyeborgs for lost vision, cochlear for lost hearing, bionic limbs for lost arm/finger mobility, pacemaker/cardioverter defibrillator implants for heart arrhythmia, etc.).

Branches of cybernetics involve (non-exhaustively) the following:

- Information and communication theory.
- Network theory.
- Decision making and game theory.
- Connectionist theory.
- Biological cybernetics.
- Economic cybernetics.

- Computational cybernetics.
- Marine cybernetics.
- Psycho-cybernetics.
- Socio-cybernetics.
- Human-machine cybernetics.

In this chapter, we have provided the fundamental ontological and epistemological issues of cybernetics, namely what is cybernetics (several views), what is the history of cybernetics, first- and second-order cybernetics, and socio-cybernetics. The concepts features and capabilities of cybernetics were presented at a fluent conceptual level suitable for both the non-expert reader and the systems and control theory professional. More detailed and complete material can be found in classical books on cybernetics (first order and second order) and its branches [51–60].

References

[1] Available at: http://en.wikipedia.org/wiki/Plato
[2] Wiener, N. (1948/1961). *Cybernetics: Or Control and Communication in the Animal and the Machine*. Cambridge, MA: The MIT Press.
[3] Kolmogorov, A. N. (1958). *Cybernetics, Bolschaya Sovetskays Entsiklopedia* 51st Supp. 149.
[4] Ashby, W. R. (1956). *An Introduction to Cybernetics*. London: Chapman and Hall, 4–5.
[5] Bateson, G. (2016). *Cybernetics and the Social Behavioral Sciences*, Narbeth Pa.com. Available at: www.anylogic.com/blog?page=post&id=18
[6] Foerster, H., and Asaro, P. M. (2007). *Heinz Foerster and the Biocomputing Movements of the 1960s*. Available at: http://cybersophe.org/writing/Asaro%20HVF%26BCL.pdf
[7] Maturana, H., and Varela, F. (1978). *Tree of Knowledge: The Biological Roots of Human Understanding*. Boulder, CO: Shambahla.
[8] Griffin, E. (2008). *A First Look at Communications Theory,* 7th edn. New York, NY: McGraw-Hill, 43–44.
[9] Greniewski, H. (1960). *Cybernetics without Mathematics*. Oxford: Pergamon.
[10] Ericson, R. F. (1972). Vision of cybernetic organizations. *Acad. Manag. J.* 15, 427.

[11] Laszlo, E. (1972). *Introduction to Systems Philosophy: Toward a New Paradigm of Contemporary Thought*. Philadelphia, PA: Gordon & Breach.
[12] Kauffman, L. (2007). *Cognitive Informatics and Wisdom Development*. (Cited in and Targowski, A) Hershey, PA: IGI Global, 68.
[13] Pask, G. (1966). "The cybernetics of human performance and learning," in *Facets of Systems Science*, ed. T. Klir (Berlin: Springer), 1989, 426.
[14] George, F. H. (1962). *The Brain as a Computer*. Oxford: Pergamon Press, 2.
[15] Helvey, T.C. (1971). *The Age of Information: An Interdisplinary Survey of Cybernetics*. Englewood Cliffs, NJ: Educational Technology Publications, 6.
[16] von Bertalanffy, L. (1968/1988). *General Systems Theory: Foundations, Development, Applications*. New York, NY: George Braziller.
[17] Couffignal, L. (2011). *Cognitive Informatic and Wisdom Development*, (Cited in A. Targowski), Hershey, PA: IGI Global, 68.
[18] von Cube, F. (1967). *Was ist Kybernetik?* Bremen: C. Schunemann Verlag.
[19] Beer, S. (1959). *Cybernetics and Management*. New York, NY: John Wiley.
[20] Ampere, A. M. (1845). *Essai sur la Philosophie de Sciences*. Paris: Chez Bachelier:
[21] Drozin, V. G. (1976). The different meaning of cybernetics. *Cybernetics Forum ASC* III, 28–31.
[22] ASC: American Society for Cybernetics, Defining 'cybernetics'. Available at: http://www.gwu.edu/~asc/cyber_definition.html
[23] Levien, R., and Maron, M. E. (1964). *Cybernetics and its development in the Soviet Union, Memo RM-4156-PR*. Santa Monica, CA: RAND Corporation.
[24] Umpleby, S. A. (2008). A short history of cybernetics in the United States. *Osterr. Z. Geschichtswiss*. 19, 28–40.
[25] von Foerster, H. (1996/1975). *The Cybernetics of Cybernetics, Biological Computer Lab*. Mineapolis, MN: Future Systems, Inc.
[26] Holloway, D. (1974). Innovation in science: the case of cybernetics in the Soviet Union. *Soc. Stud. Sci*. 4, 299–337.
[27] Dent, E. B. (1996). *The Design, Development, and Evaluation of Measures, In Survey Worldview in Organizations*. Ann Arbor, MI: University Microfilms.

[28] Umpleby, S., and Dent, E. B. (1999). The origin and purpose of several traditions in systems theory and cybernetics. *Cybern. Syst.* 30, 79–103.
[29] Valle, R. (2008). "History of cybernetics," in *Systems Science and Cybernetics*, ed. F. Parra-Luna, (Oxford: UNESCO EOLLS Encyclopedia).
[30] ASC: American Society of Cybernetics, History of Cybernetics. Available at: http://www.asc-cybernetics.org/foundations/history.htm
[31] Francois, C. (1999). Systemic and Cybernetics in a historical perspective. *Syst. Res. Behav. Sci.* 16, 203–219.
[32] Francois, C. (1999). Systemic and Cybernetics in a historical perspective. *Syst. Res. Behav. Sci.* 16, 203–219.
[33] Glanville, R. (2008). "Second order cybernetics", in *Systems Science and Cybernetics, UNESCO EOLLS Encyclopedia*, ed. F. Parra-Luna. Paris: Eolss Publishers.
[34] What is Sociocybernetics? The Research Committee on Sociocybernetics. Available at: http://mgterp.freeyellow.com/academic/whatis.html
[35] Maturana, H. (1970). "Neurophysiology of cognition," in *Cognition: A Multiple View*, ed. R. Carvin, New York, NY: Spartan Books.
[36] von Foerster, H. (1979). *Cybernetics of Cybernetics, Understanding of Understanding, Keynote Address, Essay on Cybernetics and Cognition*, Berlin: Springer.
[37] Glanville, R. (2004). The purpose of second-order cybernetics. *Kybernetes* 33, 1379–1386.
[38] Pask, G. (2016). Heinz von foerster's self-organization, the progenitor of conversation and interaction theories. *Syst. Res.* 13, 349–362.
[39] Buckley, W. (1967). *Sociology in Modern Systems Theory*. Englewood Cliffs, NJ: Prentice-Hall.
[40] Luhmann, N. (1995). *Social Systems* Translated by J. Bednarz and D. Beaker. Stanford, CA: Stanford University Press.
[41] Minuchin, S. (1974). *Families and Family Therapy*, Boston, MA: Harvard University Press.
[42] Parsons, T. (1977). *Social Systems and the Evolution of Action Theory*. New York, NY: Free-Press.
[43] Family Systems Theory-Basic Concepts/Propositions Available at: http://family.jrank.org/pages/597/Family-Systems-Theory-Basic-Concepts/propositions
[44] Hornung, B. R. (2005). "Principles of sociocybernetics," *Proceedings of 6th ESSU Symposium on Sociocybernetics*, Paris, 19–22.

[45] Ross Ashby, W. (1947). Principles of the self-organizing dynamic systems. *J. General Psychol.* 37, 125–128.
[46] Heylinghen, F. (2008). *Complexity and Self-Organization, Encyclopedia of Library and Information Sciences*, eds, M. T. Bates and M. N. Mack, London: Taylor and Francis.
[47] Lucas, C. (1997). *Self-organization, FAQ*. Available at: http://psoup.math.wisc.edu/archive/sosfaq.html
[48] Camazine, S., Deneubourg, J. L., Francis, N. R., Sneyd, J., Theraulaz, G., and Bonablau, E. (2001). *Self-organition in Biological Systems*. Princeton, NJ: Princeton University Press.
[49] Whitehead, A. N. (1927). *Symbolism: It's Meaning and Effect*. Oxford: Mac Millan.
[50] Dempster, M. B. L. (1998). *A Self-Organizing Systems Perspective on Planning for Sustainability*, Master thesis, University of Waterloo, Ontario.
[51] Tzafestas, S. G. (2018). *Energy, Information, Feedback, Adaptation, and Self-Organization: The Fundamental Elements of Life and Society*. Berlin: Springer.
[52] Dechert, C. R. (1967). *Social Impact of Cybernetics*. London: Simon & Schuster.
[53] Guilbaut, G. T. (1960). *What is Cybernetics?* New York, NY: Grove Press.
[54] Porter, A. (1969). *Cybernetics Simplified*. London: English Universities Press.
[55] Pask, G. (1961). *An Approach to Cybernetics*. London: Hutchinson & Co. Ltd.
[56] Ya Lerner, A. (1972). *Fundamentals of Cybernetics*. London: Chapman and Hall.
[57] Klir, J. and Valach, M. (1967). *Cybernetics Modeling*. London: Fourth Estate.
[58] Frank, G. H. (1960). *Automation, Cybernetics, and Society*. Melbourne, VIC: Leonard-Hills Books.
[59] Frank, G. H. (1976). *Cybernetics (Teach Yourself)*. London: Hodder & Stroughton.
[60] Maltz, M. (1989). *Psycho-Cybernetics: A New Way to Get More Living Out of Life*. New York, NY: Pocket Books.
[61] Birnbaum, R. (1991). *How Colleges Work? The Cybernetics of Academic Organization*. San Francisco, CA: Jossey-Bass.

[62] Prigozine, I., and Nicolis, G. (1996). *Self-Organization in Non-Equilibrium Systems*. New York, NY: Pocket Books.
[63] Beer, S. (1979). *The Heart of Enterprise*. Chichester, NY: John Wiley.
[64] Beer, S. (1985). *Diagnosing the System for Organizations*, Chichester, NY: John Wiley.
[65] Bale, L. S. (1995). Gregory Bateson, Cybernetics and the social/behavioral sciences. *Cybernet. Hum. Knowing* 3, 27–45.
[66] Scott, B. (2008). "The role of sociocybernetics in understanding world futures," in *Proceedings of 8th International Conference on Sociocybernetics, Complex Systems, Interdisciplinarity, and World Futures, Ceudal,* Mexico, 24–28.
[67] Williams, S. (2002). *Effective Management*, Chichester, NY: South Western College Publishing, 2002.
[68] Mc Cleland, A., and Fararo, T. J. (eds.), (2006). *Purpose, Meaning, and Action Control Systems Theories in Sociology*. New York, NY: Palgrave Macmillan, 2006.
[69] Osgood, C. H., May, H. H. and Miron, M. S. (1978). *Cross-Cultural Universals of Affective Meaning*. Urbana, NY: University of Illinois Press.
[70] Hese, D. R. (2002). "Understanding Social Interaction with Affect Control Theory," in *New Directions in Sociological Theory*, eds, J. Berger and M. Zelditch, Boulder, CO: Rowman and Littlefield.
[71] Dawkins, R. (1976). *The Selfish Gene*. Oxford: Oxford University Press.

6

Control

One of the things I've learned is to be receptive of feedback.
Ben Siebermann

The image of the world around us, which we carry in our head, is just a model.
Nobody in his head imagines all the world, government or country.
He has only selected concepts, and relationships, between them, and uses those to represent the real system.
Jay Wright Forrester

Measurement is the first step that leads to control and eventually to improvement. If you can't measure something, you can't understand it.
If you can't understand it, you can't control it.
If you can't control it, you can't improve it.
H. James Harrington

6.1 Introduction

Control theory and engineering is a vast scientific and technological field which focuses on analysis and design of systems to improve the speed of response, accuracy, and stability of systems. The term '*control*' has many meanings which often are different in different communities. Control is based on the concept of feedback and computer algorithms. It is *ubiquitous* and surrounds our everyday life with a plethora of applications in society and everyday life which for the most part humans are unaware of them. Human body is a huge collection of control systems: cells, tissues, organs all function according to specific biological, chemical or physical laws, and they are controlled on the basis of rules that obey these laws.

Actually, at its core, control is an information-based science, and employs and processes information in both analog and digital form. The two categories of control methods are (i) classical control methods, and (ii) modern control methods.

The purpose of the present chapter is to provide a review of the control systems field. Specifically, the chapter:

- Gives fundamental answers to the ontological questions 'what is feedback?', and 'what is control?'
- Provides a brief historical review of feedback and control including the periods of prehistoric and early control, pre-classical control, classical control, and modern control.
- Presents fundamental epistemological/methodological elements of classical control.
- Summarizes the epistemological/methodological elements of the basic branches of modern control, namely: state-space modeling, controllability and observability, Lyapunov stability, state-feedback control, optimal control, stochastic control, knowledge based control, neural network control, fuzzy logic control, and hybrid control.
- Discusses basic ontological aspects (features, architectures, benefits, drawbacks) of networked control systems.

6.2 Feedback and Control

6.2.1 What is Feedback?

According to *Cambridge Dictionary*, 'feedback is the return into a machine or system of part of what it produces, especially to improve what is produced. According to the *Columbia Encyclopedia* (2008 edition), 'feedback is the arrangement for the automatic self-regulation of an electrical, mechanical, or biological system by returning part of its output as input.' Extending this definition we can state that '*feedback* is any response or information about the result of a process'. In all cases 'feedback' is used to achieve and maintain a desired performance of the system or process at hand (natural or man-made, or biological, etc.). For example, in a management control system, 'feedback' is a concept equivalent to 'information about actual performance' in comparison with the planned performance, i.e. feedback is the process by which a given system or model is tested to see if it is performing as planned. Via timely feedback it is possible to exert quickly corrective action(s) when the situation goes out of hand.

Other simple examples of feedback are the following:

- If the speed of an engine exceeds a preset value, then the governor of the engine reduces the supply of the fuel, thus decreasing the speed.
- In a thermostat the actual temperature is compared with the desired temperature and the feedback enables to reduce the difference.
- In car direction (driving) control, the driver senses the actual direction of the car and compares it with the desired direction on the road. If the actual direction is to the left of the desired one, the driver turns it a little to the right. In the opposite case the turn is to the left.
- The temperature of a healthy human body is maintained in a very narrow range through the use of feedback (homeostasis).
- In business, communication feedback is the part of the receivers corrective responses communicated back to the sender.

Feedback is distinguished in *positive feedback* and *negative feedback.* The inputs are usually the outcome of the system's environment influence on the system, and the outputs are the results (product) generated by the system due to these inputs. In, other words, the inputs are the '*cause*' and the outputs are the '*effect*', which implies that there is a time separation between inputs and outputs. The inputs are applied 'before', and the outputs are obtained 'after' the elapse of some time. A system with feedback (known as *feedback system)* is pictorially shown in Figure 6.1.

In Figure 6.1 we see that there is a closed loop between input-system-output through the feedback, which is also known as *closed-loop control.* The feedback is realized through a *sensor* which measures the output and sends this information back to the input of the system.

Positive feedback occurs when a change in the output is fed back to the system input such that the output changes even more in the same direction which results in a continuing spiral of change (rheostasis). This implies that each pass around the feedback cycle magnifies the change instead of diminishing it. Positive feedback is used only in systems where the amplification of a signal is needed. For example, in electronic systems positive feedback (of a proper magnitude) is needed in order to obtain an *oscillator* which produces a sinusoidal signal of constant magnitude.

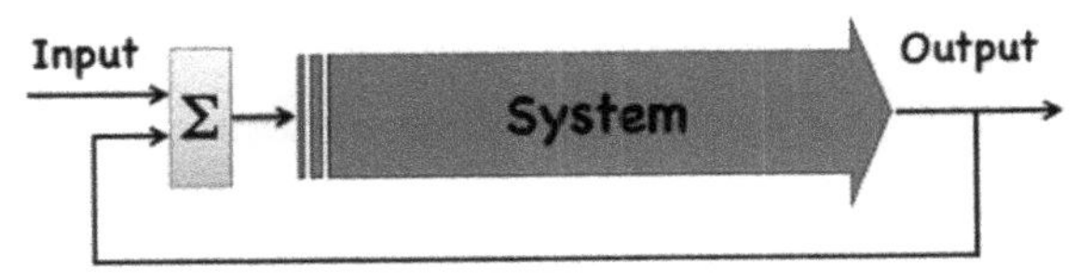

Figure 6.1 Schematic of a feedback system.

Negative feedback occurs when a corrective action is applied in order to '*damp*' or '*reduce*' the magnitude of an '*error*', so that constant conditions are maintained inside the system or biological body (known as *homeostasis*). All the examples given in Section 6.2.1 are *negative feedback (or deviation reducing feedback)* examples.

In general (except from the cases where positive feedback is balanced by an equivalent amount of negative feedback, as in the oscillator example), *positive feedback* leads to *divergent behavior* and, if a positive feedback loop is left to itself, it can lead to indefinite expansion or explosion, which may result in the destruction of the system. Some examples of positive feedback are nuclear chain reaction, proliferation of cancer cells, industrial growth, and population explosion.

Negative feedback leads to desired regulation (i.e., adaptive/goal-seeking behavior) maintaining the level of the output variable concerned (direction, speed, temperature, fluid level, material concentration, voltage level, productivity, etc.).

The goal (reference) to be sought may be:

- **Self-determined** (i.e. with no human involvement) as for example in the case of the *ecosystem* which maintains the composition of the air, or the ocean, or the level maintenance of glucose in the *human blood.*
- **Human-determined** where the goals of the technological (man-made) systems are set by the humans in order to satisfy their needs and preferences. This is actually the case of all servomechanisms and man-made automatic control systems (industrial, social, enterprise, and economic).

6.2.2 What is Control?

Control is tightly connected and inseparable from *feedback*, although in some special cases we use *feedforward (non-feedback) control* with great attention and increased care. Norbert Wiener said that '*the present time is the age of Communication and Control*' [1]. Control, like the feedback, is present in one or the other form in all physical, biological, and man-made systems. A control system involves always two subsystems:

- The subsystem or process to be *controlled* (symbolically **P**).
- The subsystem that controls the first, called *controller* (symbolically **C**).

Also, a control system is always based on two processes:

- Action.
- Perception.

The controller **C** acts on **P** and changes its state in a desired (conscious or unconscious) way. The process under control **P** informs **C** by providing to it a *perception* of the state of **P**. Therefore, control is always a kind of *action-perception* (i.e., *feedback*) *cycle*.

To illustrate in some more detail how a feedback control system operates, we will work on the operational diagram shown in Figure 6.2.

- The *perception* about the state of the system (process) under control is provided by one or more appropriate *sensors* which constitute the *feedback elements*.
- The controller receives and processes the error between the desired and the actual output (provided by an *error detector*) and sends the *control action* (signal) to the system through a suitable *actuator*.
- The *actuator* or *effector* produces the control action in accordance with the control signal and exerts it upon the system.

The goals are set by the owner, the designer or the user of the system, or by the nature in case of biological systems. In order for the control to be effective, the decisions and control actions must be exerted without or with a very small time delay. Otherwise special care is needed. The fact that the above system involves *negative feedback* is indicated by the negative sign in the feedback path, which produces the error signal $\varepsilon = x - y$. If the actual output y is greater than the desired output, then the error ε is negative, and the controller-actuator pair exerts a negative action to the system, forcing it to decrease the actual output y towards the desired output and reduce the error. If $y < x$ then $\varepsilon > 0$ and the action imposed to the system is positive so as to increase y and again reduce the error. The above type of controller, which produces a control signal proportional to the actual error,

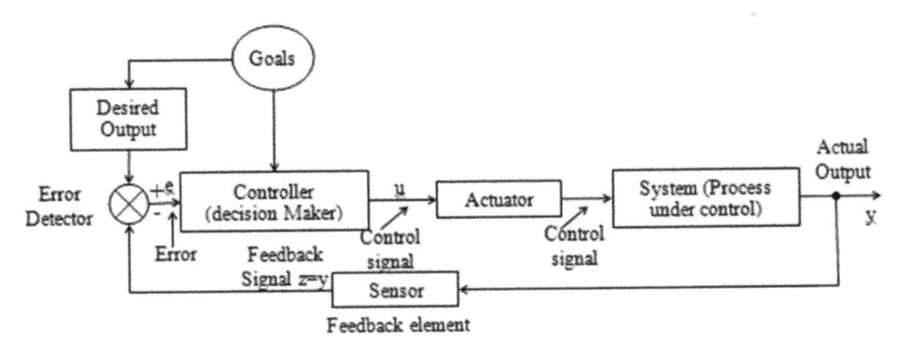

Figure 6.2 Operational diagram of a typical feedback control system.

is called *Proportional Controller* (**P**). Other types of control use either the integral of the error (*Integral Controller:* **I**) or the derivative of the error (*Derivative Controller:* **D**). In practice, we usually use combinations of the above controllers namely: **PI**, **PD**, or **PID**. A controller which is also used frequently in practice is the two-valued controller. Here, the control signal u takes a value in accordance with the *signum* of the error, which changes in a desired sequence of time instants, called *switching times*. If the two values of u are 0 and 1, the controller is *called on-off controller*, if they are -1 and $+1$ the controller is called *bang-bang controller.*

Figure 6.3 shows a typical diagram of a **PID** control system which is still at present used extensively in (physical and chemical) process control applications.

The proportional term P controls the plant/process taking into account the current value $e(t)$of the error, the integral term I controls the plant on the basis of the integral of the error over a certain time period (representing the memory time of the controller), and the derivative term D controls the systems on the basis of the derivative of the error (i.e., the slope of the function $e(t)$) and represents a future value of $e(t)$ (i.e., the predicted value of $e(t)$ at a future time). In other words, the PID controller controls the plant using the present value (*P*), the past value(s) (*I*), and the future value(s) (*D*) of the error.

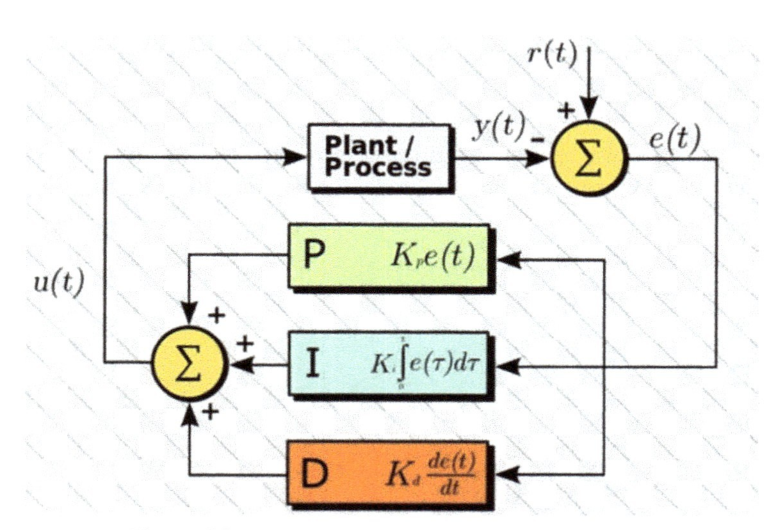

Figure 6.3 Typical structure of a PID control system.

Source: http://www.elprocus.com/wp-content/uploads/2013/12/Working-of-PID-controller.jpg

6.3 Brief Historical Review of Feedback Control

Feedback and control are inherent and endogenous in the life on Earth and during the centuries up to a certain time was in use by the human in his activity unconsciously. According to the literature [1–9] the history of automatic control goes back to more than 2500 years, and is divided in the following principal periods:

- Prehistoric and Early control period: Up to 1900.
- Preclassical Control period: 1900–1935.
- Classical Control period: 1936–1955.
- Modern Control period: 1956 till Present.

6.3.1 Prehistoric and Early Control Period

The prehistoric and Early control period essentially starts with the work of Aeschylus in the fifth century B.C. He was the first to use the term machine *(μηχανή)* with its modern meaning. Investigators in this period include:

Ktesibios of Alexandria (285–222 BC) His work (water clock, etc.) marks the beginning of conscious control system design.

Heron of Alexandria (1st Century BC) In his work '*Pneumatica*' describes several primitive control systems (or automata) for gate opening in temples, water supply, automatic dispensing of wine, etc.

William Salmon He constructed in 1630 a steam engine which used a floating value to control the water level in the boiler. In 1784, Sulton Thomas Word has constructed in England a similar engine based on the ideas of Heron.

Edmund Lee He constructed in 1745 the first automatic wind mill.

James Watt In 1769 he designed the '*fly-ball governor*' (called the *Watt's governor)*, later improved by many engineers (Figure 6.4).

James Clerk Maxwell (1831–1879) In 1868 he published his celebrated paper entitled 'On Governors', where he showed that the stability of second-, third-, and fourth-order systems can be determined examining the coefficients of the differential equations (or characteristic polynomials) deriving the necessary and sufficient conditions).

Edward Routh (1831–1907) Based on Maxwell's stability results, he derived in 1877 the algebraic stability criterion, known today as *Routh criterion.*

Figure 6.4 Watt's fly-ball governor. It was designed to control the speed of a steam engine.
Source: http://www.usciencecompendium.blogspot.com (2014/12/mechanical-governors-working-principle.html).

Adolf Hurwitz (1831–1907) derived an alternative algebraic stability criterion (Hurwitz criterion). The two alternative equivalent/formulations are today known as *Routh-Hurwitz criterion*.

Alexandr M. Lyapunov (1857–1918) He developed the so-called Lyapunov stability criterion (in 1892) using a generalized energy function.

O. Heaviside He developed (during 1892–1898) the operational calculus and studied the transient behavior of systems through a concept similar to what is presently called '*transfer function*'.

6.3.2 Pre-Classical Control Period

Pre-classical control period starts at about 1900 with the application of feedback controllers to electrical, mechanical, hydraulic, pneumatic, thermal, and other man-made systems. However, during the first two decades of the 20th century these devices were designed without good knowledge of the dynamics and stability problems involved. An exception to this was the automatic ship steering system designed by E. Sperry in 1910.

Elmer Sperry He invented the *gyroscope* which originally (1911) was used to the ship steering feedback control, and later in aircraft control.

N. Minorsky (1885–1970) In 1922 he made one of the first applications in nonlinear control for steering ships through the use of PID control (3-term control), which was introduced for the first time by him. He developed this control law by observing how a helmsman steered a ship. Minorsky's work became widely known after his publications in the journal '*The Engineer*', in late 1930s.

Harold Stephen Black (1898–1983) He was the first who showed and demonstrated (in 1927) that by using negative feedback amplifiers the noise amplification could be reduced. This technique started to be used within AT&T in 1931.

Harry Nyquist (1889–1976) He was a cooperator of Black and in 1932 he published a paper called '*Regeneration Theory*' which has presented the frequency-domain stability criterion, called now *Nyquist criterion which is* based on the polar plot of a complex function (transfer function).

Clesson E. Mason In 1928 Mason started experiments with feedback using the flapper-nozzle amplifier invented by Edgar H. Bristol at Foxboro Company. He developed in 1930 a feedback circuit which linearized the value operation enabling the introduction of integral (reset) component in the controller.

Harold Locke Hazen (1901–1980) Hazen designed a complete servo system and studied deeply the performance of servomechanisms. His 1934 publication '*Theory of Servomechanisms*' is considered to be the starting point for the next period of control, called '*classical control period*'. He coined the term '*servomechanism*' from the words 'servant' and 'mechanism'.

6.3.3 Classical Control Period

The previous control period was marked and ended by the three developments already mentioned, namely:

- Negative feedback electronic amplifier.
- Linearized pneumatic controller.
- Design and study of servomechanisms.

The first part of the classical control period (1935–1940) is marked by intensive work and new advances made in USA and Europe. The key workers of this initial period are the following:

- **Hendryk Bode:** In 1938 he used the magnitude and phase frequency plots (in terms of frequency logarithm) and investigated the stability of the closed-loop system using the concepts of *gain margin* and *phase margin*. He adopted the point (−1,0) as the critical point for measuring these margins.
- **J. G. Ziegler and N. B. Nichols:** They developed the now called '*Ziegler-Nichols*' tuning methods for selecting the parameters of PI and PID controllers. Their results were published in the 1942 book '*Network Analysis and Feedback Amplifier Design*'. Nichols in 1947 developed this *Nichols Chart* for the design of feedback control systems.
- **W. R. Evans:** In 1948 developed his *root locus* technique, which provided a direct way to determine the closed-loop pole locations in the s-plane.
- **Norbert Wiener:** He studied stochastic systems through the frequency domain stochastic analysis and developed the celebrated *Wiener statistical optimal filter* (in 1949) for the case of continuous-time signals which improved the signal-to-noise ratio of communication systems. The case of discrete-time stochastic processes was studied by the Russian scientist A.N. Kolmogorov (1941).
- **Donald McDondald:** His 1950 book 'Non-Linear Techniques for Improving Servo Performance' inspired during the 1950s extensive research work on the time-optimal control problem of single-input single-output systems using a saturating control. The work pursued on this topic was briefly described by Oldenburger in his 1966 book entitled 'Optimal Control'. Workers contributed in this area include Bushaw (1952), Bellman (1956) and J. P. LaSalle (1960) who showed that the time-optimal control is of the *bang-bang* type.

The achievements of the classical control period (1935–1955) were disseminated through many textbooks. These include in chronological order the books of the following authors: Bode (1940); Kolmogorov (1941); Smith (1942); Gatder and Barnes (1942); Bode (1945); MacColl (1945); James, Nichols and Philips (1947); Laner, Lesnik and Matson (1947); Brown and Campbell (1948); Wiener (1949); Porter (1950); Chestnut and Mayer (1951, 1955); Tustin (1952); Thaler and Brown (1953); Nixon (1953); Evans (1954); Bruns and Sauders (1954); Truxal (1954); and Thaler (1955).

6.3.4 Modern Control Period

The *Modern Control Period* starts with the works of R. Bellmann, L. S. Pontryagin, R. Kalman, and M. Athans.

Richard Bellman: During the period 1948–1953 Bellman developed the '*dynamic programming*' via his '*principle of optimality*'. The word 'programming' was selected as more proper term than 'planning'. In 1957 he applied dynamic programming to the optimal control problem of discrete-time (sampled-data) systems and derived the optimal control algorithm in backward time as a multistage decision making process [10]. The main drawback of dynamic programming is the so-called '*curse of dimensionality*', which was partially faced by Bellman and his co-worker Stuart Dreyfus via the development of carefully designed numerical solution computer programs. The dimensionality problem is still an obstacle despite the considerable increased computer power of present day supercomputers.

Lev S. Pontryagin: By 1958, Pontryagin has developed his *maximum principle* for solving optimal control problems on the basis of the *calculus of variations* developed in the middle seventies by L. Euler. The optimal controller of Pontryagin was of the on/off relay type [11].

Rudolf E. Kalman: In the Moscow IFAC Conference (1960), Kalman presented his paper: 'On the General Theory of Control Systems' where he coined the duality between multivariable feedback control and multivariable feedback filtering, thus showing that solving the control problem one has also the solution of the filtering problem, and vice versa [13]. In the same year (1960) he published his seminal paper on the optimal filtering problem for discrete time systems in the state space [14], and his joint paper with Bertram on the analysis and design of control systems via Lyapunov's second method [15]. The Lyapunov method was extended in 1960 by Lasalle [12]. The continuous-time version of the Kalman filter was published in 1961 jointly with Bucy [16].

Michael Athans: He published in 1966 one of the first comprehensive books on optimal control jointly with Falb [17] which played a major role in the research and education of the field. He published since then numerous publications with important results on optimal and robust multivariable control, many of which are included, along with other new results in [18].

V. M. Popov: In 1961 he developed the so-called circle criterion for the stability of nonlinear systems. His work was extended in the years that followed by many researchers including I.W. Sandberg in 1964, C.A. Desoer in 1965, and others [19, 60–69].

Further important developments on modern control theory were made by many control scientists including: J. Ackerman, B. D. O. Anderson,

D. P. Atherton, J. S. Baras, T. Basar, D. P. Bertsekas, R. W. Brockett, R. S. Bucy, J. P. Cruz, Jr., E. J. Davison, P. Eykoff, G. F Franklin, K. Glover, A. H. Haddad, I. Horowitz, C. H. Houpis, R. Isermann, E. J. Jury, T. Kailath, P. V. Kokotovic, H. Kwakernaak, I. D. Landau, R.Larson, F. L. Lewis, D. G. Luenberger, A. G. I. MacFarlane, J. M. Maciejwosky, Mayne, D. Q., Meditch, J. S., Mendel, J. M., Meystel, A., Narendra, K. S., Sage, A. P., Saridis, G. N., Seinfeld, J., Sheridan, T. B., Singh, M., Slotine, J. J., Spong, M. W., Titli, A., Tou, J. T., Unbehauen, H., Utkin, V. L., Vamos, T., Yurkovich, S., Wolovich, W. A., Wonham, W. M., Zadeh, L., Zames, G., Ziegler, B. P.

Looking at this literature one can see that what is collectively called '*modern control*' has been evolved along two main avenues:

- Optimal, stochastic, and adaptive control avenue.
- Algebraic and frequency-domain control avenue.

All techniques of these avenues are based on the assumption that a mathematical model of the system to be controlled is available, and in overall they are called '*model-based*' control techniques. Another family of control techniques does not use any mathematical model; instead they employ what they are called '*intelligent*' processes to specify how the system works and bypass the lack of a model. These techniques are named '*model-free*' control techniques [20–27]. Two important key concepts in the whole modern control theory are '*observability and controllability*' which reveal the exact relationships between transfer functions and state variable representations. The basic tool in optimal control theory is the celebrated matrix *Riccati* differential equation which provides the time-varying feedback gains in a linear-quadratic control system cell. The fundamental concept upon which the multivariable frequency methodology is based are the '*return ratio*' and '*return difference*' quantities of Bode. These concepts, together with the '*system matrix*' concept of Rosenbrock, were used to develop the multivariable (vectorial) Nyquist-theory as well as the algebraic modal (pole shifting or placement) control theory. Rosenbrock has coined several new concepts such as the diagonal dominance, characteristic locus and inverse Nyquist array [28, 29]. The theory of sampled-data control systems was firmly developed by G. Franklin, J. Ragazzini, and E. Jury who coined the '*Jury' stability criterion* [30–33]. The field of robust control for systems with uncertainties has been initiated by Horowitz [34–36] leading to the so-called '*quantitative feedback theory*' (QFT). Important contributions in this area were made by C. Houpis, A. MacFarlane, I. Postlethwaite, M. Safonov [35–42], and others. The stochastic control area after the seminal work of Kalman was promoted by K. Astrom, B. Wittenmark, H. Kwakernaak and J. Meditch [43–52]. The adaptive control area, especially the model-reference approach, was initiated by Landau [53–55]. The modern non-linear control theory has been evolved, after the works of Popov and Lyapunov, by G. Zames, K. Narendra, J.-J. Slotine [56–62], and others [63–65]. In the following section, the classical and modern analysis and design methodologies will be shortly discussed for both continuous-time and discrete-time (sampled-data) systems.

6.4 Classical Control Epistemology

In this section we will present the fundamental epistemological/ methodological elements of classical control at a level and detail appropriate for the purpose of this book. These elements are:

- Transfer function.
- Stability.
- Closed-loop controlled system.
- Performance specifications.
- Root-locus method.
- Frequency response methods (Nyquist, Bode, and Nichols).
- Compensators (phase lead, phase lag, phase lead-lag).
- Discrete-time control systems.

6.4.1 Transfer Functions and Stability

Transfer function

A general *linear-time-invariant* **(LTI)** *single-input/single-output* **(SISO)** dynamic system is described by a linear differential equation, with constant coefficients, of the form:

$$a_n \frac{d^n y(t)}{dt^n} + a_{n-1} \frac{d^{n-1} y(t)}{dt^{n-1}} + \cdots + a_2 \frac{d^2 y(t)}{dt^2} + a_1 \frac{dy(t)}{dt} + a_0$$

$$= b_m \frac{d^m u(t)}{dt^m} + b_{m-1} \frac{d^{m-1} u(t)}{dt^{m-1}} + \cdots + b_1 \frac{du(t)}{dt} + b_0 \quad (6.1)$$

where $a_k\,(k = 0, 1, \ldots, n)\,, b_k\,(k = 0, 1, \ldots, m)$ are constant coefficients, and $a_n \neq 0$ (typically assumed $a_n = 1$). For realizability of the system, the orders m and n must be $n > m$

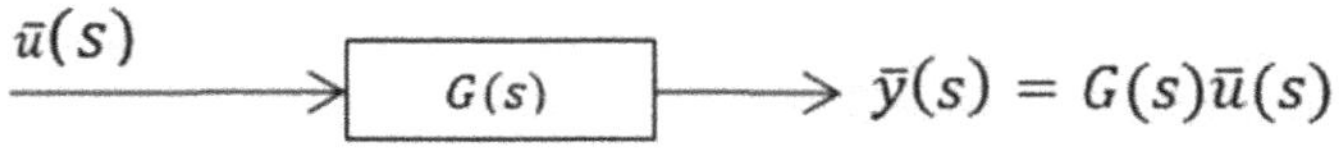

$\bar{u}(s)$ → $G(s)$ → $\bar{y}(s) = G(s)\bar{u}(s)$

Figure 6.5 Block diagram of a general SISO system.

Application of the Laplace transform

$$\overline{y}(s) = \int_0^\infty y(t)\, e^{-st}\mathrm{d}t \quad \text{and} \quad \overline{u}(s) = \int_0^\infty u(t)\, e^{-st}\mathrm{d}t$$

where $s = a + j\omega$ is the complex frequency, on both sides of (6.1), with zero initial conditions $u(0) = 0, y(0) = 0$, gives:

$$\overline{y}(s) = G(s)\,\overline{u}(s) \tag{6.2}$$

where $G(s) = \overline{y}(s)/\overline{u}(s) = B_m(s)/A_n(s)$ and

$$A_n(s) = a_n s^n + a_{n-1}s^{n-1} + \cdots + a_1 s + a_0$$

$$B_m(s) = b_m s^m + b_{m-1}s^{m-1} + \cdots + b_1 s + b_0$$

Equation (6.2) has the block diagram of Figure 6.5.

The polynomial $A_n(s)$ is called the output (characteristic) polynomial, and $B_m(s)$ is called the input polynomial. The function G(s) given by (6.2) is called by definition the *transfer function* of the system. The roots of $A_n(s) = 0$ are called the *poles* (or characteristic values), and the roots of the equation $B_m(s) = 0$, are called the *zeros* of the system. The value $G(s)]_{s=0} = [B_m(s)/A_n(s)]_{s=0}$ is, called the *dc* (direct current) *gain* of the system.

Stability

An input-free (autonomous) system is called to be absolutely stable if for any initial conditions, the output of the system:

- Is bounded for all time.
- The system returns to its equilibrium state (position) when the time tends to infinity $t \to \infty$.
- A system is said to be '*asymptotically stable*' if its impulse response has finite absolute value and goes to zero as $t \to \infty$, i.e., $|h(t)| < \infty$ for any $t \geq 0$ or $\lim |h(t)| = 0$.
- A system with input $u(t)$ is called to be bounded input-bounded output (BIBO) stable, if for any bounded input the resulting output is also bounded, i.e.:

'If $|u(t)| \leq u_{\max} < +\infty$ for any $t > 0$
Then $|y(t)| \leq y_{\max} < +\infty$ for any $t > 0$'

On the basis of the above it follows that:

'A continuous LTI system H(s) is BIBO stable (and absolutely stable) if and only if the poles of its transfer function H(s) lie strictly in the left hand semi-plane s, i.e., if and only if all of its poles have negative real part'.

The stability of a SISO system can be checked by examining its impulse response (defined as the inverse Laplace transform of its transfer function H(s)). If the impulse response tends to zero as time tends to infinity, i.e., if:

. $|y_\delta(t| \rightarrow 0$ as $t \rightarrow \infty$, then the system is asymptotically stable. Direct algebraic criteria for checking stability by using the coefficients of the characteristic polynomial (directly) are:

- ***Routh criterion:***

Given the system's characteristic polynomial $A_n(s)$ in (6.2) we construct the Routh's table as follows

$$
\begin{array}{c|cccc}
s^n & a_n & a_{n-2} & a_{n-4} & \cdots \\
s^{n-1} & a_{n-1} & a_{n-3} & a_{n-5} & \cdots \\
s^{n-2} & \beta_1 & \beta_2 & \beta_3 & \cdots \\
s^{n-3} & \gamma_1 & \gamma_2 & \gamma_3 & \cdots \\
\cdots & \cdots\cdot & \cdots\cdot & \cdots & \cdots \\
s^o & \cdots & \cdots & \cdots & \cdots
\end{array}
$$

where the coefficients $\beta_1, \gamma_1, \beta_2, \gamma_2$, etc. are given by:
$\beta_1 = \frac{a_{n-1}a_{n-2} - a_n a_{n-3}}{a_{n-1}}, \gamma_1 = \frac{\beta_1 a_{n-3} - a_{n-1}\beta_2}{\beta_1}$, etc. The table is completed up to the row that corresponds to s^0. All the other rows are zero. *Routh's stability criterion* states that:

'For $a_n > 0$ all the poles of the system lie on the left-hand semi-plane s, i.e., the system is stable, if and only if all the elements of the first column of the Routh table are positive. If this is not so the number of sign changes gives the number of poles that lie on the right-hand semi-plane s.' Some special cases need special treatment. This means that for stability:

$$a_n > 0,\ a_{n-1} > 0,\ \beta_1 > 0,\ \gamma_1 > 0, \ldots \tag{6.3}$$

If $A_n(s)$ involves any varying parameter (e.g., a gain K), the stability conditions (6.3) provide the values of K for which the system is stable.

- ***Hurwitz criterion:***

The Hurwitz criterion (which is equivalent to the Routh criterion) is stated by using the following determinants (Hurwitz determinants):

$$\Delta_1 = \alpha_{n-1}\left(a_n > 0\right), \quad \Delta_2 = \begin{vmatrix} a_{n-1} & a_{n-3} \\ a_n & a_{n-2} \end{vmatrix}$$

$$\Delta_3 = \begin{vmatrix} a_{n-1} & a_{n-3} & a_{n-5} \\ a_n & a_{n-2} & a_{n-4} \\ 0 & a_{n-1} & a_{n-3} \end{vmatrix}$$

$$= a_{n-1}a_{n-2}a_{n-3} + a_n a_{n-1} a_{n-5} - a_n a_{n-3}^2 - a_{n-1} a_{n-1}^2$$

The Hurwitz criterion states that the system is stable if and only if $\Delta_k > 0$ for all $k = 1, 2, ..., n-1$ (under the assumption that $a_n > 0$). Again the conditions $\Delta_k > 0$, $k = 1, 2, ..., n-1$ can be used to determine the values of K for which the system is stable.

6.4.2 Closed-Loop Controlled Systems and Performance Specifications

Closed-loop controlled systems Consider the closed-loop negative feedback system of Figure 6.6.

Assuming zero disturbance, the controlled system is the combination of the actuator and the system and is symbolized as $G(s)$, $G_c(s)$ is the series controller (compensator), $F(s)$ is the feedback element/controller), $y_f(s)$ is the feedback signal, and $e(s) = \overline{u}(s) - \overline{y_f}(s)$ is the error signal, where u is the input/command c.

Working on Figure 6.6, with zero disturbance, we find the equations:

$$\frac{\overline{y_c}(s)}{\overline{e}(s)} = G_c(s)\,G(s)$$
$$\overline{e}(s) = \overline{u}(s) - F(s)\,\overline{y_c}(s)$$

from which we get:

$$\mathrm{H}_c(s) = \frac{\overline{y_c}(s)}{\overline{u}(s)} = \frac{G_c(s)\,G(s)}{1 + G_c(s)\,G(s)\,F(s)} \tag{6.4}$$

The function $\mathrm{H}_c(s)$ is by definition the *closed-loop transfer function* of the system where $\overline{y_c}(s)$ is the closed-loop output of the system. The roots of $G_c(s)\,G(s) = 0$ are the zeros, and the roots of $1 + G_c(s)\,G(s)\,F(s) = 0$ are the poles of the closed-loop system (the polynomial $1 + G_c(s)G(s)F(s)$

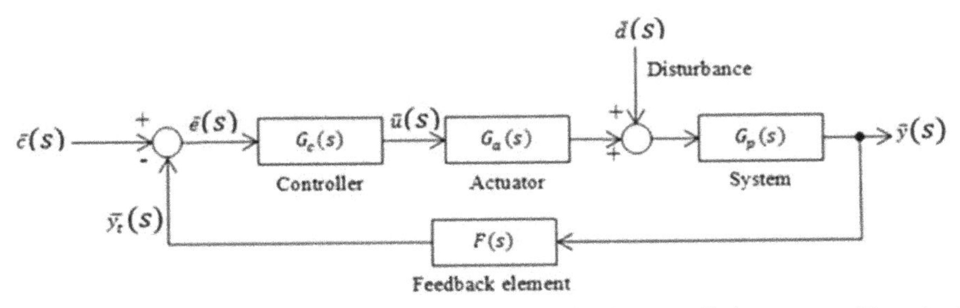

Figure 6.6 Basic control loop for a negative feedback controlled system with output disturbance ($G_\alpha(s)G_p(s) = G(s)$).

is the *closed-loop characteristic polynomial).* If $u(t) = \delta(t)$ *(impulse function:* $\overline{\delta}(s) = 1$) the output $y_d(t)$ is called *impulse response* of the system. If $u(t) =$ unit step function, in which case $u(s) = 1/s$, then $y_u(t)$ is called the *step response* of the system.

Performance specifications The performance specifications of control systems are distinguished in:

- *Transient response specifications.*
- *Steady-state response specifications.*

The transient response performance specifications (parameters) of a system are defined using the unit step time response of a second-order system:

$$\left(a_2 s^2 + a_1 s + a_0\right) y(t) = u(t)$$

which has transfer function:

$$\frac{\overline{y}(s)}{\overline{u}(s)} = \frac{1}{a_2 s^2 + a_1 s + a_0} \tag{6.5}$$

A mechanical example of second-order system is a mass-spring system with linear friction (proportional to velocity) (Figure 6.7).

The characteristic polynomial of the system can be written in the canonical form:

$a_2 s^2 + a_1 s + a_0 = ms^2 + \beta s + k = m\left(s^2 + 2\zeta\omega_n s + \omega_n^2\right)$ where ζ (the damping ratio), and ω_n (the undamped natural cyclic frequency) are defined as:

$$2\zeta\omega_n = \beta/m, \quad \omega^2 = k/m \tag{6.6}$$

Assuming that $\zeta = \beta/2m\sqrt{k/m} = \beta/2\sqrt{km} < 1$ *(underdamping)*, the roots of the characteristic equation $s^2 + 2\zeta\omega_n s + \omega_n^2 = 0$ (i.e., the poles of the system) are:

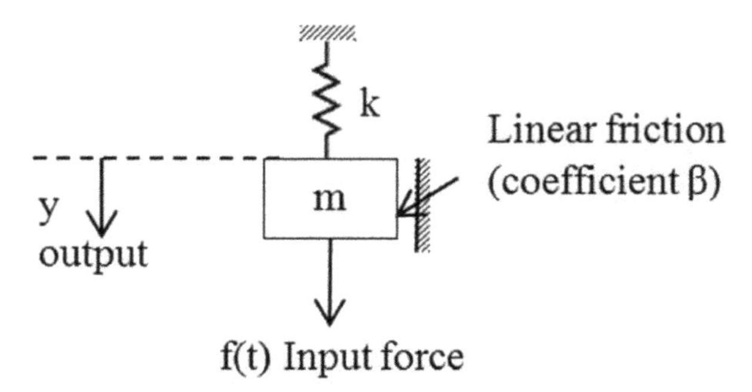

Figure 6.7 Mass–spring system with linear friction (damping).

$$\pi_1 = -a + j\omega_0 \quad \text{and} \quad \pi_1 = -a - j\omega_0$$

where $a = \zeta\omega_n$ (system damping) and $\omega_0 = \omega_n\sqrt{1-\zeta^2}$ (natural cyclic frequency under damping).

If the force $f(t)$ is a *unit step function* $u(t)$, $(\overline{u}(s) = 1/s)$ then $\overline{y}(s)$ is given by

$$\overline{y_u}(s) = 1/s\,(s-\pi_1)(s-\pi_2)\,.$$

The step response is found to be:

$$y_u(t) = L^{-1}y_u(s) = 1 - (\omega_n/\omega_0)\,e^{-\alpha t}\sin(\omega_0 t + \Psi) \tag{6.7}$$

here L^{-1} is the inverse Laplace transform and tan $\Psi = \omega_0/\alpha$. Graphically the two conjugate poles π_1 and π_2 on the s-plane have the positions shown in Figure 6.8.

From this figure we find that

$$\cos\Psi = \alpha/\omega_n = \zeta, \tag{6.8}$$

where the angle Ψ is as shown in Figure 6.8.

The typical unit step response of a damped system $(\zeta > 0)$ has the form of Figure 6.9.

Working on Equation (6.7) we find that the first overshoot h and the corresponding time (see Figure 6.9) are given by:

$$h = e^{-\zeta\pi/\sqrt{1-\zeta^2}} \quad \text{and} \quad t_h = \pi/\omega_n\sqrt{1-\zeta^2} \tag{6.9}$$

The settling time T_s for accuracy $\eta = \pm 2\%$ is found by using the equation

$$\left[e^{-\alpha t}\right]_{t=T_s} = \left[e^{-\zeta\omega_n/t}\right]_{t=T_s} = \eta$$

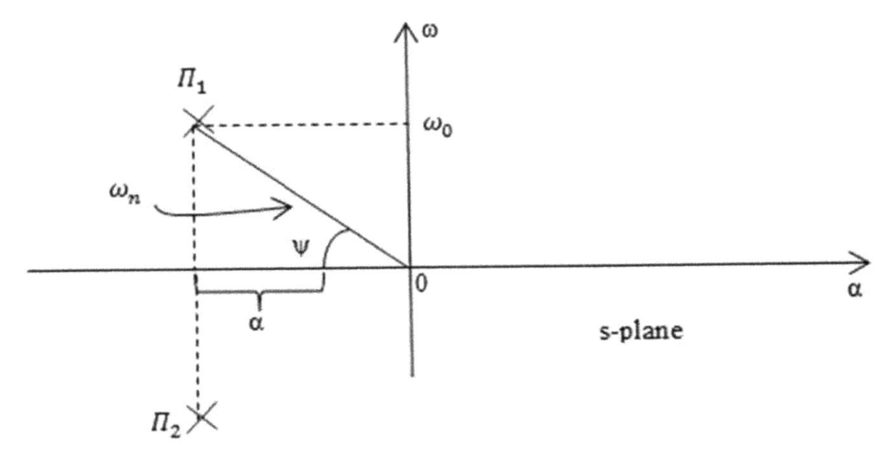

Figure 6.8 A pair of conjugate complex poles on the s-plane.

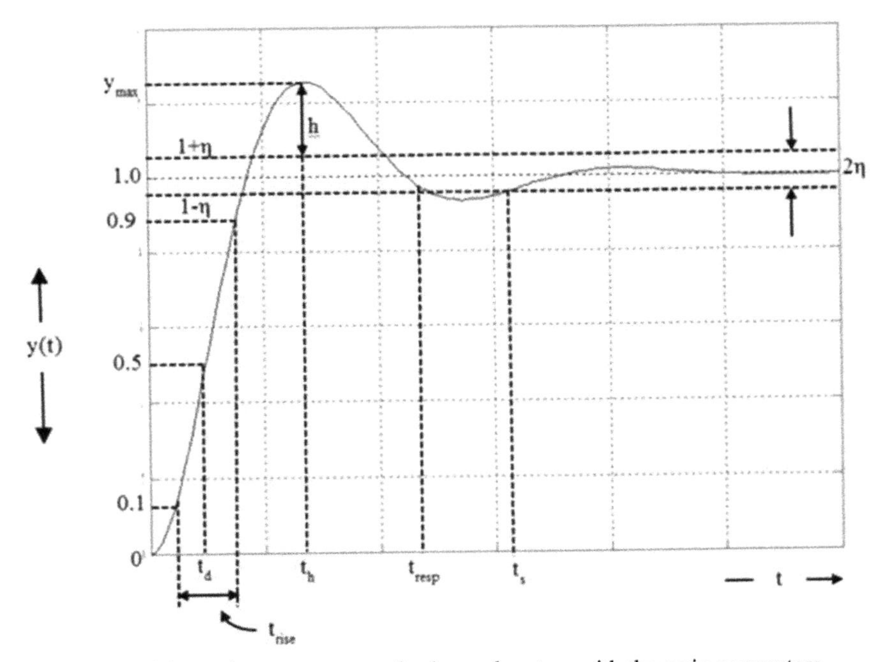

Figure 6.9 Unit step response of a damped system with the main parameters.

This gives $\zeta\omega_n \mathrm{T}_s = 4$, i.e.:

$$\mathrm{T}_s = \frac{4}{\zeta\omega_n} \quad \text{for error} \quad \eta = 0.02 \tag{6.10}$$

- In general (not only for second-order systems) the *overshoot* h is defined as:

$$h = 100 \times \frac{y_{\max} - y_{\mathrm{ss}}}{y_{\mathrm{ss}}} \%$$

where $y_{\max}$ is the maximum value of the output, and y_{ss} the steady-state value of the output.
The expressions (6.7), (6.8) and (6.9) give the values of h, t_h and T_s in terms of the system parameters ζ and ω_n (or equivalently in terms of the original system parameters (a_2, a_1, a_0) or (m, β, k)). For systems of higher order than two we use a second-order approximate model that has the dominant pair of conjugate poles, i.e., the poles with the smaller damping $a_{\mathrm{d}} = \zeta\omega_n$ which lie nearer to the imaginary axis. This approximation is very good if all the other poles have a damping α that satisfies the inequality:

$$|a| \geq 10\,|a_{\mathrm{d}}| = 10\,|\zeta\omega_n|$$

The dominant time constant τ_d is equal to $\tau = 1/\,|a|$.
- The *rise time* t_{rise} is the time needed for response $y\,(t)$ to go from 10% to 90% of y_{ss}. The *response time* t_{resp}. is the time at which the step response goes back to y_{ss} after the (first) overshoot. The *dominant time constant* t_{dom} is defined as the time needed by the exponential envelope to go at the 63% of its steady-state value.
- The *settling time* T_s is defined to be the minimum time needed by for the response to enter the window $\eta = \pm 2\%$ (or 5%) centered at the steady-state value.

Steady-state response specifications

Consider the feedback system of Figure 6.6. The system is said to be a *type-p system* if it involves p pure integrations (i.e., $\pi_0 = \pi_1 = \ldots = \pi_p = 0$).

The steady state error e_{ss} of the closed-loop system (without $G_{\mathrm{c}}\,(s)$) under the assumption that it is stable is given by:

$$e_{\mathrm{ss}} = e\,(\infty) = \lim_{s\to 0} s\bar{e}\,(s) = \lim_{s\to 0} \frac{s\bar{u}(s)}{1 + G\,(s)\,F\,(s)}$$

where $\overline{u}(s)$ is the applied input. For the steady state error we have the following:

- *Position error* e_p This error is obtained if the input is a unit step function $(u(t)) = 1$ for $t > 0, u(t) = 0$ for $t \leq 0$), i.e.,

$$e_p = \lim_{s \to 0} \frac{s}{1 + G(s) F(s)} \frac{1}{s} = \frac{1}{1 + K_p}$$

  where K_p is called the position error constant, is equal to:
  $K_p = \lim_{s \to 0} G(s) F(s) = G(0) F(0) \neq 0 \quad$ for $p = 0$, and
  $K_p = \infty$ for $p \geq 1$.
- *Velocity error* e_v This error is obtained when $u(t)$ is a unit ramp $\left(u(t) = t, \quad t \geq 0, \quad \overline{u}(s) = 1/s^2\right)$, i.e.:

$$e_v = \lim_{s \to 0} \frac{s}{1 + G(s) F(s)} \frac{1}{s^2} = \frac{1}{K_v}$$

  where K_v, called the *velocity error constant*, is given by

$$K_v = \lim_{s \to 0} sG(s) F(s) = 0 \quad \text{for } p = 0$$

$$K_v = G(0) F(0) \neq 0 \quad \text{for } p = 1$$

$$K_v = \infty \quad \text{for } p \geq 2$$

- *Acceleration error* e_α This error is obtained when $u(t)$ is a unit parabolic input $\left(u(t) = t^2/2, t \geq 0, \overline{u}(s) = 1/s^3\right)$, i.e.:

$$e_\alpha = \lim_{s \to 0} \frac{s}{1 + G(s) F(s)} \frac{1}{s^3} = \frac{1}{K_\alpha}$$

  where K_α, called the *acceleration error constant*, is given by

$$K_\alpha = \lim_{s \to 0} s^2 G(s) F(s) = 0 \quad \text{for } p = 0, 1$$

$$K_\alpha = G(0) F(0) \neq 0 \quad \text{for } p = 2$$

$$K_\alpha = \infty \quad \text{for } p \geq 3$$

6.4.3 Root-Locus and Frequency Response Methods

These methods are used both for control systems analysis (i.e., determination of response and system specifications) and control systems design (i.e., for a given system design of compensators to achieve desired performance specifications).

Root-Locus Method

This method provides an algebraic-geometric way for determining the positions of the closed-loop poles for different values of the open loop gain K, i.e., the parameter in the expression:

$$\begin{aligned} G(s)\,F(s) &= \mathrm{K}Q_1(s)\,/Q_2(s) \\ &= \mathrm{K}(s-\mu_1)(s-\mu_2)\ldots(s-\mu_m)\,/\,(s-\pi_1)(s-\pi_2)\ldots(s-\pi_n) \end{aligned}$$

where $m < n$. The closed-loop transfer function poles are the roots of the closed-loop characteristic equation $1 + G(s)\,F(s)$ or

$$Q_2(s) + KQ_1(s) = 0 \tag{6.11}$$

The Root Locus is defined to be the locus (line) on which the closed-loop poles lie (move) when K changes from $\mathrm{K} = 0$ to $\mathrm{K} \to \infty$. For $\mathrm{K} = 0$ the closed-loop poles coincide with the open loop poles. This means that the root locus begins from the open loop poles. For $\mathrm{K} \to \infty$ the closed loop poles tend to the positions of the open loop zeros (i.e., to the roots of $Q_1(s) = 0$). In general, for the root locus to passed through a point s_0 of the complex plane s, s_0 must be the root of the characteristic equation $Q_2(s_0) + \mathrm{K}Q_1(s_0) = 0$ or equivalently:

$$G(s_0)\,F(s_0) = KQ_1(s_0)\,/Q_2(s_0) = -1 \tag{6.12}$$

For (6.12) to hold the following two conditions must hold:

$$|G(s_0)\,F(s_0)| = |\mathrm{K}|\,|Q_1(s_0)\,/Q_2(s_0)| = 1 \tag{6.13a}$$

and

$$\underline{/G(s_0)\,F(s_0)} = \underline{/\mathrm{K}Q_1(s_0)\,/Q_2(s_0)} = 180° + p360°,\ p = 0,\ \pm 1,\ \pm 2, \ldots. \tag{6.13b}$$

Condition (6.13(a)) is called the *magnitude condition*, and (6.13(b)) is called the *phase condition*. The magnitude and phase of $G(s)\,F(s)$ can be determined with the aid of a number of rules. Today there are available several software packages (e.g., MATLAB, Matrix-X, Control-C, etc.) that can be used to draw the root locus.

Using the root-locus we can determine the following parameters:

- The critical gain K_c and *critical frequency* ω_c for which the system passes from stability to instability (i.e., the root locus crosses the imaginary axis).

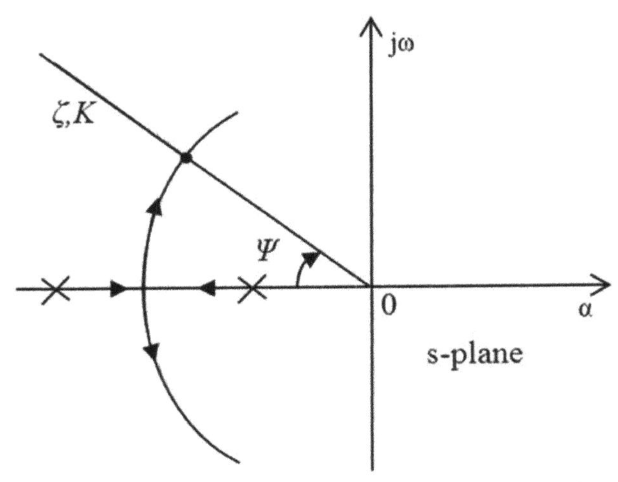

Figure 6.10 Determination of the gain K needed for a given $\zeta = a/\omega \quad (s = \alpha + j\omega)$, where $\cos \Psi = \zeta$.

- The gain margin and phase margin.
- The gain K needed for the damping factor ζ to have a desired value (see Figure 6.10).
- The system transfer function (and the time response).

It is remarked that if the number of open-loop poles exceed the number of (finite) open-loop zeros by 3 or more then there always exist a critical value K_c of the gain.

Frequency Response Methods

These methods include the Nyquist, Bode and Nichols methods, which are based on the Nyquist (polar), Bode, and Nichols plots.

Nyquist Plot

Given a transfer function $G(s)$, by setting $s = j\omega$ we get the frequency domain function $G(j\omega)$ which is a complex function of the real variable ω (the frequency), i.e.:

$G(j\omega) = \mathrm{A}(\omega) + j\mathrm{B}(\omega)$ Cartesian form, or

$G(j\omega) = |G(j\omega)| \underline{/\phi(\omega)}$ Polar form, or

$G(j\omega) = |G(j\omega)] [\cos\phi(\omega) + j\phi(\omega)]$ Euler form

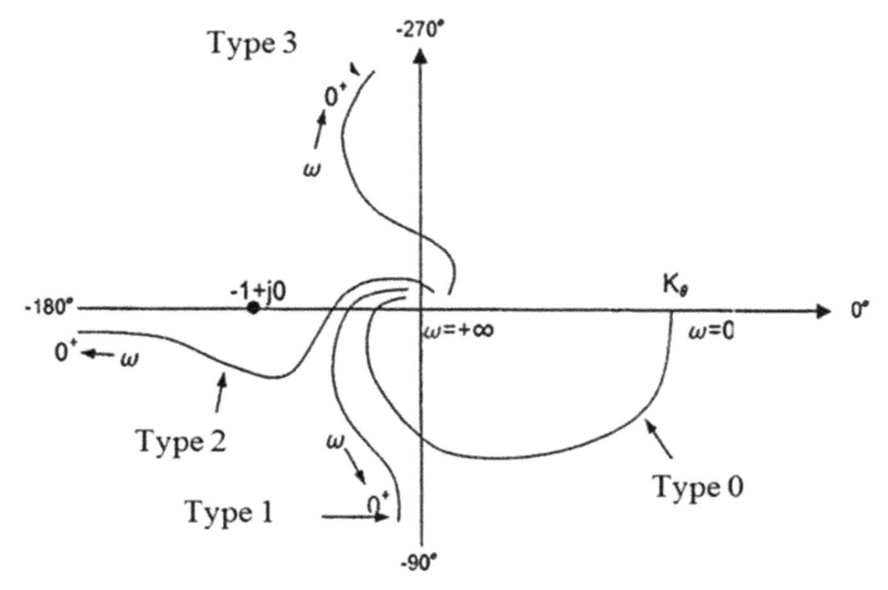

Figure 6.11 General Form of polar plots for systems of types 0 up to 3.

where $\mathrm{A}(\omega) = \mathrm{Re}[G(j\omega)]$ (real part), and $\mathrm{B}(\omega) = \mathrm{Im}[G(j\omega)]$ (imaginary part), and $\tan\phi = \mathrm{B}(\omega)/\mathrm{A}(\omega)$ (phase), $\mathrm{M}(\omega) = |G(j\omega)| = \sqrt{\mathrm{A}(\omega)^2 + \mathrm{B}(\omega)^2}$ (magnitude).

Direct polar diagram (plot) or simply *polar plot* of $G(j\omega)$ is its graphical representation on the plane $G(j\omega)$ for values of ω from $\omega = 0$ up to $\omega = +\infty$ (i.e., when ω traverses the positive imaginary axis). The polar plot of $G(j\omega)$ is the same no matter which representation form is used (Cartesian, Polar, Euler).

The form of polar plots of a general system with:

$$G(j\omega) = \frac{\mathrm{K}(1 + j\omega\tau_a)\ldots(1 + j\omega\tau_p)}{(j\omega)^n(1 + j\omega\tau_1)\ldots(1 + j\omega\tau_n)},$$

for $n = 0$ (type 0), $n = 1$ (type 1), $n = 2$ (type 2) and $n = 3$ (type 3) is given in Figure 6.11.

The polar plot of a simple RL circuit with transfer function:

$$G(j\omega) = R/(R + j\omega L) = 1/(1 + j\omega\tau), \tau = L/R$$

has the circular form of Figure 6.12:

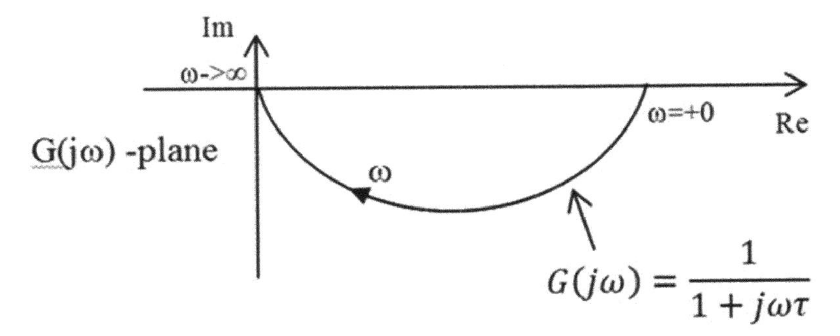

Figure 6.12 Polar plot of $1/(1 + j\omega\tau)$.

Bode Plots

The Nyquist plot describes a control system (magnitude and phase) via the unique curve $G(j\omega)$ in the plane $G(j\omega)$ with parameter the frequency ω. On the contrary, *Bode plots* describe the system via two district curves: the curve for the magnitude $|G(j\omega)|$ and the curve for the phase $\underline{|G(j\omega)}$. Because, the transfer function (can almost always) be expressed as products of simpler transfer functions, Bode has coined the use of logarithms and introduced the following definitions.

- *Magnitude Bode plot* is the plot of $20\log_{10}|G(j\omega)|\, db$ (decibels) in terms of $\log_{10}\omega$.
- *Phase Bode plot* is the plot of the phase $\underline{|G(j\omega)}$ in terms of $\log_{10}\omega$.

In general, the following relation holds:

$$\begin{aligned}\log_{10}[G(j\omega)] &= \log_{10}|G(j\omega)|\, e^{j\phi(\omega)} = \log_{10}|G(j\omega)| + \log_{10} e^{j\phi(\omega)} \\ &= \log_{10}|G(j\omega)| + j0.434\phi(\omega)\end{aligned} \tag{6.14}$$

which implies that

$$\begin{aligned}\mathrm{Re}[\log_{10} G(j\omega)] &= \log_{10}|G(j\omega)| \\ \mathrm{Im}[\log_{10} G(j\omega)] &= 0.434\phi(\omega),\end{aligned}$$

where $\mathrm{Re}[z]$ and $\mathrm{Im}[z]$ represent the *real part and imaginary part* of the complex number z.

Obviously, if:

$$G(j\omega) = F_1(j\omega)\, F_2(j\omega) / [H_1(j\omega)\, H_2(j\omega)]$$

Then:

$$20\log_{10}|G(j\omega)| = 20\log_{10}|F_1(j\omega)| + 20\log_{10}|F_2(j\omega)| \\ -20\log_{10}|H_1(j\omega)| - 20\log_{10}|H_2(j\omega)| \tag{6.15}$$

This means that the Bode plot of the magnitude of $G(j\omega)$ can be found by the algebraic sum of the magnitude Bode plots of the factors:

$$F_1(j\omega), F_2(j\omega), H_1(j\omega) \quad \text{and} \quad H_2(j\omega).$$

The same is true for the phase plot, i.e.:

$$\underline{|G(j\omega)} = \underline{|F_1(j\omega)} + \underline{|F_2(j\omega)} - \underline{|H_1(j\omega)} - \underline{|H_2(j\omega)}. \tag{6.16}$$

Therefore, in order to draw the Bode plots for the magnitude and phase of a general linear system of the form:

$$G(j\omega) = \frac{\mathrm{K}(1+j\omega\tau_1)(1+j\omega\tau_2)\ldots}{(j\omega)^p(1+j\omega_1\mathrm{T}_1)(1+2\zeta/\omega_n)j\omega+(1/\omega_n^2)(j\omega)^2\ldots}$$

We separately plot and add the magnitude Bode plots of its factors in the numerator and denominator. In practice, it is simpler and usually adequate to draw and use the so-called, asymptotic Bode diagrams, which are linear approximations of the exact Bode diagrams.

Performance specifications in the frequency domain

As seen in Section 6.4.2, the second-order system plays a key role in control systems analysis and design because it provides simple formulas for ζ and ω_n. In the frequency domain the system specifications corresponding to ζ and ω_n are the *phase margin* and the *bandwidth* B of the system which are approximately related to ζ and ω_n as:

$$\zeta = \phi^0_{\text{margin}}/100 \text{ for } \begin{matrix} \zeta < 1/\sqrt{2} = 0.707 \\ \overline{B = \omega_n\sqrt{1-2\zeta^2+\sqrt{2-4\zeta^2+4\zeta^4}} = \omega_n} \end{matrix} \tag{6.17}$$

for $\zeta = 1/\sqrt{2} = 0.707$.

Equivalent frequency domain parameters are the *peak frequency* ω_p and *peak amplitude value* M_p of the conventional frequency plot of the system's magnitude:

$$\omega_\mathrm{p} = \omega_n\sqrt{1-2\zeta^2} \quad (0<\zeta<1/\sqrt{2}) \\ M_\mathrm{p} = 1/2\zeta\sqrt{1-\zeta^2} \quad (0<\zeta<1/\sqrt{2}) \tag{6.18}$$

The *bandwidth B* is related to the *rise time* t_{rise} as:

$$t_{\text{rise}} = \pi / B$$

All the above relations can be properly used in the design of compensators and controllers. Two other frequency response specifications are the *gain margin* K_{margin} and *phase margin* Φ_{margin}. The *gain margin* is computed at the frequency ω_{ϕ} at which the phase of the system G(jω) F(jω) is $\pm 180°$, and is defined as 'how many times the gain at ω_{ϕ} can be increased so that $|G\,(j\omega)\;F(j\omega)|_{\omega=\omega\varphi} = 1$ or 0 db. The phase margin Φ_{margin} is computed at the frequency ω_{α} where

$|G\,(j\omega_a)\,F\,(j\omega_a)| = 1,$ or 0 db and is defined as how much the phase of the system at ω_a can be increased until $\underline{|G\,(j\omega)\,F\,(j\omega)}\,]_{\omega=\omega_a} = \pm 180°$.

The feedback control system is stable if:

$$K_{\text{margin}} > 1 \text{ or } 0\,\text{db} \quad \text{and} \quad \Phi_{\text{margin}} > 0°$$

and unstable if:

$$K_{\text{margin}} < 1 \text{ or } 0 \text{ db} \quad \text{and} \quad \Phi_{\text{margin}} < 0°$$

Using the *Bode* plots K_{margin} and Φ_{margin} are illustrated as shown in Figure 6.13.

Using polar (Nyquist) plots, the above stability conditions are equivalent to: '*The system is stable if the polar plot does not encircle the Nyquist point* $-1 + jo$ *(leaves this point to the left), and is unstable if the plot encircles the point* $-1 + jo$.

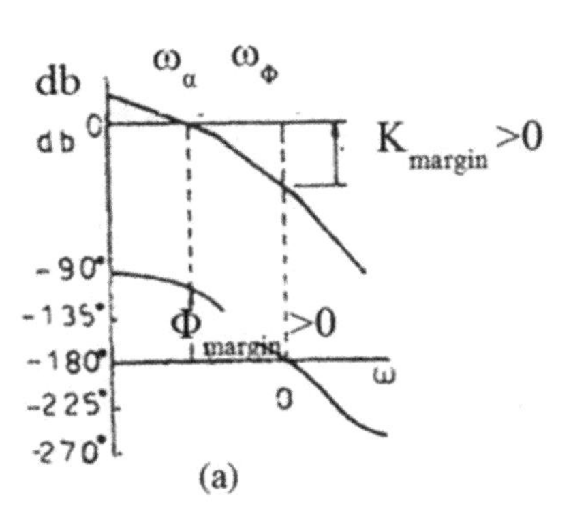

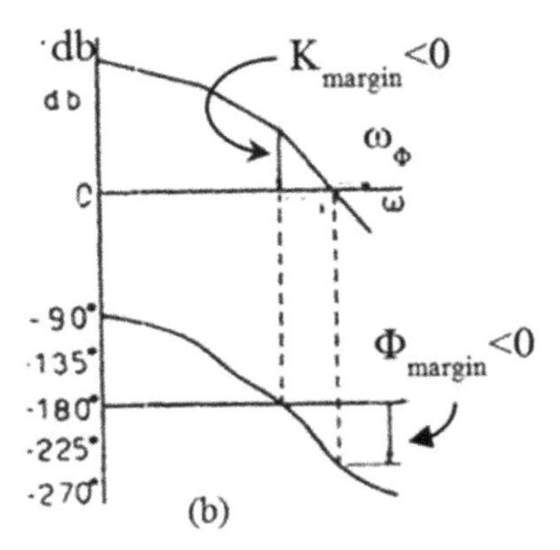

Figure 6.13 Gain and phase margins illustrated via Bode plots (a) $K_{\text{margin}} > 0$ db, $\Phi_{\text{margin}} > 0°$ and (b) $K_{\text{margin}} < 0$ db, $\Phi_{\text{margin}} < 0$.

Nichols Plot

This is the plot of the magnitude $|G(j\omega)F(j\omega)|$ of the system in db, i.e., $20\log_{10}|G(j\omega)F(j\omega)|$, and the phase $\underline{/G(j\omega)F(j\omega)}$ (in degrees), in orthogonal axes. The Nichols plot of the system:

$$G(j\omega)F(j\omega) = \frac{4(1 + j0.5\omega)}{j\omega(1 + j2\omega)\left[1 + j0.05\omega + (j\omega)^2/8\right]}$$

has the form shown in Figure 6.14, where the gain margin K_{margin}; at $\underline{/F(j\omega)G(j\omega)} = -180°$, and Φ_{margin} at $|F(j\omega)G(j\omega)| = 0$ db, are shown. We see that $K_{\text{margin}} > 0$ and $\Phi_{\text{margin}} > 0°$. Therefore the system is *stable*.

A change of the system gain implies a translation of the plot *upwards* (if the gain is *increased)* or *downwards* (if the gain is *decreased)* without any change of the phase. Clearly, if the Nichols plot is moved *upwards* the gain margin is *decreased,* whereas if it is moved *downwards* the gain margin is *increased.* In general, in terms of the Nichols (or *L*) plot, the Nyquist criterion is stated as '*A closed-loop minimum phase system with open-loop transfer function $G(s)F(s)$ is* **stable** *if its Nichols plot* $\left(20\log_{10}|G(j\omega)F(j\omega)|, \underline{/G(j\omega)F(j\omega)}\right)$ *traversed in the direction of increasing ω, leaves the Nichols–Nyquist point* (0 db, −180°) on **its right**'.

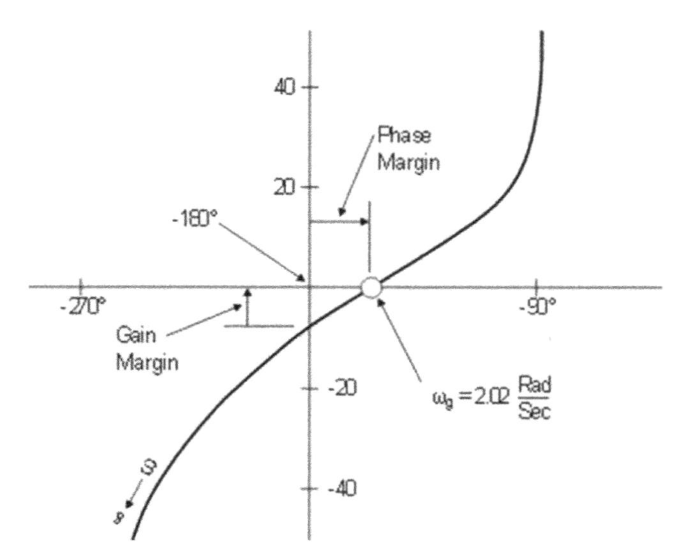

Figure 6.14 The Nichols plot of a stable system.

6.4.4 Compensators

Compensators (or controllers) are distinguished in:

- *Series compensators* (if they are placed in the forward path, Figure 6.15(a)).
- *Feedback compensators* (if they are placed in the feedback path, Figure 6.15(b)).

In many cases a combined series compensator and feedback compensator are employed. Compensators are designed such that if applied to a given system, lead to an overall closed-loop system that possesses desired specifications of transient and steady-state performance.

The compensator design can be performed using any one of the methods described thus far, i.e.:

- Design via root locus.
- Design via Nyquist plots.
- Design Bode plots.
- Design via Nichols plots.

A special design method for series compensation is *Ziegler–Nichols* PID compensator. Our purpose here is to provide a tour to the root-locus method. Compensator design techniques using polar, Bode and Nichols plots are fully studied in D'Azzo and Houpis, Dorf, and DiStefano [65, 73, 74].

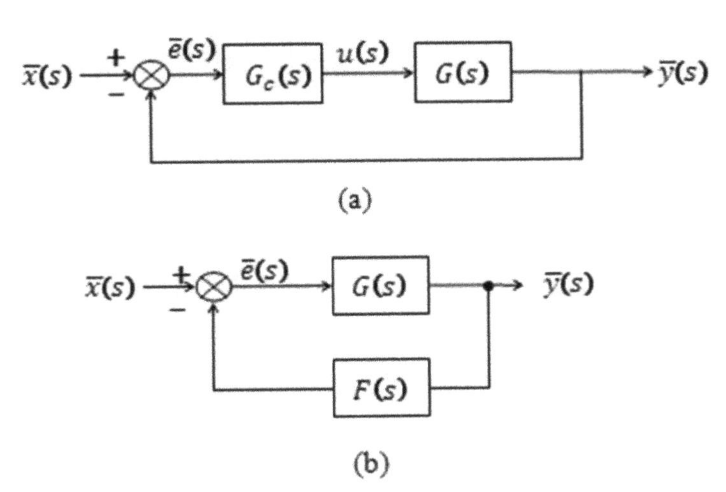

Figure 6.15 Compensators: (a) series compensator $G_c(s)$ and (b) feedback compensator $F(s)$.

Root Locus: Series Compensator Design The root locus can be used for selecting the value of the gain Ksuch that the closed-loop poles are moved to positions that correspond to the desired specifications $\zeta,\ \omega_n,\ h,\ \tau_{dom}$, etc. In many cases this is not possible by mere *gain-control*. In these cases the root locus must be properly modified. The *gain* is selected such that to achieve the desired steady-state performance (errors). Here we distinguish the following cases:

- The transient response is acceptable, but the steady-error is large. In this case we can increase the gain without any change of the root locus (pure proportional control).
- The system is stable but the transient response is not acceptable. In this case the root locus must be shifted to the left, i.e., at a greater distance from the imaginary axis.
- The system is stable, but both the steady-state performance (steady-state errors) and the transient performance are not adaptable. Here, both again increase and a shift of the root locus to the left are needed.
- The system is unstable for all the values of the gain. Thus the root locus should be modified such that some part of every branch lies on the left semi-plane *s*. In this way the system will become conditionally stable.

Actually, the use of a compensator introduces new poles and new zeros in the open-loop transfer function. Typically, the input used in all cases is the unit step function. If the desired response is *subcritical*, the compensator is usually selected such that to obtain a pair of dominant complex poles, so that the contributions of all the other poles are negligible. In general one can apply the following.

Phase-lag compensation This is done when the transient response is good, but the steady-state error is very high. We actually increase the type of the system with minimal change of the closed-loop poles. The ideal phase-lag compensator has a transfer function:

$$G_c\left(s\right) = \left(s - \mu\right)/s, \quad \mu < 0$$

i.e., a pole at $s = 0$ and a zero in the left semi-plane s, but very near to $s = 0$ to secure that the actual positions of the closed-loop poles remain the same, and so the transient response remains practically unchanged. The above compensator is actually a *proportional plus integral controller*. This is verified by writing $G_c\left(s\right)$ as:

$$G_c\left(s\right) = 1 + K_i/s, \quad K_i = -\mu > 0,$$

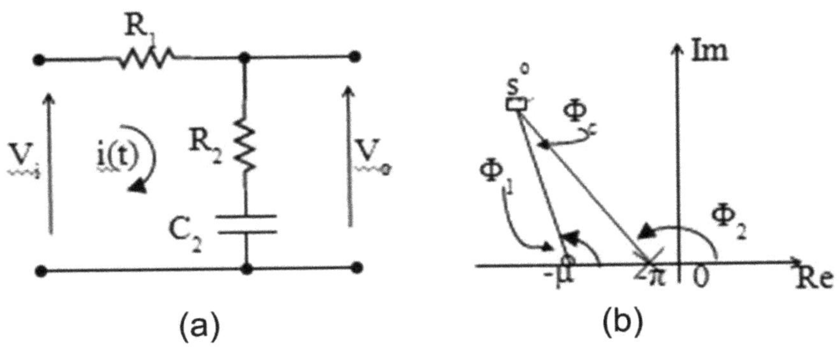

Figure 6.16 Phase-lag compensation. (a) Compensator circuit and (b) Criterion of 3–5° compensator angle $\phi_c = \phi_1 - \phi_2$.

where K_i is the gain of the *integral term.* The RC circuit of the phase-lag compensator typically used in practice has the transfer function (Figure 6.16(a)):

$$\frac{\overline{v}_o(s)}{\overline{v}_i(s)} = \frac{1 + \mathrm{T}_2 s}{1 + a\mathrm{T}_2 s} = \frac{s - \mu}{s - \pi},$$

where $\mathrm{T}_2 = R_2 C_2$, $a = 1 + R_1/R_2 > 1$, $\mu = -1/\mathrm{T}_2$, $\pi = -1/a\mathrm{T}_2$ and $|\pi| < |\mu|$, i.e., the lag compensator's pole is to the right of the zero.

To maintain the root locus of the system under compensation unchanged as much as possible (since it meets the transient specifications), the angle of the compensator must be as small as possible, less than 5°). Thus, we first draw the constant damping line which passes via the desired closed-loop pole, and select the pole of the compensator at the origin of the s-plane. Then we select the zero of the compensator at an angle 3°, and move the pole from the origin to a position π such as $|\mu/\pi| = a$ where a is the gain needed for achieving the desired steady-state error (see Figure 6.16(b)).

Phase-lead compensation This is done when the steady-state error is acceptable, but the transient response is not acceptable. The transient response is improved if we move the root locus to the left of its original position. The ideal phase lead compensator has transfer function ($\mu < 0$):

$$G_\mathrm{c}(s) = s - \mu = 1 + sK_\mathrm{d}, \quad K_\mathrm{d} = -1/\mu > 0$$

This compensator adds a zero to the forward path, which in practice can be obtained using a differentiator. We observe that this controller is actually a proportional plus derivative controller. Practically a lead compensator is implemented by an appropriate RC circuit (Figure 6.17(a)).

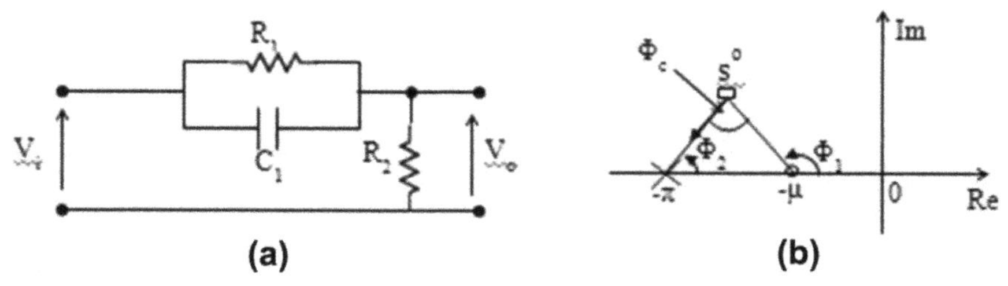

Figure 6.17 (a) Phase-lead RC circuit. (b) The phase lead at s^0 is $\phi_c = \phi_1 - \phi_2$.

This circuit has the transfer function:

$$\frac{\overline{v}_o(s)}{\overline{v}_i(s)} = a\frac{1 + \mathrm{T}_1 s}{1 + a\mathrm{T}_1 s} = \frac{s - \mu}{s - \pi}$$

Where $\alpha = R_2/(R_1 + R_2) < 1$, $\mathrm{T}_1 = R_1 C_1$, $\mu = -1/\mathrm{T}_1$, and $\pi = -1/\alpha\mathrm{T}_1$. Here, $|\mu| < |\pi|$, i.e., the lead compensator's zero is to

The right-side of the pole, here, we will use Dorf's method which is the simplest one. Other methods include the bisection method and the constant phase circles method. In all cases the desired specifications of the transient response $(h, \zeta, \omega_\mathrm{n}, \tau_\mathrm{d}, \mathrm{T_s}, \mathrm{etc.})$ are converted to desired positions of the dominant pair of poles of the closed-loop system.

Phase lead-lag compensation This type of compensation is needed when both the steady-state and the transient performances are not satisfactory. We can separately design a lag compensator to achieve the steady-state specifications and a lead compensator for the transient specifications, and then combine them in series with a buffer circuit between them (to avoid the loading of the first circuit by the second circuit). Four parameters will be selected in this case; the two zeros and two poles of the circuits. It is, however, more convenient to use a circuit where only three parameters define the compensator (Figure 6.18).

It is remarked that the *PID controller* of Figure 6.3 is a kind of phase lead–lag compensator having the time-domain equation

$$u(t) = K_\mathrm{a}\left[e(t) + \tau_\mathrm{d}\frac{\mathrm{d}e(t)}{\mathrm{d}t} + \frac{1}{\tau_i}\int_0^t e(t')\,\mathrm{d}t\right],$$

where K_a is the controller gain, τ_d is the time constant of the derivative term, and τ_i is the time constant of the integral term (usually called 'reset time').

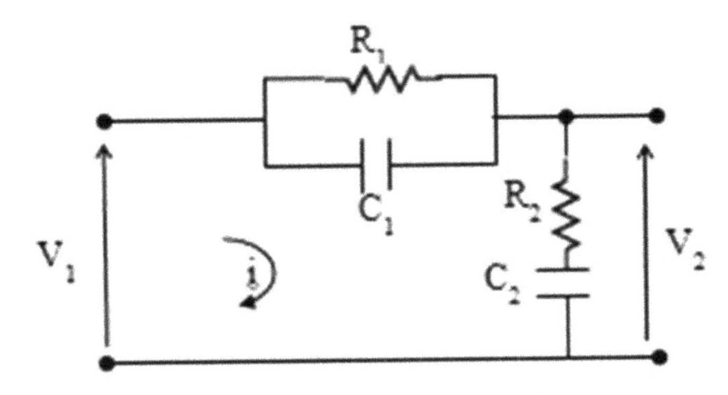

Figure 6.18 Phase lead–lag circuit.

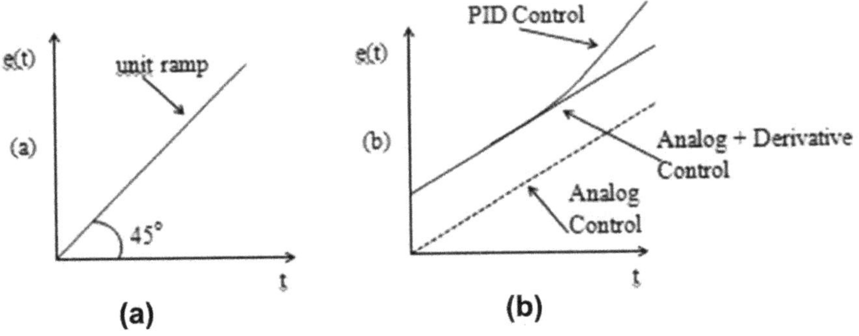

Figure 6.19 Unit ramp response of PID controller: (a) Unit ramp error and (b) PID control signal corresponding to (a).

The transfer function $G_{\mathrm{c}}(s)$ of the PID compensator is found to be

$$G_{\mathrm{c}}(s) = \frac{\overline{u}(s)}{\overline{e}(s)} = K_a \left[1 + s\tau_{\mathrm{d}} + \frac{1}{s\tau_i}\right]$$

If the error $e(t)$ is a unit ramp function (Figure 6.19(a)), then the control signal (the output of the compensator) has the form of Figure 6.19(b)).

Obviously, to design (tune) the controller we have here to select the three parameters K_{d}, τ_{d} and τ_i. The most popular PID parameter tuning method is the Ziegler–Nichols method (1942), a popular variant of which is based on the stability limits of the closed-loop system, and involves the following three steps:

Step 1: Disconnect the derivative and integral terms (i.e., use only the proportional term).

Step 2: Increase the gain K_{a} until the stability limit is reached and oscillations are started. Let $\mathrm{T_o}$ be the oscillations' period and K_{c} the critical value of the gain.

Step 3: Select the parameters K_{a}, τ_i and τ_{d} as follows:

- For proportional control: $K_{\mathrm{a}} = K_{\mathrm{c}}/2$.
- For PI control $K_{\mathrm{a}} = 0.45K_{\mathrm{c}},\ \tau_i = 0.8\mathrm{T_o}$.
- For PID control: $K_{\mathrm{a}} = 0.6K_{\mathrm{c}}, \tau_i = \mathrm{T_o}/2$ and $\tau_{\mathrm{d}} = \tau_i/4$.

The performance achieved by these values in typical process control systems is quite acceptable (giving about 10–20% overshoot). The PID tuning can also be done by other methods, e.g., root locus method, Bode method, and control parameter optimization based on a certain performance index, etc.

6.4.5 Discrete-Time Control Systems

The concepts involved in Sections 6.4.1–6.4.5 have analogous counterparts for the case of discrete-time systems or sampled-data systems which are described by discrete-time (difference) equations. In this case the transfer function (called pulse transfer function) model (Definition 6.1) has the form:

$$G(z) = \frac{Y(z)}{X(z)} = \frac{\mathrm{B}(z)}{\mathrm{A}(z)} \tag{6.19}$$

where $\mathrm{A}(z)$ and $\mathrm{B}(z)$ are polynomials in z, the discrete-time complex frequency. In analogy to the closed-loop transfer function (6.2), here we have under certain conditions (i.e., a sampler after the error $\overline{e}(s)$ and a sampler after $G(s)$ the closed-loop system of Figure 6.20).

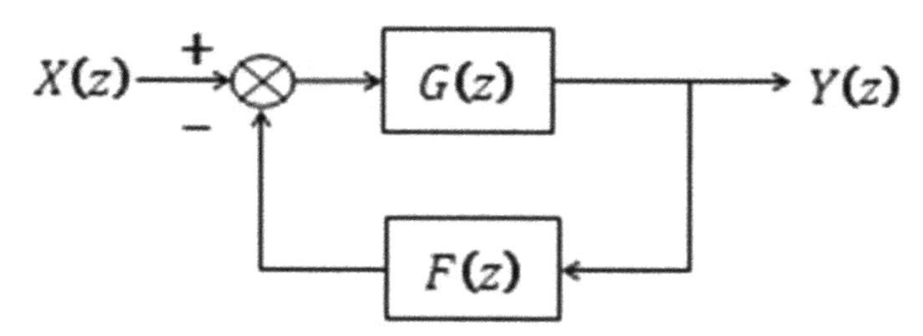

Figure 6.20 The basic typical form of closed-loop discrete-time systems.

The closed-loop discrete-time transfer function is:

$$\frac{Y(z)}{X(z)} = \frac{G(z)}{1 + G(z)F(z)} \tag{6.20}$$

Here, the role of the imaginary axis $j\omega$ of continuous-time systems, plays the unit circle of the z-plane. If the poles in the z-plane of the closed loop system lie inside the unit z-circle the system is stable. If they lie outside of the unit z-circle the system it is unstable.

Analogous plots corresponding to root-locus, polar plot, and Bode plots can be drawn for discrete time systems, and correspondingly analogous formulas can be derived for the performance specifications $\zeta, \omega_n, t_h, \mathrm{T}_s$, etc. of discrete-time systems. The constant damping line for an under-damping system is a spiral inside the unit circle starting at $z = 1$.

The Routh, Hurwitz, Nyquist, Bode, and Nichols stability criteria can be formulated for discrete-time systems as well.

6.5 Modern Control Epistemology

In this section, the following epistemological/methodological elements of modern control will be briefly discussed:

- State-space modeling.
- Controllability and observability.
- Lyapunov stability.
- State feedback control.
- Optimal and stochastic control.
- Model-free control.

6.5.1 State Space Modeling, Controllability, and Observability

State-space modeling Control systems modeling with state-space equations is applicable to time-varying, linear/nonlinear multivariable systems, distributed-parameter systems in both continuous-time and discrete-time representations.

The key concept is the state vector $\mathbf{x}(t)$ which is the minimum-dimensionality vector (with components called state variables) which at $t = t_0$ together with the input(s) $\mathbf{u}(t)$ for $t \geq t_0$ determine completely the system behavior for any time $t \geq t_0$. In practice, not all state variables are measurable physical quantities. Because state-feedback control needs all of them, the non-measurable ones are estimated, under certain conditions, by proper estimates (e.g., Luenberger observer, Kalman filter).

Given a system described by (6.1) with $m = n$ where the coefficients may be time invariant or time-varying we define $D = \mathrm{d}/\mathrm{d}t,\ D^2 = \mathrm{d}^2/\mathrm{d}t^2$, Then, solving for

$y(t)$ in terms of D, we find:

$$y(t) = b_0 u(t) + \frac{1}{D}(b_1 u - a_1 y) + \ldots + \frac{1}{D^n}(b_n u - a_n y)$$

where the ordering of the coefficients has been reversed. Now, defining the state variables $x_1, x_2, \ldots, x_n$ as:

$$x_1 = \tfrac{1}{D}(-a_1 y + b_1 u) + x_2, \quad x_2 = \tfrac{1}{D}(-a_2 y + b_2 u) + x_3$$
$$\ldots . x_n = \tfrac{1}{D}(-a_n y + b_n u)$$

we find the state-space model:

$$\begin{aligned} &\mathrm{d}x_1/\mathrm{d}t = -a_1 y + x_2 + b_1 u \\ &\ldots\ldots. \\ &\mathrm{d}x_{n-1}/\mathrm{d}t = -a_{n-1} y + x_n + b_{n-1} u \\ &\mathrm{d}x_n/\mathrm{d}t = -a_n y + b_n u \\ &y = x_1 + b_0 u \end{aligned}$$

which can be written in the compact form

$$\frac{\mathrm{d}\mathbf{x}}{\mathrm{d}t} = \mathbf{Ax} + \mathbf{Bu}, \quad y = \mathbf{Cx} + \mathbf{Du} \tag{6.21}$$

where $\mathbf{x} = [x_1, x_2, \ldots, x_n]^{\mathrm{T}}$, $\mathbf{C} = \left[\begin{array}{ccccc} 1 & 0 & \ldots & 0 \end{array}\right]$, $\mathbf{D} = [b_o]$, and $\mathbf{u} = [u]$, and

$$\mathbf{A} = \left[\begin{array}{ccccc} -a_1 & 1 & 0 & \ldots & 0 \\ -a_2 & 0 & 1 & \ldots & 0 \\ \ldots & \ldots & \ldots & \ldots & \ldots \\ -a_{n-1} & 0 & 0 & \ldots & 1 \\ -a_n & 0 & 0 & \ldots & 0 \end{array}\right], \quad \mathbf{B} = \left[\begin{array}{c} b_1 - a_1 b_0 \\ b_2 - a_2 b_0 \\ \ldots \\ b_n - a_n b_0 \end{array}\right] \tag{6.22}$$

The model (6.21) with matrices (6.22) is known as *observable canonical* model. Other canonical state-space models are the following:

Controllable canonical model

$$\mathbf{x}=\begin{bmatrix} x_1 \\ x_2 \\ \vdots \\ x_n \end{bmatrix},\ \mathbf{A}=\begin{bmatrix} 0 & 1 & 0 & \cdots & \cdots & 0 \\ 0 & 0 & 1 & & & 0 \\ \cdots \quad \cdots & \cdots \quad \cdots & \cdots & \cdots & \cdots & \cdots \\ 0 & 0 & 0 & \cdots & \cdots & 1 \\ -a_n & -a_{n-1} & -a_{n-2} & \cdots & \cdots & -a_1 \end{bmatrix},$$

$$\mathbf{B}=\begin{bmatrix} 0 \\ 0 \\ \vdots \\ 0 \\ 1 \end{bmatrix},\ \mathbf{u}=u \tag{6.23}$$

Modal canonical model

$$\mathbf{A}=\begin{bmatrix} \lambda_1 & & & 0 \\ & \lambda_2 & & \\ & & \ddots & \\ 0 & & & \lambda_n \end{bmatrix},\ \mathbf{B}=\begin{bmatrix} 1 \\ 1 \\ \vdots \\ 1 \end{bmatrix},\ \mathbf{C}=[\rho_1,\rho_2,...,\rho_n],\ \mathbf{D}=[b_0] \tag{6.24}$$

where the poles (eigenvalues) were assumed distinct $(\lambda_1 \neq \lambda_2 \neq ... \neq \lambda_n)$, and the state variables are defined as:

$$\overline{x_1}(s)=\frac{1}{s-\lambda_1}\overline{u}(s),\ \ \overline{x_2}(s)=\frac{1}{s-\lambda_2}\overline{u}(s),...,\overline{x_n}(s)=\frac{1}{s-\lambda_n}\overline{u}(s)$$

We see that in these cases the system matrix A is diagonal. If some eigenvalues are multiple, then the matrix A is block-diagonal with submatrices the so-called *Jordan blocks*. Given a system with arbitrary matrices **A, B**, we can convert it to a desired canonical form by using a suitable state similarity transformation.

The block diagram of the state space model (6.21), with canonical matrices or not, has the form shown in Figure 6.21.

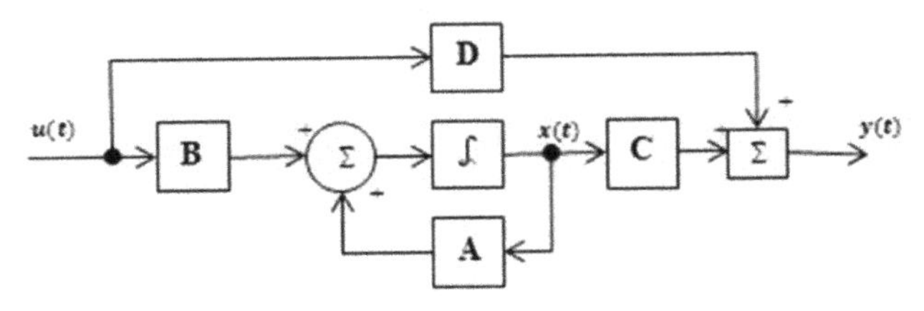

Figure 6.21 Block diagram of a general linear state-space model.

Controllability and Observability

Controllability

The ultimate goal of any automatic control system is to improve (and if possible to optimize) the performance of the physical process under control. The question here is whether a satisfactory controller can actually be designed that provide this improvement. In many cases the control input affects the entire system and so such a proper controller exists. But in many other cases (especially in MIMO systems) some control inputs affect only part of the dynamic performance. The concept of *controllability* has been developed exactly to study the ability of a controller to alter the performance of the system in an arbitrary desired way. A state $\mathbf{x}_0$ of a system is called *totally controllable* if it can be driven to a final state $\mathbf{x}_f$ as quickly as desired independently of the initial time t_0. A system is said to be *totally controllable* if all of its states are totally controllable'. Intuitively, we can see that if some state variables do not depend on the control input $\mathbf{u}(t)$, no way exists that can drive them to some other desired states. Thus, these states are called '*non-controllable states*'. If a system has at least one non-controllable state it is said to be non-totally controllable or, simply, non-controllable. The above controllability concept refers to the states of a system and so it is characterized as *state controllability*. If the controllability is referred to the outputs of a system then we have the so-called *output controllability*. In general, state controllability and output controllability are not the same. A linear time-invariant system: $\dot{\mathbf{x}} = \mathbf{Ax} + \mathbf{Bu}$, $\mathbf{u} = [u_1, u_2, .., u_m]^{\mathrm{T}}$ is state controllable if and only if the $n \times nm$ matrix:

$$\mathbf{P}_c = \left[\mathbf{B} \vdots \mathbf{AB} \vdots \mathbf{A}^2\mathbf{B} \vdots \cdots \vdots \mathbf{A}^{n-1}\mathbf{B}\right],$$

Called *controllability matrix*, has rank n, where n is the dimensionality of the state vector $\mathbf{x}$, i.e., if and only if

$$rank\mathbf{P}_c = n$$

An analogous controllability criterion can be formulated for discrete-time systems $\mathbf{x}(k+1) = \mathbf{Ax}(k) + \mathbf{Bu}$, for which the controllability matrix has exactly the same form:

$$\mathbf{P}_c = \left[\mathbf{B} \vdots \mathbf{AB} \vdots \mathbf{A}^2\mathbf{B} \vdots \cdots \vdots \mathbf{A}^{n-1}\mathbf{B}\right]$$

and the discrete-time system is totally state controllable if and only if

$$rank\mathbf{P}_c = n$$

Observability

Observability theory deals with the problem of determining whether the state variables of a system can be measured or estimated using only the input/output relation of the system and the measured output history from the initial time to the present. In practice not all state variables may be physically accessible (measurable). *Observability* is a dual concept to *controllability*, defined as follows:

(a) The state $\mathbf{x}(t)$ of a system is said to be observable at some time instant t, if the knowledge of the input $\mathbf{u}(t)$ and the output $\mathbf{y}(\tau)$ for some finite interval $t_0 \leq \tau \leq t$ determines completely $\mathbf{x}(t)$.
(b) A system is said to be totally observable if all states $\mathbf{x}(t)$ are observable.

A similar definition of observability holds for the discrete-time system $\mathbf{x}(k+1) = \mathbf{A}\mathbf{x}(k) + \mathbf{B}(k)\mathbf{u}(k)$, $\mathbf{y}(k) = \mathbf{C}(k)\mathbf{x}(k)$, with $k \geq k_0$ ($\mathbf{x}$ is an n-vector, $\mathbf{u}$ is an r vector, and $\mathbf{y}$ is an m-vector).

In analogy to the controllability, a linear time-invariant system $\dot{\mathbf{x}} = \mathbf{A}\mathbf{x} + \mathbf{B}\mathbf{u}$, $\mathbf{y} = \mathbf{C}\mathbf{x}$ is *state observable* if and only if the *observability matrix:*

$$\mathbf{S}_o = \begin{bmatrix} \mathbf{C} \\ \mathbf{CA} \\ \vdots \\ \mathbf{CA}^{n-1} \end{bmatrix}$$

has rank n ($rank\mathbf{S} = n$) where n is the state vector dimensionality.

The state vector x of any system S can be decomposed in four parts as:

$$\mathbf{x}^{\mathrm{T}} = \left[\mathbf{x}^{1^{\mathrm{T}}}, \mathbf{x}^{2^{\mathrm{T}}}, \mathbf{x}^{3^{\mathrm{T}}}, \mathbf{x}^{4^{\mathrm{T}}}\right]$$

where:

$\mathbf{x}^1$: Controllable and observable part (S_1)

$\mathbf{x}^2$: Uncontrollable and observable part (S_2)

$\mathbf{x}^3$: Controllable and unobservable part (S_3)

$\mathbf{x}^4$: Uncontrollable and unobservable part (S_4)

This decomposition is known as *Kalman decomposition*.

6.5.2 Lyapunov Stability

The Lyapunov stability theory has contributed substantially in the development of modern control.

The algebraic stability criteria as well as the stability criteria of Nyquist, Bode, and Nichols are applicable only to linear time-invariant systems. Lyapunov's stability method can also be applied to time-varying and non-linear systems. Lyapunov has introduced a generalized notion of *energy* (called *Lyapunov function*) and studied dynamic systems without external input. Combining Lyapunov's theory with the concept of **BIBO** stability (Section 6.4.1) we can derive stability conditions for *input-to-state* stability **(ISS)**. Lyapunov's criterion can be applied to both continuous-time and discrete-time systems. Lyapunov has introduced two stability methods. The first method requires the availability of the system's time response (i.e., the solution of the differential equations). The second method, also called *direct Lyapunov method*, does not require the knowledge of the system's time response. A brief description of this method is as follows.

Definition 1 The equilibrium state $\mathbf{x} = \mathbf{0}$ of the free system $\dot{\mathbf{x}} = \mathbf{A}(t)\mathbf{x}$ is stable in the Lyapunov sense *(L-stable)* if for every initial time t_0 and every real number $\varepsilon > 0$, there exists some number $\delta > 0$ as small as desired, that depends on t_0 and ε, such that:

If $\|\mathbf{x}_0\| < \delta$ then $\|\mathbf{x}(t)\| < \varepsilon$ for all $t \geq t_0$ where $\|\cdot\|$ denotes the norm of the vector $\mathbf{x}$, i.e., $\|\mathbf{x}\| = \left(x_1^2 + x_2^2 + \ldots + x_n^2\right)^{1/2}$.

Theorem 1 The transition matrix $\mathbf{\Phi}(t, t_0)$ of a linear system is bounded by $\|\mathbf{\Phi}(t, t_0)\| < k(t_0)$ for all $t \geq t_0$ if and only if the equilibrium state $\mathbf{x} = 0$ of $\dot{\mathbf{x}} = \mathbf{A}(t)\mathbf{x}$ is L-stable.

The bound of $\|\mathbf{x}(t)\|$ of a linear system does not depend on $\mathbf{x}_0$. In general, if the system stability (of any kind) does not depend on $\mathbf{x}_0$ we say that we have *global (total) stability or stability in the large*. If stability depends on $\mathbf{x}_0$ then it is called *local stability*. Clearly, local stability of a linear system implies also total stability.

Definition 2 The equilibrium state $\mathbf{x} = \mathbf{0}$ is asymptotically stable if:

(i) It is L-stable
(ii) For every t_0 and $\mathbf{x} = \mathbf{0}$, the condition $\mathbf{x}(t) \to \mathbf{0}$, for $t \to \infty$, holds.

Definition 3 If the parameter δ (in Definition 1) does not depend on t_0, then we have uniform L-stability.

Definition 4 If the system $\dot{\mathbf{x}}(t) = \mathbf{A}(t)\mathbf{x}$ is uniformly L-stable and for all t_0 and for arbitrarily large ρ, the relation $\|\mathbf{x}_0\| < \rho$ implies $\mathbf{x}(t) \to 0$ for $t \to \infty$, then the system is called *uniformly asymptotically stable*.

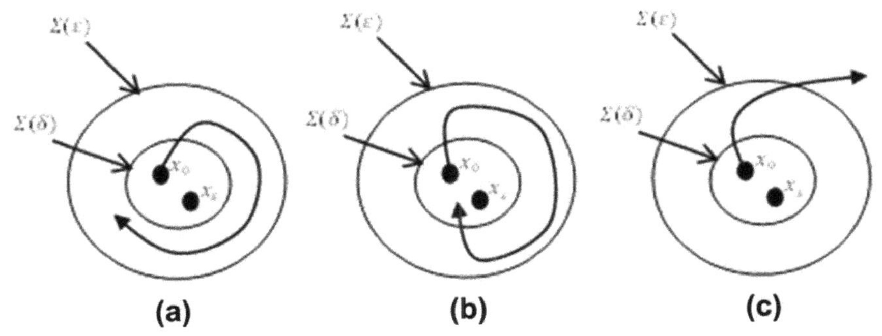

Figure 6.22 Illustration of L-stability (a), L-asymptotic stability (b), and instability (c). $\Sigma\left(\varepsilon\right)$ and $\Sigma\left(\delta\right)$ symbolize n-dimensional balls (spheres) with radii ε and δ respectively.

Theorem 2 The linear system $\dot{\mathbf{x}} = \mathbf{A}\left(t\right)\mathbf{x}$ is uniformly asymptotically stable if and only if there exist two constant parameters k_1 and k_2 such that: $\left\|\mathbf{\Phi}\left(t, t_0\right)\right\| \leq k_1 e^{-k_1(t-t_0)}$ for all t_0 and all $t \geq t_0$.

Definition 5 The equilibrium state $\mathbf{x} = \mathbf{0}$ of $\dot{\mathbf{x}} = \mathbf{A}\left(t\right)\mathbf{x}$ is said to be *unstable* if for some real number $\varepsilon > 0$, some $t_1 > t_0$ and any real number δ arbitrarily small, there always exist an initial stable $\left\|\mathbf{x}_0\right\| < \delta$ such that $\left\|\mathbf{x}\left(t\right)\right\| > \varepsilon$ for $t \geq t_1$.

Figure 6.22 illustrates geometrically the concepts of L-stability, L-asymptotic stability, and instability.

Direct Lyapunov Method

Let $d\left(\mathbf{x}\left(t\right), \mathbf{0}\right)$ be the distance of the state $\mathbf{x}\left(t\right)$ from the origin $\mathbf{x} = \mathbf{0}$ (defined using any valid norm). If we find some distance $d\left(\mathbf{x}\left(t\right), \mathbf{0}\right)$ which tends to zero for $t \to \infty$, then we conclude that the system is asymptotically stable. To show that a system is asymptotically stable using Lyapunov's direct method, we don't need to find such a *distance* (norm), but a *Lyapunov function* which is actually a generalized energy function.

Definition 6 *Time-invariant Lyapunov function* is called any scalar function $V\left(\mathbf{x}\right)$ of $\mathbf{x}$ which for all $t \geq t_0$ and $\mathbf{x}$ in the vicinity of the origin satisfies the following four conditions:

1. $V\left(x\right)$ is continuous and has continuous derivatives
2. $V\left(\mathbf{0}\right) = 0$
3. $V\left(\mathbf{x}\right) > 0$ for all $\mathbf{x} \neq 0$

4. $\frac{dV(\mathbf{x})}{dt} = \left[\frac{\partial V(\mathbf{x})}{\partial \mathbf{x}}\right]^{\mathrm{T}} \frac{d\mathbf{x}}{dt} < 0$ for $\mathbf{x} \neq \mathbf{0}$

Theorem 3 If a Lyapunov function $V(\mathbf{x})$ can be found for the state of a nonlinear or linear system $\dot{\mathbf{x}}(t) = \mathbf{f}(\mathbf{x}(t), t)$ where $\mathbf{f}(\mathbf{0}, t) = \mathbf{0}$ (**f** is a general function), then the state $\mathbf{x} = \mathbf{0}$ is asymptotically stable.

Remarks

1. If the definition 5 holds for all t_0, then we have 'uniformly asymptotic stability'.
2. If the system is linear, or we replace in the definition 5 the condition '**x** *in the vicinity of the origin*', by the condition '*for* **x** *everywhere*', then we have 'total asymptotic stability:
3. If the condition 4 of the definition 6 becomes $dV(\mathbf{x})/dt \leq 0$, then we have (simple) L-stability.

Clearly, to establish L-stability of a system we must find a Lyapunov function. Unfortunately, there does not exist a general methodology for this. The above definitions and results hold also for discrete-time systems, for which we have the following.

Definition 7 Discrete-time Lyapunov function (time-invariant) is any scalar function $V(\mathbf{x}(k))$ of $\mathbf{x}(k)$ which satisfies the following conditions:

1. $V(\mathbf{x})$ is continuous with respect to $\boldsymbol{x}$
2. $V(\mathbf{0}) = V(\mathbf{x}(k) = \mathbf{0}) = 0$
3. $V(\mathbf{x}) > 0$ for $\mathbf{x}(k) \neq \mathbf{0}$
4. $\Delta V(\mathbf{x}) < \mathbf{0}$ for $\mathbf{x}(k) \neq \mathbf{0}$ where $\Delta V(\mathbf{x}) = \Delta V(\mathbf{x}(k)) = V(\mathbf{x}(k+1)) - V(\mathbf{x}(k))$

Theorem 4 If a Lyapunov function $V(\mathbf{x}(k))$can be found for the state of a discrete-time system $\mathbf{x}(k+1) = \mathbf{f}(\mathbf{x}(k))$ with $\mathbf{f}(\mathbf{x}(k)) = 0$ for all k, then the equilibrium state $\mathbf{x}(k) = 0$ (for all *k*) is globally asymptotically stable.

Analogous results hold for the case of time-varying Lyapunov functions $V(\mathbf{x}(t), t)$ and $V(\mathbf{x}(k), k)$.

6.5.3 State Feedback Control

State-feedback control is more powerful than classical control because it can be applied to **MIMO** systems, performing the control in a unified way for all loops simultaneously, and not serially one loop after the other which does not guarantee the overall stability and robustness. State feedback controllers are distinguished in:

- Eigenvalue placement controller.
- Non interacting (decoupling) controller.
- Model matching controller.

Let a **SISO** system:

$$\dot{\mathbf{x}}(t) = \mathbf{A}\mathbf{x}(t) + \mathbf{B}u(t), \ y(t) = \mathbf{C}\mathbf{x}(t) + Du(t)$$

where **A** is an $n \times n$ constant matrix, **B** is an $n \times 1$ constant matrix (column vector), **C** is an $1 \times n$ matrix (row vector), u is a scalar input, and D is a scalar constant. In this case, a state feedback controller has the form:

$$u(t) = \mathbf{F}\mathbf{x}(t) + v(t)$$

where $v(t)$ is a new control input and **F** is an n-dimensional constant row vector:

$\mathbf{F} = [f_1, f_2, ..., f_n]$. Introducing this control law into the system we get the state equations of the closed-loop (feedback) system:

$$\dot{\mathbf{x}}(t) = (\mathbf{A} + \mathbf{BF})\mathbf{x}(t) + \mathbf{B}v(t), \ y(t) = (\mathbf{C} + \mathbf{DF})\mathbf{x}(t) + Dv(t)$$

A similar state-feedback controller can also be formulated for discrete time systems:

$$\begin{aligned}
&\mathbf{x}(k+1) = \mathbf{A}\mathbf{x}(k) + \mathbf{B}\mathbf{u}(k), \ y(k) = \mathbf{C}\mathbf{x}(k) + Du(k) \\
&\mathbf{u}(k) = \mathbf{F}\mathbf{x}(k) + \boldsymbol{v}(k), \ \mathbf{F} = [f_1, f_2, ..., f_n] \\
&\mathbf{x}(k+1) = (\mathbf{A} + \mathbf{BF})\mathbf{x}(k) + \mathbf{B}\boldsymbol{v}(k)
\end{aligned}$$

Eigenvalue placement controller

Here the problem is to select the controller gain matrix **F** such that the eigenvalues of the closed-loop matrix $\mathbf{A} + \mathbf{BF}$ are placed at desired positions $\lambda_1, \lambda_2, ..., \lambda_n$. It can be shown that this can be done (i.e., the system eigenvalues are controllable by state feedback) if and only if the system $(\mathbf{A}, \mathbf{B})$ is totally controllable. Three techniques by which the feedback matrix **F** can be selected are:

- Use of the controllable canonical form.
- Equating the characteristic polynomials.
- Ackerman technique.

Decoupling controller

Consider a MIMO system with m inputs and m outputs. We say that the system is input-output decoupled (or non-interacting) if each output is affected by one only input and each input affects only one output. The transfer matrix $\mathbf{G}(s)$ of an input-output decoupled system is diagonal, i.e.:

$$\mathbf{G}(s) = diag\left[g_{11}(s), ..., g_{ii}(s), ..., g_{mm}(s)\right]$$

in which case the outputs $y_i(s)\,(i = 1, 2, ..., m)$ are given by $y_i(s) = g_{ii}(s)\,u_i(s)$, where $u_i(s)\,(i = 1, 2, ..., m)$ are the inputs of the system.

Falb and Wolovich have shown that a system: $\dot{\mathbf{x}} = \mathbf{A}\mathbf{x} + \mathbf{B}\mathbf{u}$, $\mathbf{y} = \mathbf{C}\mathbf{x}$ ($\mathbf{u}$ is an m-vector, $\mathbf{y}$ is an m-vector) that has not a diagonal transfer matrix $\mathbf{G}(s)$ with a $|\mathbf{G}(s)| \neq 0$, can be decoupled by a static state feedback controller:

$$\mathbf{u}(t) = \mathbf{F}\mathbf{x}(t) + \mathbf{G}\boldsymbol{v}(t),$$

if and only if the matrix (called decoupling matrix):

$$\mathbf{B}^* = \begin{bmatrix} \mathbf{c}_1\mathbf{A}^{d_1}\mathbf{B} \\ \mathbf{c}_2\mathbf{A}^{d_2}\mathbf{B} \\ \vdots \\ \mathbf{c}_m\mathbf{A}^{d_m}\mathbf{B} \end{bmatrix} \tag{6.25}$$

is nonsingular $|\mathbf{B}^*| \neq 0$. A solution $\mathbf{F}$ and $\mathbf{G}$ that decouples the system is:

$$\mathbf{F} = -\left(\mathbf{B}^*\right)^{-1}\mathbf{A}^*, \quad \mathbf{G} = \left(\mathbf{B}^*\right)^{-1} \tag{6.26}$$

where

$$\mathbf{A}^* = \begin{bmatrix} c_1\mathbf{A}^{d_1+1} \\ \vdots \\ c_m\mathbf{A}^{d_m+1} \end{bmatrix}$$

and the indexes $d_i\,(i = 1, 2, ..., m)$ are equal to:

$$d_i = \begin{cases} \min\left\{j : \mathbf{c}_i\mathbf{A}^j\mathbf{B} \neq \mathbf{0}, j = 0, 1, .., n-1\right\} \\ \quad n-1 \text{ if } \mathbf{c}_i\mathbf{A}^j\mathbf{B} = 0 \text{ for all } j \end{cases}$$

Model matching controller

In this case the problem is to find a state feedback controller:

$$\mathbf{u} = \mathbf{F}\mathbf{x} + \mathbf{G}\boldsymbol{v}$$

which, when applied to the system $\dot{\mathbf{x}} = \mathbf{Ax} + \mathbf{Bu}$, $\mathbf{y} = \mathbf{Cx} + \mathbf{Du}$, leads to a closed-loop system that matches the transfer function of a desired model:

$$\dot{\hat{\mathbf{x}}} = \hat{\mathbf{A}}\hat{\mathbf{x}} + \hat{\mathbf{B}}\hat{\upsilon}, \ \hat{\mathbf{y}} = \hat{\mathbf{C}}\hat{\mathbf{x}} + \hat{\mathbf{D}}\hat{\upsilon}$$

This means that it is desired to match the zeros, the poles and the D.C. gains of the closed-loop system to those of the desired model. A suitable technique is to use the controllable canonical form of the system under control ($\mathbf{A}$, $\mathbf{B}$; $\mathbf{C}$, $\mathbf{D}$). Under certain conditions it is possible to have *exact model matching*. If not, then one may obtain *approximate model matching*, depending on the approximation criterion used. In the general case the derivation of the controller is somehow complicated.

6.5.4 Optimal and Stochastic Control

Optimal control This type of control is obtained by optimizing a certain performance (cost) function. There are two principles for doing this:

- Bellman's principle of optimality.
- Pontryagin's minimum principle.

These principles were derived independently, but in later years they were shown to be equivalent. The principle of optimality says:

> *'An optimal policy (or optimal control policy) has the property that for every initial state and initial decision, the remaining decisions constitute an optimal policy with respect to the state that results from the initial decision'*.

This principle will be applied to the following general discrete-time optimal control problem: '*Given* a discrete-time system:

$$\mathbf{x}_{k+1} = \mathbf{F}_k\left(\mathbf{x}_k, \mathbf{u}_k\right), k = 1, 2, ..., N$$

Find the control sequence $\{\mathbf{u}_k\} = \{\mathbf{u}_k : k = 1, 2, ..., N\}$ which minimizes the cost function

$$J_N = \sum_{k=1}^{N} L_k\left(\mathbf{x}_k, \mathbf{u}_k\right),$$

According to the principle of optimality J_N can be written as:

$$J_N = L_1\left(x_1, u_1\right) + J_{N-1}\left(x_2\right) = L_1\left(x_1, u_1\right) + J_{N-1}\left\{F\left(x_1, u_1\right)\right\}$$

where the first term is the initial cost and the second term is the optimal cost resulting from the next $N-1$ decisions. Thus the optimal cost $J_N^0(x_1)$ is given by:

$$J_N^0(x_1) = \min_{u_1}\left[L_1(x_1,u_1) + J_{N-1}\{F(x_1,u_1)\}\right], N \geq 2$$

For $N=1$ we have $J_1^0(x_1) = \min_{u_1} L_1(x_1,u_1)$. Thus for $N = N, N-1, N-2, ..., 2, 1$ we get:

$$J_N^0(x_1) = \min_{u_1}\left[L_1(x_1,u_1) + J_{N-1}^0(x_2)\right], x_2 = F_1(x_1,u_1)$$
$$J_{N-1}^0(x_2) = \min_{u_2}\left[L_1(x_2,u_2) + J_{N-2}^0(x_3)\right], x_3 = F_2(x_2,u_2)$$

....

....

$$J_2^0(x_{N-1}) = \min_{u_{N-1}}\left[L_{N-1}(x_{N-1},u_{N-1}) + J_1^0(x_N)\right],$$
$$x_N = F_{N-1}(x_{N-1},u_{N-1})$$
$$J_1^0(x_N) = \min_{u_N}\left[L_N(x_N,u_N)\right]$$

Consequently, starting from $J_1^0(x_N)$ we compute u_N, then we compute u_{N-1} from $J_2^0(x_{N-1})$, and so on. This means that applying the principle of optimality we find an optimal control policy in which the current control action depends only on the current state and the current time, and does not depend on previous states or control actions. A policy of this type is known as *Markovian policy*.

This principle can be applied to linear and nonlinear systems in both continuous-time and discrete-time representations.

Continuous time

Let the system:

$$\dot{\mathbf{x}}(t) = \mathbf{f}(\mathbf{x},\mathbf{u},t), \mathbf{x}(t_0) = \mathbf{x}_0 \tag{6.27}$$

The problem is to determine the control input $\mathbf{u}(t), t_0 \leq t \leq t_f$ such that to minimize the cost function:

$$J(\mathbf{u}) = \int_{t_0}^{t_f} L(\mathbf{x},\mathbf{u},t)\,dt \tag{6.28}$$

Defining the optimal cost $J^o(\mathbf{x},t)$ as:

$$J^o(\mathbf{x},t) = \min_{\mathbf{u}(\tau),\ \tau\in[t,t_f]} \int_t^{t_f} L(\mathbf{x},\mathbf{u},\tau)\,d\tau,$$

the solution $\mathbf{u}(t) : t \in [t_0, t_f]$ is given by the solution of the equation:

$$-\frac{\partial J^o}{\partial t} = \min_{u(t)} H(\mathbf{x}, \mathbf{u}, t) \tag{6.29}$$

where $H(\mathbf{x}, \mathbf{u}, t)$ is the system Hamilton-Jacobi-Bellman:

$$H = L(\mathbf{x}, \mathbf{u}, t) + \frac{\partial J^{o\mathrm{T}}(\mathbf{x}, t)}{\partial x}\mathbf{f}(\mathbf{x}, \mathbf{u}, t) \tag{6.30}$$

Let the system be linear:

$$\dot{\mathbf{x}} = \mathbf{A}(t)\mathbf{x} + \mathbf{B}(t)\mathbf{u}, \;\; \mathbf{x}(t_0) = \mathbf{x}_o \tag{6.31}$$

and the cost function quadratic:

$$J = \int_{t_0}^{t_f} L dt, \;\; L = \frac{1}{2}\mathbf{x}^{\mathrm{T}}\mathbf{Q}\mathbf{x} + \frac{1}{2}\mathbf{u}^{\mathrm{T}}\mathbf{R}(t)\mathbf{u}, \;\; |\mathbf{R}(t)| \neq 0 \tag{6.32}$$

We define the optimal cost J^o as

$$J^o(\mathbf{x}, t) = \frac{1}{2}\mathbf{x}^{\mathrm{T}}(t)\mathbf{P}(t)\mathbf{x}(t) \tag{6.33}$$

where $\mathbf{P}(t)$ is a symmetric positive definite matrix.

Then, the optimal Hamiltonian is found to be:

$$H^o = \min_{\mathbf{u}(t)} H = \frac{1}{2}\mathbf{x}^{\mathrm{T}}\left(\mathbf{Q} + \mathbf{A}^{\mathrm{T}}\mathbf{P} + \mathbf{P}\mathbf{A} - \mathbf{P}\mathbf{B}\mathbf{R}^{-1}\mathbf{B}^{\mathrm{T}}\mathbf{P}\right)\mathbf{x}$$

and the solution of the Hamilton-Jacobi-Bellman equation is found to be given by:

$$\mathbf{u}^o(t) = -\mathbf{R}^{-1}(t)\mathbf{B}^{\mathrm{T}}(t)\mathbf{P}(t)\mathbf{x}(t) \tag{6.34}$$

with

$$-d\mathbf{P}(t)/dt = \mathbf{A}^{\mathrm{T}}\mathbf{P}(t) + \mathbf{P}(t)\mathbf{A} + \mathbf{Q} - \mathbf{P}(t)\mathbf{B}\mathbf{R}^{-1}\mathbf{B}^{\mathrm{T}}\mathbf{P}(t) \;\; \text{and}\; \mathbf{P}(t_f) = 0. \tag{6.35}$$

The weight matrix $\mathbf{P}(t)$can be found by solving in reverse (backward) time, the matrix Riccati Equation (6.35), and then use it in forward time to compute $\mathbf{u}^o(t)$ for $t_0 \leq t \leq t_f$. The resulting closed-loop optimal control system is pictorially shown in Figure 6.23.

The linear controller (6.34)–(6.35) constitutes one of the cornerstones of modern control and has found numerous applications in important applications of society.

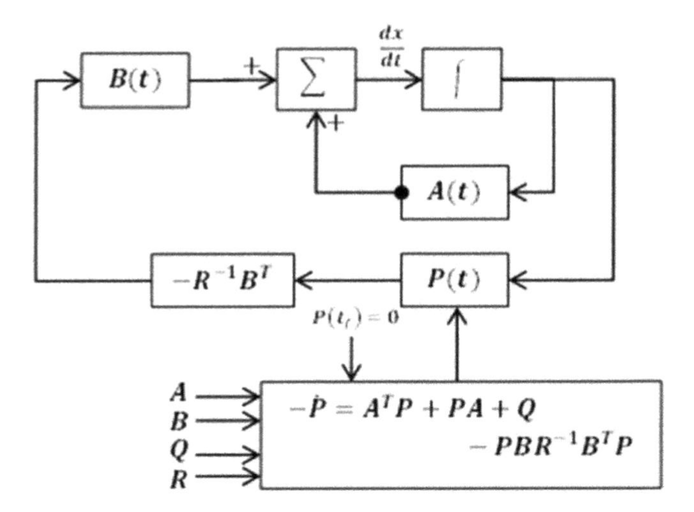

Figure 6.23 Closed-loop optimal control system.

Pontryagin Minimum Principle

This principle (originally formulated by Pontryagin as *maximum principle*) is applicable to the cases where the control signal is subject to constraints specified by a lower and an upper bound. By definition, the optimal control signal $\mathbf{u}^0$ corresponds to a local minimum of the cost function if: $J(\mathbf{u}) - J(\mathbf{u}^0) = \Delta J \geq 0$ for all the allowable signals $\mathbf{u}$ close to $\mathbf{u}^0$. Let $\mathbf{u} = \mathbf{u}^0 + \delta\mathbf{u}$. Then:

$$\Delta J\left(\mathbf{u}^0, \delta\mathbf{u}\right) = \delta J\left(\mathbf{u}^0, \delta\mathbf{u}\right) + (higher + order\ terms)$$

The variation δJ is a linear function of $\delta\mathbf{u}$ and the higher-order terms tend to zero for $\|\delta\mathbf{u}\| \to 0$. If the control signal is free (i.e., if it is not subject to some constraint), then the control variation $\delta\mathbf{u}$ can take any arbitrary value, and the necessary condition for $\mathbf{u}^0$ to minimize $J(\mathbf{u})$ is:

$$\delta J\left(\mathbf{u}^0, \delta\mathbf{u}\right) = 0 \text{ for } \|\boldsymbol{\delta}\| \text{ sufficiently small.}$$

But, if the control signal is subject to constraints, the control variation $\delta\mathbf{u}$ is arbitrary only if the total $\mathbf{u}$ lies in the interior of the permissible control region for all $t \in [t_0, t_f]$. As long as this is valid, the constraints do not have any effect on the solution of the problem. However, if $\mathbf{u}$ falls at the boundary of the allowable region, at least for some time instants $t \in [t_1, t_2]\,, t_0 \leq t_1 \leq t_2 \leq t_f$, then there exist allowable variations $\delta\mathbf{u}$ for which their opposite variations $-\delta\mathbf{u}$ are not allowable. If we consider only these variations the necessary condition for $\mathbf{u}^0$ to minimize J is:

$\delta J\left(\mathbf{u}^0, \delta\hat{\mathbf{u}}\right) \geq 0$. Therefore, the necessary condition for $\mathbf{u}^0$ to minimize J is:

$$\delta J\left(\mathbf{u}^0, \delta\mathbf{u}\right) \geq 0 \tag{6.36}$$

where $\|\delta\mathbf{u}\|$ is sufficiently small such that the sign of ΔJ to be specified by the sign of ΔJ, and the signal $\mathbf{u} = \mathbf{u}^0 + \delta\mathbf{u}$ to be *allowable* (i.e., it does not go outside the boundary of the allowable control region). This is the *minimum principle of Pontryagin* (see [11]). A useful practical optimal control problem which was solved by Pontryagin principle is the time optimal control problem of choosing $u\,(t)$such that to minimize the time of control under the constraint $u_L \leq |u\,(t)| \leq u_M$.

Stochastic Optimal Control

Here the system is stochastic. In the continuous-time case it is described by the *Gauss-Markov* model:

$$\begin{aligned} &\dot{\mathbf{x}}\,(t) = \mathbf{A}\,(t)\,\mathbf{x}\,(t) + \boldsymbol{\Gamma}\,(t)\,\mathbf{w}\,(t) \\ &\mathbf{y}\,(t) = \mathbf{C}\,(t)\,\mathbf{x}\,(t) + \mathbf{v}\,(t) \qquad \mathbf{x}\,(t_0) = \mathbf{x}_0,\ t \geq t_0 \end{aligned} \tag{6.37}$$

where all stochastic processes $w\,(t)\,, v\,(t)$ and $\mathbf{x}\,(t_0)$ have Gaussian distributions with known mean values and covariance's (Figure 6.24).

The cost function to be minimized is

$$\begin{aligned} J &= \mathrm{E}\left[\int_{t_0}^{t_f} L dt + J_f\right] \\ L &= \frac{1}{2}\mathbf{x}^{\mathrm{T}}\mathbf{Q}\,(t)\,\mathbf{x} + \frac{1}{2}\mathbf{u}^{\mathrm{T}}\mathbf{R}\,(t)\,\mathbf{u},\ J_f = \frac{1}{2}\mathbf{x}^{\mathrm{T}}\,(t_f)\,\mathbf{Q}_f\mathbf{x}\,(t_f)\,, \end{aligned}$$

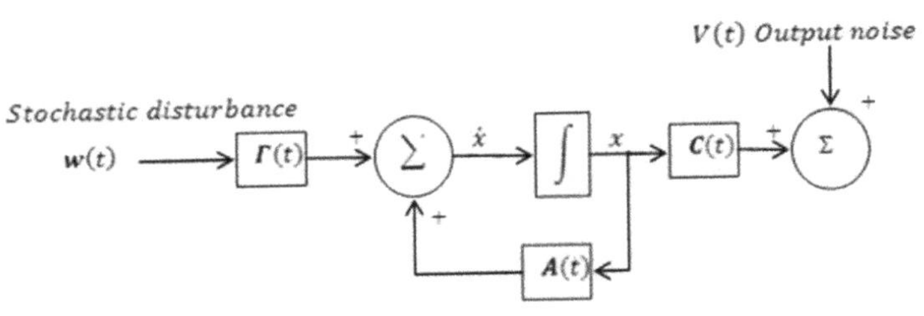

Figure 6.24 Pictorial representation of Gauss-Markov model. The processes $\mathbf{w}\,(t)\mathbf{v}\,(t)$ and $\mathbf{x}_0$ have mean values $\bar{\mathbf{w}}\,(t)\,, \bar{\mathbf{v}}(t)$; and covariance matrices $Q\,(t)\,, R\,(t)\,, t \geq t_0$, and $\bar{x}\,(t_0)\,, \Sigma\,(t_0)$ positive definite covariance matrix.

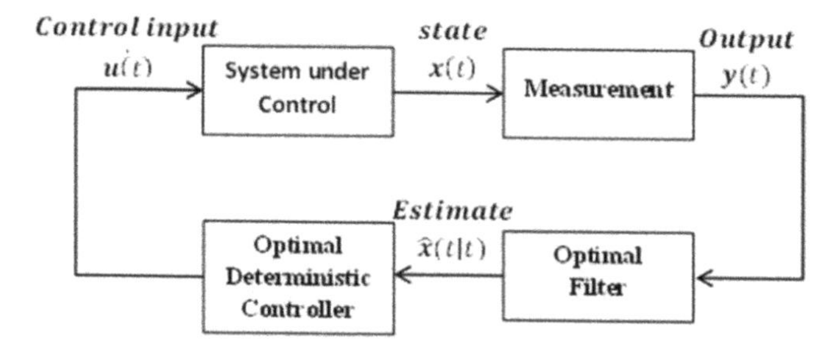

Figure 6.25 Block diagram representation of separation between estimation and control.

where $\mathrm{E}[\cdot]$ is the expectation operator. Applying Bellman's principle we find that the solution is:

$$\hat{\mathbf{u}}^0(t) = -\mathbf{R}^{-1}(t)\,\mathbf{B}^{\mathrm{T}}(t)\,\mathbf{P}(t)\,\hat{\mathbf{x}}(t\,(t)),\ t \geq t_0, \tag{6.38}$$

where $\hat{\mathbf{x}}(t|\,t)$ is the optimal estimate of $\mathbf{x}(t)$ provided by the Kalman filter, [14], and $\mathbf{P}(t)$ is the solution of the control Riccati Equation (6.35). Here, the Kalman principle of '*separation between optimal state estimation and optimal control*' holds, and so the overall optimal stochastic control problem solution has the pictorial form shown in Figure 6.25.

6.5.5 Model-Free Control

This kind of control includes all approaches that develop methods of control which do not require the availability of a mathematical model of the system under control. These methods are also collectively called '*intelligent control methods*' and include:

- Knowledge based control **(KBC)**.
- Behavior-based control (**BBC**).
- Fuzzy logic-based control **(FLC)**.
- Neural network-based control (**NNC**).
- Genetic algorithm-based control (**GAC**).
- Hybrid control (Neuro-fuzzy: NF, fuzzy genetic: FG, etc.).

Fuzzy sets and *fuzzy logic* were initiated by Lofti Zadeh in 1965, and provide the means for the design of control systems than can operate and make inferences in a linguistic way under conditions of uncertainty as it is done by experts in every-day (*approximate reasoning*) [81–83].

Neural networks (NN) were initiated by *McCulloch and Pitts* in 1943 using a model of the basic cell of the human brain, presently known as '*artificial neuron*'. The next big step in NNs was made by Hebb in 1949 who coined the concept of learning through the updating of synaptic weights. Fifteen years after the introduction of McCulloch and Pitts '*neuron*', Rosenblatt has developed the concept of *Perceptron* as a solution to the identification problem. The big progress in NNs was made in the 1980s (*Grossberg* (1980), *Broamhead* radial basis functions (RBF, 1988) [84–85].

Genetic algorithms (*GA*) were introduced and expanded by John Holland in the 1960's and 1970's, and have exploited the 'adaptation' phenomenon as it occurs in nature. The first book on GA was Holland's book: 'Adaptation in Natural and Artificial Systems' where the operators of '*mutation*', *and* '*inversion*' were used in a population of computational chromosomes (string of 'genes', 0's and 1's) [85–87]. Figure 6.26 shows the flow chart of a fundamental GA.

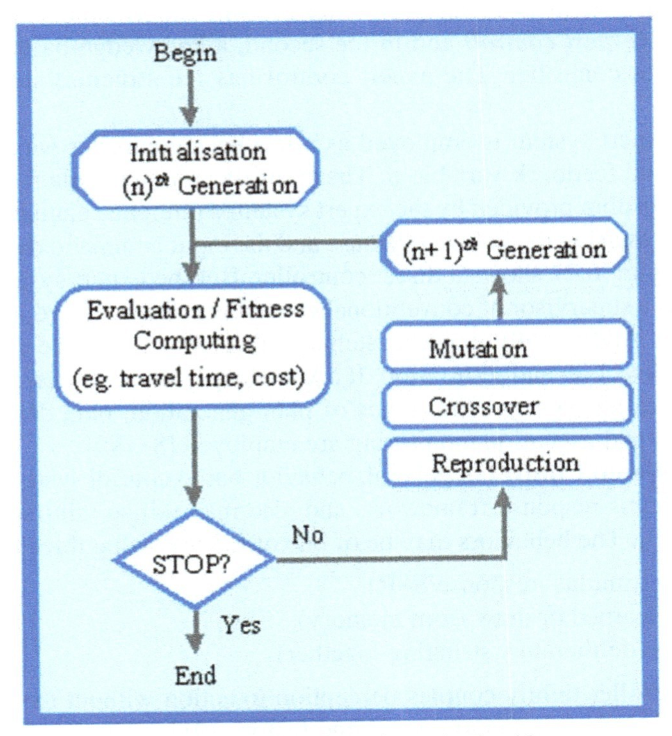

Figure 6.26 Flow chart of genetic algorithm.

Source: www.business-fundas.com (Genetic Algorithm).

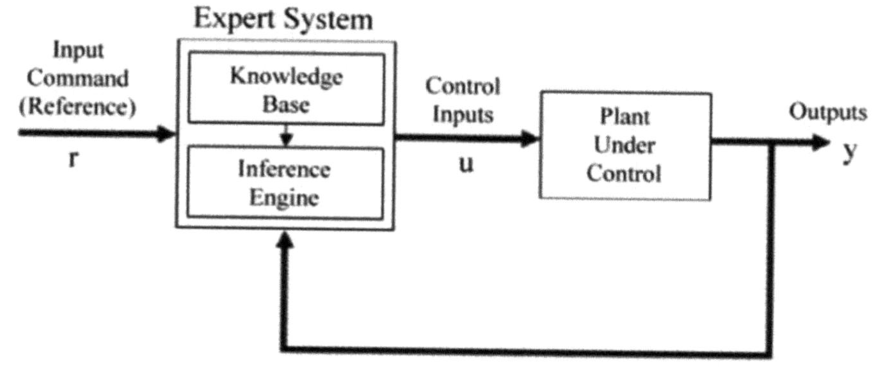

Figure 6.27 Basic general structure of expert control.

Knowledge-Based Control In general, knowledge-based (or expert) control is distinguished in two classes. In the first an expert system is used as a controller *(expert control)* and in the second, a knowledge-based planner is used as a controller. The expert control has the structure shown in Figure 6.27.

Here the expert system is employed as a feedback controller with reference input r, and feedback variable y. The controller's output (plant control input) is an algorithm provided by the expert system's inference engine which makes inferences using the knowledge base and the input command data. The structure of Figure 6.27 shows a direct controller. But the expert system can also be used as a supervisor of conventional controllers (the so-called *indirect expert control*). Additionally, expert systems themselves can also be used for fault (failure) detection and restoration. If a knowledge-based planner is used instead of the expert systems the stages of plan generation, plan decisions, plan execution, and execution monitoring are employed [88–90].

Behavior-Based Control In general, behavior-based control systems rely on a tight stimulus-response framework, and also may utilize minimal state information [99]. The behaviors may be of one of the following three types:

- *Reflexive* (stimulus–response/**S–R**).
- *Reactive* (learned or draw from memory).
- *Conscious* (deliberately stringing together).

A reactive controller tightly couples perception to action without the use of intervening abstract representations or time history [91]. The two principal influential architectures are:

- Subsumption architecture (Brooks, 1986).
- Motor schemas architecture (Arkin, 1987).

According to subsumption philosophy modules should be grouped into layers of competence, and at a higher level can override or subsume behaviors in the next lower level (this can be done through *suppression*, i.e., substitute input going to a module or *inhibition*, i.e., turn off output from a module). The subsumption architecture does not involve internal state of a local persistent representation similar to a world model.

Conscious (deliberative planning) behavior can be merged with reactive behavior in three ways namely [91]:

- Hierarchical integration of planning and reaction (Figure 6.28(a)).
- Planning to guide reaction (i.e., selecting and setting parameters for the reactive control) (Figure 6.28(b)).
- Couple planning and reaction (which allows the two concurrent activities, each guiding the other) (Figure 6.28(c)).

Behavior-based control finds application to mobile robot obstacle-avoiding path planning control. One of the first robot control schemes that were designed using the hybrid deliberative planning and reactive (schema-based) architecture is the *Autonomous Robot Architecture (AuRA)* [91].

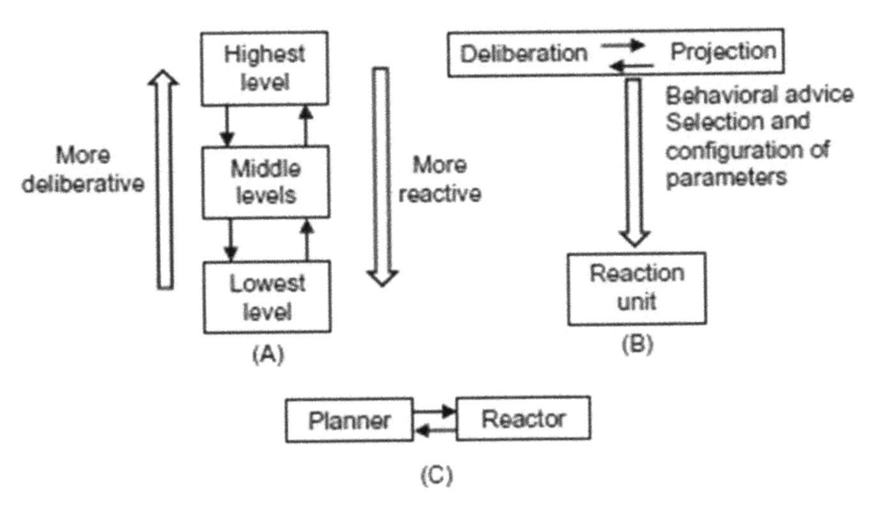

Figure 6.28 (a) Hierarchical hybrid deliberative-reactive structure, (b) Planning to guide reaction scheme, and (c) Coupled planning and reacting scheme.

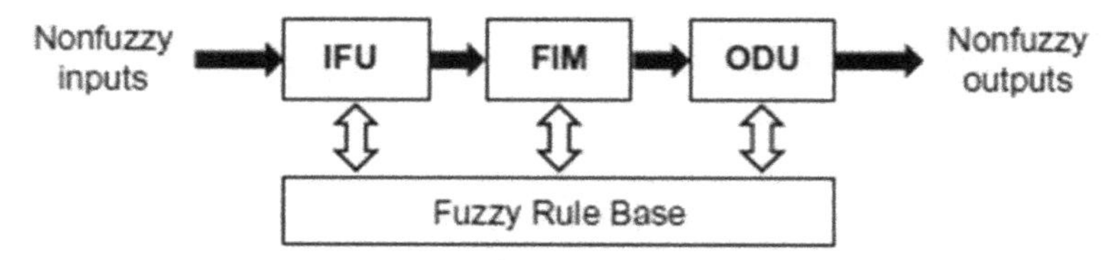

Figure 6.29 General form of FLA/FLC.

Fuzzy Logic-based Control Fuzzy-logic, or linguistic, algorithms (FLAs) and controllers (FLCs) have found particular applications in process control, robotics and management. Robotic and management systems are complex systems that cannot be modeled precisely even under specific assumptions and approximations. In many cases the control of robotic and management systems needs intervention of human operators who employ experimental rules than can cast into the fuzzy logic framework [82, 83].

The general structure of any FLA/FLC controller is shown in Figure 6.29 and involves four principal units [82]:

- Fuzzification interface (**FI**), or input fuzzification unit (IFU).
- Knowledge-base/Fuzzy rule base (**KB/FRB**).
- Fuzzy inference mechanism (**FIM**).
- Output defuzzification interface/Unit (**DI/ODU**).

The fuzzification interface performs the following operations:

- Measures the values of inputs.
- Maps the input range of values to suitable universes of discourse.
- Fuzzifies the incoming data (i.e., converts them to appropriate fuzzy form).

The knowledge base involves a numeric 'data base' section and fuzzy (linguistic) rule-base section. The data base section involves all numeric information required to specify the fuzzy control rules and fuzzy reasoning. The fuzzy rule base involves the specification of the control goals and control strategies in linguistic form.

The fuzzy inference mechanism, which constitutes the core of the FLC, involves the required decision making logic (fuzzy reasoning) such as generalized modus ponens, Zadeh's max-min composition, etc.

The defuzzification interface carries out the following tasks:

- Maps the range of output variables into corresponding universes of discourse.

- Defuzzifies the result of FIM (i.e., converts to nonfuzzy form the fuzzy control action received from FIM).

The fuzzy rule base contains the rules that are to be used for the control of the process. These rules are usually the result of interviews with the expert operators (very rarely the rules come out of mathematical analysis or simulations) and have the IF-THEN form. In the general case the rules have many inputs and many outputs (MIMO). However, it can be shown that a set of MIMO rules can be transformed to a set of MISO (multi-input-single-output) rules.

The structure of the closed-loop fuzzy-logic based system (plant) has the form of Figure 6.27 where in place of the *Expert System* an FLA/FLC controller of the form of Figure 6.29 is used.

Neural Network-Based Control The basic (artificial) neuron model is based on the McCulloch-Pitts model which has the form shown in Figure 6.30. The neuron has a basic processing unit which consists of three elements [84].

1. A set of connection branches (synapses).
2. A linear summation node.
3. An activation (nonlinear) function $\sigma(\cdot)$.

Each connection branch has a weight (strength) which is positive if it has an *excitatory* role and negative if it has an *inhibitory* role. The summation node sums the input signals multiplied by the respective synaptic weights. Finally, the *activation function* (or as otherwise called) *squashing function* limits the allowable amplitude of the output signal to some finite value, typically in the normalized interval [0,1] or, alternatively, in the interval [−1,1]. The neuron model has also a threshold θ which is applied externally, and practically lowers the net input to the activation function.

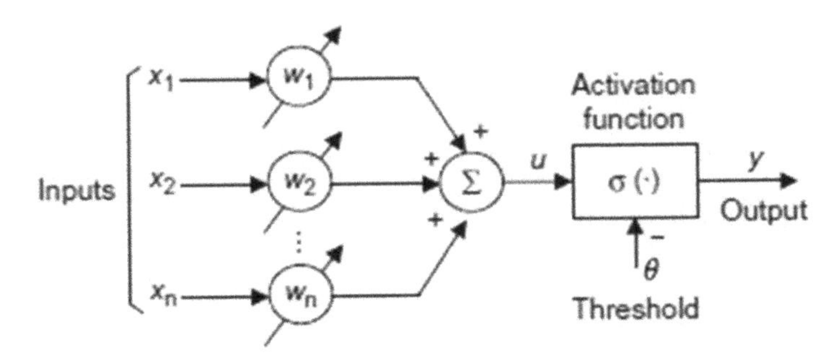

Figure 6.30 The basic artificial neuron model (The threshold θ is represented by the input $x_0 = -1$ and the weight $w_0 = \theta$).

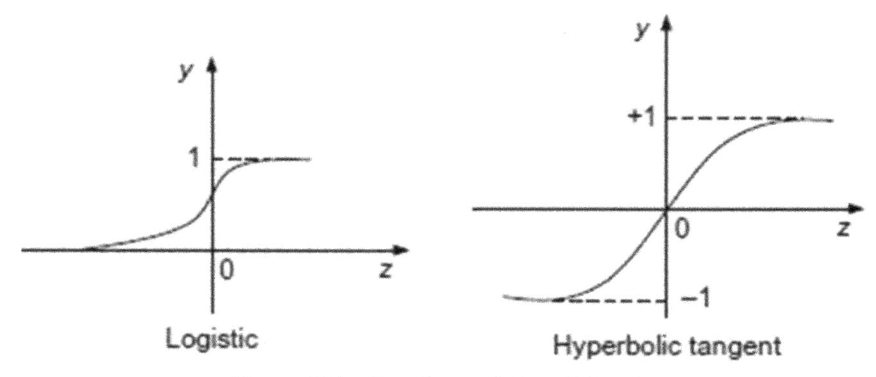

Figure 6.31 Two forms of sigmoid function.

From Figure 6.30 it follows that the neuron is described by the equation:

$$y = \sigma\left(z\right), \; z = \sum_{i=0}^{n} w_i x_i$$

The nonlinear activation function $\sigma\left(x\right)$ can be of the *on-off* or the *saturation* function type, or of the *sigmoid* function type with values either in the interval [0,1] or in the interval [−1,1], as shown in Figure 6.31.

The first sigmoid function is the *logistic function:*

$$y = \sigma\left(z\right) = \frac{1}{1+e^{-z}}, \; y \in [0,1]$$

and the second is the *hyperbolic tangential* function:

$$y = \sigma\left(z\right) = \tan\left(\frac{z}{2}\right) = \frac{1-e^{-z}}{1+e^{-z}}, \; y \in [-1,1]$$

as shown in Figure 6.31.

The two principal models of artificial neural networks are the *multilayer perceptron* (**MLP**) and the *radial-basis function* (**RBF**) networks [84].

The Multilayer Perceptron: The MLP NN has been developed by Rosemblat (1958) and has the structure shown in Figure 6.32. It involves the input layer of nodes, the output layer of nodes, and a number of intermediate (hidden) layers of nodes. It is noted that even one only hidden node layer is sufficient for the MLP NN to perform operations that can be achieved by many hidden layers. This comes from the universal approximation theorem of Kolmogorov.

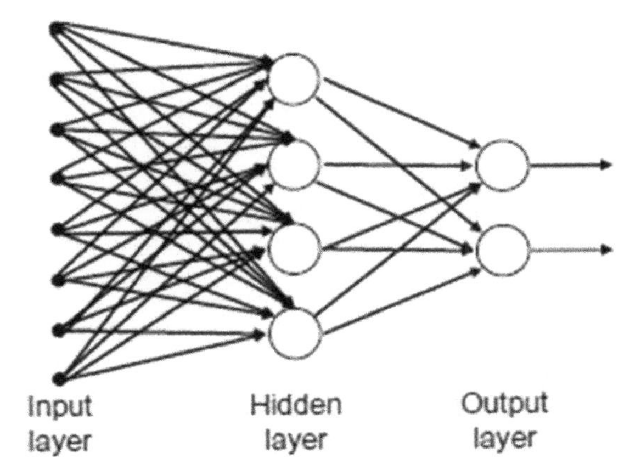

Figure 6.32 A single-hidden layer MLP with eight input nodes, four hidden nodes, and two output nodes.

In this case, the output of the NN (with m output neurons and L neurons in the hidden layer) is given by the relation:

$$y_i = \sum_{j=1}^{L} \left[v_{ij} \sigma \left(\sum_{k=0}^{n} w_{jk} x_k \right) \right], \quad i = 1, 2, ..., m, \tag{6.39}$$

where x_k $(k = 0, 1, 2, ...n)$ are the NN inputs (including the thresholds), w_{jk} are the input-to-hidden-layer interconnection weights, and v_{ij} are the hidden-to-output-layer interconnection weights.

In compact form, Equation (6.39) can be written as:

$$\mathbf{y} = \mathbf{V}^{\mathrm{T}} \boldsymbol{\sigma} \left(\mathbf{W}^{\mathrm{T}} \mathbf{x} \right), \tag{6.40}$$

where:

$$\mathbf{x} = \left[x_0, x_1, ..., x_n \right]^{\mathrm{T}}, \quad \mathbf{y} = \left[y_1, y_2, ..., y_m \right]^{\mathrm{T}}$$

The synaptic weights of the MLP are updated by the **BP** algorithm which adjusts (adapts) the weights such that to minimize the mean square error between the desired and actual outputs after the presentation of each input vector (pattern) at the NN input layer.

The RBF Network: An RBF network approximates an input-output mapping by employing a linear combination of radially symmetric functions (Figure 6.33). The k-th output y_k is given by:

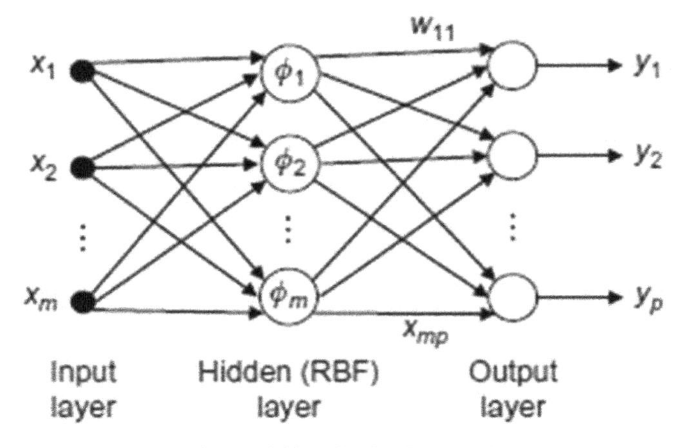

Figure 6.33 The RBF network.

$$y_k(\mathbf{x}) = \sum_{i=1}^{m} w_{ki}\phi_i(\mathbf{x}) \quad (k = 1, 2, ..., p) \tag{6.41}$$

where:

$$\phi_i(\mathbf{x}) = \phi(\|\mathbf{x} - \mathbf{c}_i\|) = \phi(r_i) = \exp\left(-\frac{r_i^2}{2\sigma_i^2}\right), \quad r_i \geq 0, \ \sigma_i \geq 0 \tag{6.42}$$

The RBF networks have always one hidden layer of computational nodes with no monotonic transfer functions $\phi(\cdot)$. Theoretical studies have shown that the choice of $\phi(\cdot)$ is not very crucial for the effectiveness of the network. In most cases, the Gaussian RBF given by Equation (6.42) is used, where $\mathbf{c}_i$ and σ_i $(i = 1, 2, ..., m)$ are selected centers and widths, respectively.

The selection of the centers c_i is done using some clustering algorithm on the training patterns, and the widths are selected by least squares.

Neural control schemes: Neural control uses 'well-defined' neural networks for the generation of desired control signals. Neural Networks have the ability to learn and generalize, from examples, nonlinear mappings (i.e., they are universal approximators), and so they are suitable for solving complex nonlinear control systems, like robotic systems, with high speed [92].

Neural control can be classified in the same way as NNs, that is:

- Neural control with supervised learning.
- Neural control with unsupervised learning.
- Neural control with reinforcement learning.

In each case, the proper NN should be used. Here, the case of neural control with supervised learning, which is very popular for its simplicity, will be considered. The structure of supervised learning neurocontrol is as shown in Figure 6.34.

The teacher trains the neurocontroller via the presentation of control signal examples that can control the robot successfully. The teacher can be either a human controller or any classical, adaptive, or intelligent technological controller. The outputs or states are measured and sent to the teacher as well as to neurocontroller. During the control period by the teacher, the control signal and outputs/states of the system are sampled and stored for training of the neural network. After the training period, the neurocontroller takes the control actions, and the teacher is disconnected from the system.

The most popular type of supervised neurocontrol is the direct inverse neurocontrol in which the NN learns successfully the system (robot) inverse dynamics and is used directrly as controller as shown in Figure 6.35.

The most general type of neurocontrol involves two NNs: the first is used as FFC and the second as FBC. The structure of this control scheme which is known as *feedback error learning neurocontroller* is shown in Figure 6.36.

A typical structure for tuning the gains (parameters) K_{p}, K_{d} and K_{i} of a PID controller, by a NN is shown in Figure 6.37.

Hybrid Control: Hybrid control is obtained by proper combination of neural and fuzzy control which in many cases is enhanced with **GAs** for facilitating the optimization of the NN synaptic weights and the membership functions of the fuzzy sets [83, 85, 92]. A typical NF controller architecture is shown

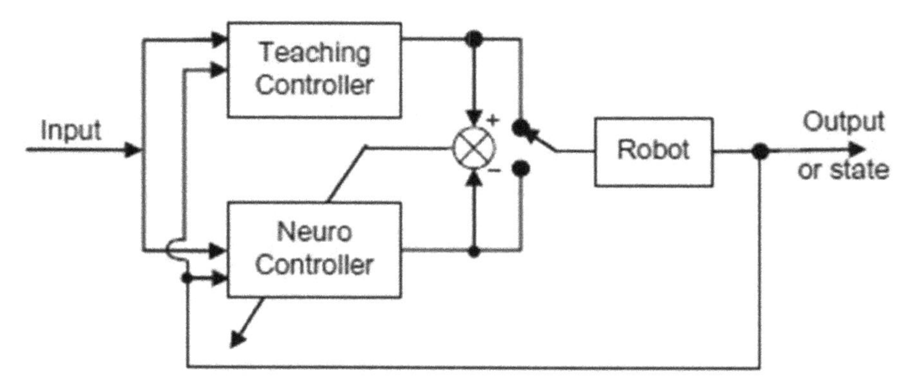

Figure 6.34 Structure of neurocontrolled robot with supervised learning.

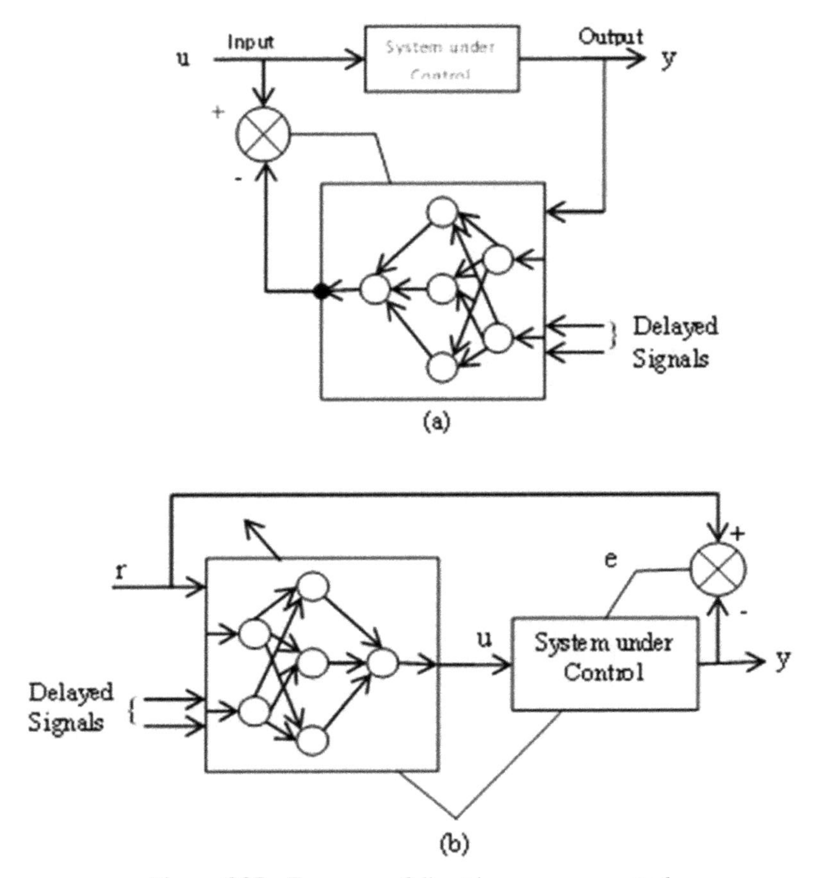

Figure 6.35 Two ways of direct inverse neurocontrol.

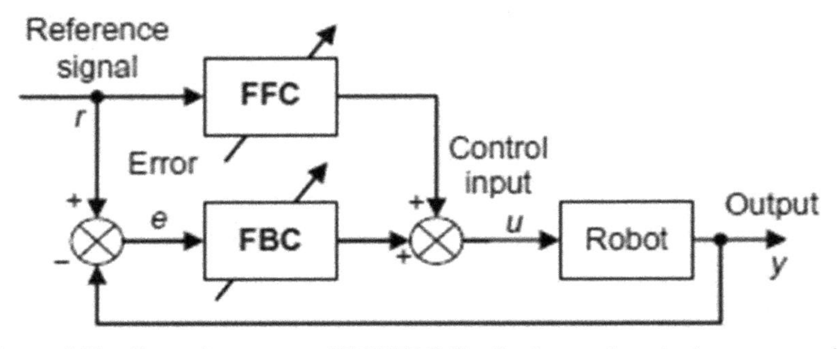

Figure 6.36 General structure of FFC-FBC (feedback error learning) neurocontrol.

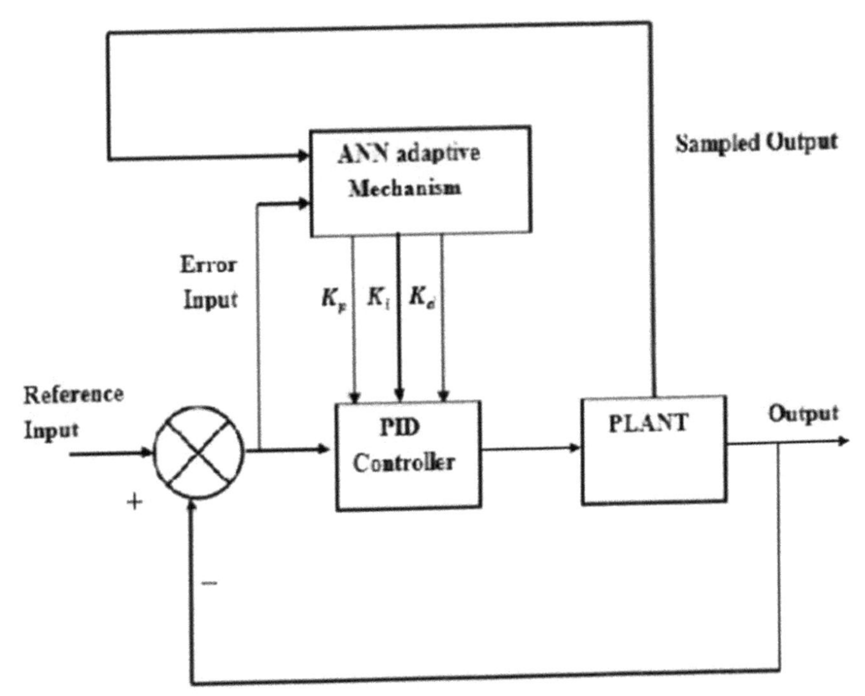

Figure 6.37 PID optimal parameter tuning using an ANN.

Source: https://rroij.com (open-access/optimized-speed-control-for-bldc-motor.php?aid=51339).

in Figure 6.38, where the $a_i^{'}s$ $(i = 1, 2, ..., p)$ represent the input basis functions (fuzzy sets) that can be Gaussian basis functions, etc.

Very broadly NF systems are distinguished in:

- *Cooperative NF systems:* The NN determines some sub-blocks of the FS, e.g., fuzzy rules which are then used without the presence of NN.
- *Concurrent NF systems:* The NN and FS work continuously together, where the NN pre-processes the inputs, or post-processes the output of the fuzzy system.

Figure 6.39 shows the general structure of a neuro-genetic (NG) controller [93].

Figure 6.40 shows the structure of a fuzzy controller of an induction motor where its elements (fuzzy sets and fuzzy rules) are adjusted (evolved) by a GA (evolutionary algorithm; see also [94]).

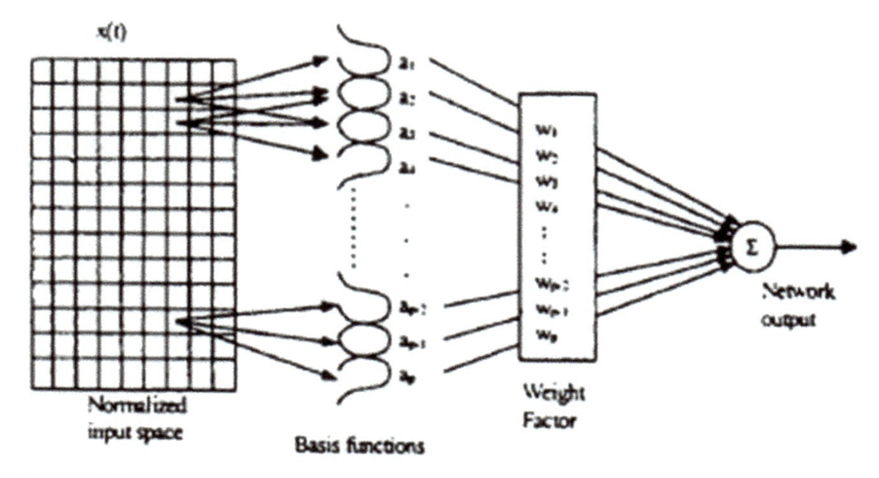

Figure 6.38 General structure of a NF controller. (similar to fuzzy associative memory).

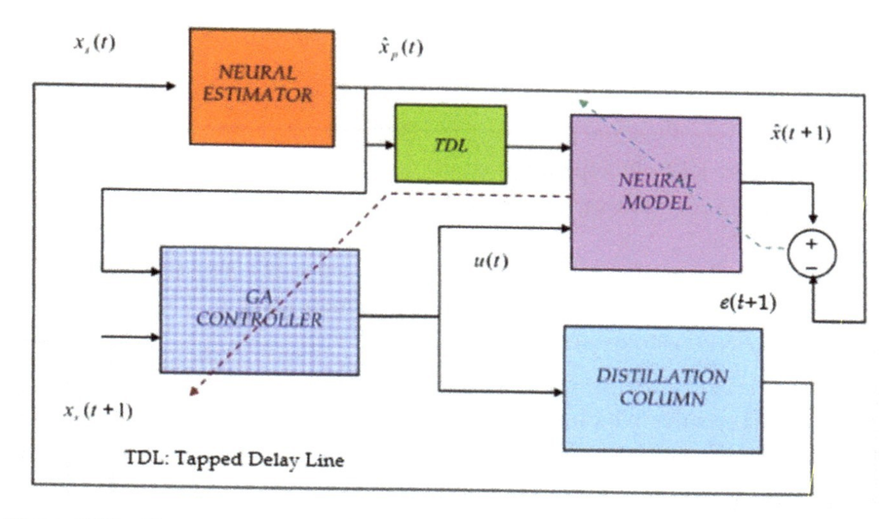

Figure 6.39 Neuro-genetic controller that uses a tapped delay line (The weights of the neural estimator are genetically evolved by an optimizer).

6.6 Networked Control Systems

An important and challenging class of control systems with a plethora of applications is the class of **NCSs**. A networked control system is a control system in which the feedback loops are realized using a shared

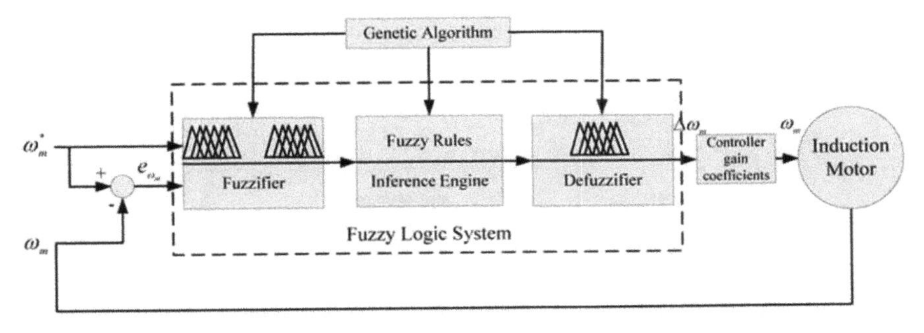

Figure 6.40 General structure of a fuzzy-genetic controller.

Source: www.intechopen.com (N. Saad, et al., An optimized hybrid fuzzy–fuzzy controller).

computer network. Control and feedback signals are exchanged among the system components in the form of information packages through a network. As shown in Figure 6.41(a,b), a NCS involves, besides the standard elements of control systems (sensors that get information, controllers that generate commands and decisions, actuators that perform the control commands), a *communication network* that enables the transfer and exchange of information between sensors, controllers and actuators.

Communication networks used in NCSs include:

- *Wide area networks* (**WAN**) for connection between different cities or countries.
- *Local area networks* (**LAN**) for connection between different rooms of a building or different nearby buildings.
- *Metropolitan area networks* (**MAN**) for connection of computers in a metropolitan area.
- *Fieldbuses* (CAN, LON, etc.).
- *Ethernet* used in LAN, MAN and WAN (e.g., Ethernet IEEE 802 Standard).

NCSs eliminate unnecessary wiring and thus reduce the complexity and total design and implementation cost of the control system. However, it is generally known that the employment of a shared communication network very often introduces uncertainties that can degrade performance and sometimes lead to closed-loop instability. Examples of such network-induced effects are:

- Varying or unknown transmission time delays.
- Varying sampling and transmission intervals.
- Quantization.
- Packet loss.

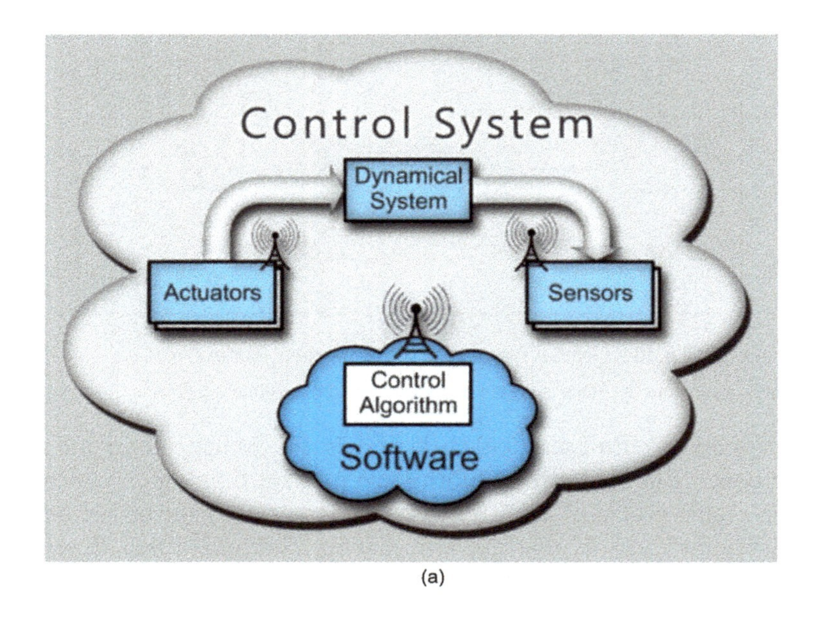

(a)

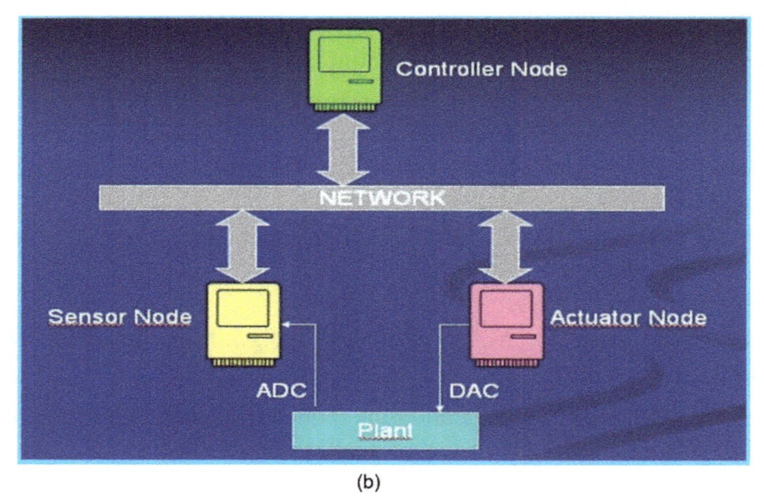

(b)

Figure 6.41 (a) Components of networked control systems, (b) A possible architecture of NCSs in which sensors, controllers, and actuators are designed as separate nodes connected by a communication network (ADC stands for analog to digital converter, and DAC stands for digital to analog converter).

Source: (a) https://heemels.tue.nl (/research/networked-control). (b) http://people.sabanciuniv.edu (/~onat/Files/NCS_intro.htm).

Over the years many techniques were developed for eliminating or reducing the consequences of the above effects, and assure better exploitation of the features and the potential benefits of NCSs [96–98]. Packet loss (or packed drop out) takes place when one or more data packets travelling across the computer network do not arrive at their destination. It is usually the result of congestion, and measured in a percentage of packets lost with respect to total packets sent. Packet loss is detected by **TCP** and corrected by retransmission of the data. It is closely related to **QoS** considerations.

Figure 6.42 shows the block diagram of a networked feedback control system that indicates the three types of delays occurring in the communication network and the controller. These delays are:

- $\tau^{sc}{}_k$: Delay from sensor to controller.
- $\tau^{ca}{}_k$: Delay from controller to actuator.
- $\tau^{c}{}_k$: Input/output delay of controller.

Time delays in control systems can be handled using several forms of delay estimators and predictors (e.g., Smith predictor, neural estimator, self-tuning estimator, maximum likelihood estimator, etc.) [96–98].

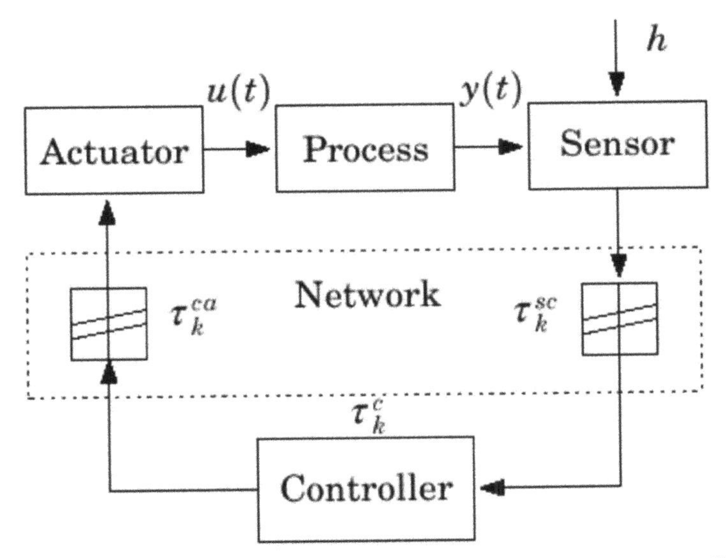

Figure 6.42 The three types of delay in networked control systems. The controller can be of any type (classical, modern, multivariable, MBPC, neural fuzzy, distributed, decentralized, embedded, etc.). The index k denotes discrete time ($t = kt$ where T is the sampling period).

Source: [99].

Applications of NCS include:

- Industrial process control.
- Factory automation.
- Remote fault diagnosis.
- Telerobotics/telesurgery.
- Space, sea, and terrestrial exploration.
- Treatment of hazardous materials in remote places.
- Robotic rescue and other human life saving processes, etc.

The benefits of NCS include the following [95]:

- *Flexibility* (NCSs enable to extend and share the network and its infrastructure).
- *Data sharing* (NCSs enable efficient data sharing as data available to the network).
- *Extendibility* (More sub-control systems can be easily added, and the network can be easily extended to more nodes whenever required. The system extension can be performed without massive changes to the layout of the system.).
- *Remote control* (NCSs allow the control to be exerted from remote sides, and make the vulnerable sites less sensitive thus protecting operators).
- *Reduced complexity* (Embedding the control in a network usually leads to reduction of the system complexity).
- *Cloud connection* (NCSs connect cloud/cyber to the physical system which makes easy to control the site from distance).

The drawbacks of NCS include the following [95]:

- *Bandwidth issue* (Extending the network to include more nodes may lead to bandwidth insufficiency problems. To face these problems high-bandwidth infrastructure is needed).
- *Data scheduling* (To achieve the desired NCS performance under traffic congestion conditions, one should use proper scheduling algorithms).
- *Network security issue* (Data over the network may be available to anyone connected to the network which may have implications for the NCS security. To face this potential problem, special cautions should be taken, e.g., use of special *cypher algorithms*.).
- *Network delays* (As mentioned above, network delays due to data congestion in the network may cause closed-loop instability. Therefore, proper delay compensation techniques should be used).

Figure 6.43 shows a hypothetical NCS which involves a variety of sub-systems and components that potentially are used in most cases, namely:

- Telecommunications network.
- Internet computer network.
- Business computer network.
- Control center (operator, computers, HMI)
- Controller units (PLC, RTC, DCS, PID, etc.).
- Business and maintenance computers with MODEMs.
- Laptop with wireless communication links.
- Physical controlled plan units (solution tanks, valves, motors, etc.).

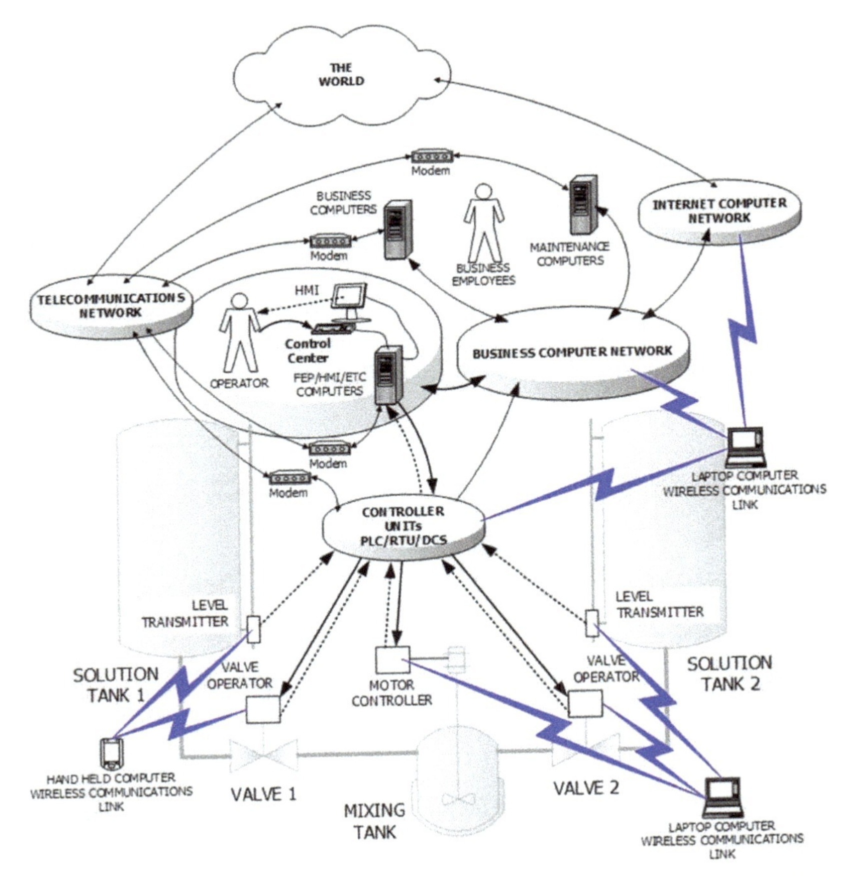

Figure 6.43 An overall hypothetical NCS with its sub-systems and components.

Source: https://ics-cert.us-cert.gov (content/overview-cyber-vulnerability).

To analyze network security, NCSs should be analyzed with regard to actual and potential vulnerabilities. This can be done by using:

- Diagrams of network architecture.
- Configurations of network components (switches, firewalls, routers, etc.).
- Configurations of host devices.
- Access control strategies.
- Software and firmware implementations.

Figure 6.44 shows the structure of a typical double-firewall NCS that covers the business/corporate network and the production control system network.

In this system:

- The controller units are connected to a data acquisition server via several communication protocols that perform data packaging.
- The collection, processing, and storage of system data are performed by a master database server.
- The operator/dispatcher monitors and controls the system via a proper human-machine interface (HMI) which includes graphical displays for monitoring the status of equipment, alarms, and events.
- The control system network is connected to the business/corporate/factory office network to enable real-time transfer of data from the control network to the elements of that office.

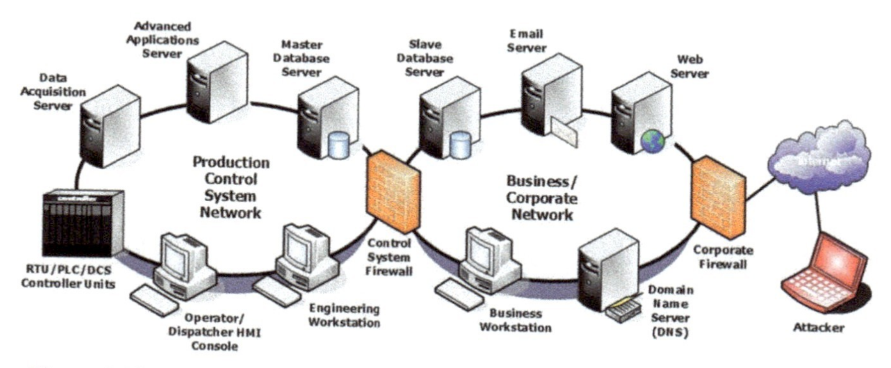

Figure 6.44 Double firewall (corporate firewall, control system firewall) architecture.
Source: https://ics-cert.us-cert.gov (content/overview-cyber-vulnerability).

Attackers attempt to:

- Get access to the LAN of the control system.
- Understand the functions of the system via discovery efforts.
- Gain the control of the system, etc.

In the following we outline the levels of networked control of an enterprise or automation company. From top to bottom these control levels are [99]:

- **Level 1**: *Corporate management level* (WAN connections to various cities or countries).
- **Level 2**: *Plant management level* (Workstations).
- **Level 3**: *Supervisory control level* (LAN: Ethernet, TCP/IP, MAP, Modbus/TCP).
- **Level 4**: *Cell control level* (Fieldbus: PLC, PID, CNC, DSC.PC).
- **Level 5**: *Sensor/actuator level* (Sensors and actuators).

The principles of NCSs are also used in web-based or other forms of distance education. A review and several case studies of control and robotics education over the web are provided in [96]. A possible architecture for a web-based laboratory where students can perform remote physical control/ robotic experiments using the Internet is shown in Figure 6.45 [96, Chapter 1]. Using web labs minimizes the need to have available many copies of the same equipment, and allow students to work on the experiment any time they want provided that the experiment is not occupied by another student.

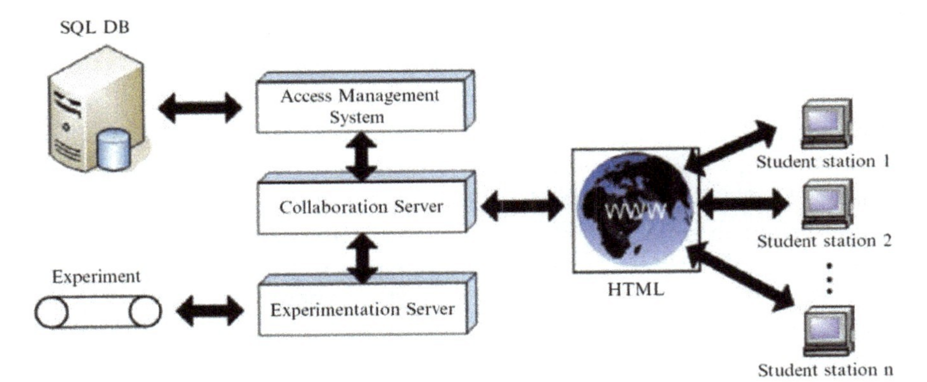

Figure 6.45 Web-based remote lab architecture. Access of students to the lab is controlled by the access management system. The same architecture holds also for virtual labs, where, in the place of the physical experiment, there is a simulated/virtual model of it.

6.7 Conclusions

Control is the process of assuring that the output (product) or state variable of a system possesses desired quality features. Typically, we have available a reference or command input variable and an undesired input variable (disturbance, noise), and the goal of control is the system output to follow precisely the reference variable and to be unaffected by the disturbance/noise as much as possible. If the command is fixed, we have the case of regulation (regulatory control), and if the command variable is changing, the control is called tracking or trajectory following control. Automatic (automated) control systems do not include (or minimize) the involvement of manual control, and so, as a result, lower consumption of energy or lower cost of process, and increased quality and productivity are achieved.

Modern control theory includes a wide repertory of techniques and implementations. This chapter has discussed a number of fundamental concepts and techniques at a conceptual level keeping the mathematical details at a minimum. The chapter has covered ontological/ and epistemological issues referring to both model-based and model-free control theory, also including a discussion of networked control systems.

Full presentations of model-free methods are provided in [20, 21, 24, 27, 81–92]. Further system classes that can be efficiently controlled by modern control techniques include: LSS, MD systems, DPSs, TDSs, FSA, and DESs [33, 75–80].

References

[1] Mayr, O. (1970). *The Origins of Feedback Control*. Cambridge, MA: MIT Press.

[2] Bennett, S. (1996). A brief history of automatic control. *IEEE Control Syst. Mag*. 6, 17–25.

[3] Bissel, C. C. (1991). Secondary sources for the history of control engineering: an annotated bibliography. *Intl. J. Control* 54, 517–528.

[4] Khramoi, A. V. (1969). History of Automation in Russia before 1917, Moscow 1956, English Translation, Jerusalem, 1969.

[5] Mayr, O. (1971). James clerk maxwell and the origins of cybernetics. *ISIS* 62, 425–444.

[6] Lewis, F. L. (1992). "A brief history of feedback control," in *Applied Optimal Control and Estimation*, Chap. 1, (Upper Saddle River, NJ: Prentice-Hall).

[7] Bennet, S. (1993). *A History of Control Engineering 1930–1955*. Stevenage: Peter Peregrinus.

[8] Bennet, S. (1979). *A History of Control Engineering 1800–1930*. Stevenage: Peter Peregrinus. (Reprinted 1986).

[9] Valavanis, K. P., Vachtsevanos, G. J., and Antsaklis, P. J. (2014). Technology and autonomous mechanisms in the mediterranean: from ancient greece to by zantium. *IEEE Control Syst. Mag*. 34, 110–119.

[10] Belmann, R. (1957). *Dynamic Programming*. Princeton, NJ: Princeton University Press.

[11] Pontryagin, L. S., Boltyansky, V. G., Gamkrelidze, R. V., and Mishchenko, E. F. (1962). *The Mathematical Theory of Optimal Processes*. New York, NY: J. Wiley.

[12] Lasalle, J. P. (1960). Some extensions of Lyapunov's second method. *IRE Trans. Circ. Theor*. 7, 520.

[13] Kalman, R. E. (1960). "On the general theory of control systems," in *Proceedings of the 1st IFAC Congress, Moscow*, Vol. 1 (London: Butteworth), 481–492.

[14] Kalman, R. E. (1960). A new approach to linear filtering and prediction problems. *ASME J. Basic Eng*. 82, 34–45.

[15] Kalman, R. E., and Bertram, J. E. (1960). Control system analysis and design via the 'second method' of Lyapunov, (i) continuous-time systems. *Trans. ASME J. Basic Eng*. 371–393.

[16] Kalman, R. E., and Bucy, R. S. (1961). New results in linear filtering and prediction theory. *ASME J. Basic Eng*. 80, 193–196.

[17] Athans, M., and Falb, P. (1966). *Optimal Control*. New York, NY: Mc Graw-Hill.

[18] Asl, F. S., Athans, M., and Pascoal, A. (2006). Issues, progress and new results in robust adaptive control, *J. Adap. Control Sig. Process*. 20, 519–579.

[19] Popov, V. M. (1961). Absolute stability of nonlinear system of automatic control. *Autom. Remote Control* 22, 857–875.

[20] Fu, K. S. (1970). Learning control systems-review and outlook. *IEEE Trans. Autom. Control* AC-15, 210–221.

[21] Fu, K. S. (1971). Learning control systems and intelligent control systems: an intersection of artificial intelligence and automatic control. *IEEE Trans Autom. Control* AC-16, 70–72.

[22] Fu, K. S. (1982). *Syntactic Pattern Recognition and Applications*. Englewood Cliffs, NJ: Prentice-Hall.

[23] Haddad, H., and Lee, G. C. (2002). GoS guided bandwidth management in differentiated time scales. *J. Optimiz. Theor. Appl.* 115, 517–547.

[24] Harris, C. I., Moore, C. G., and Brown, M. (1993). *Intelligent Control: Aspects of Fuzzy Logic and Neural Nets*. Singapore: World Scientific.

[25] Brown, M., and Harris, C. J. (1984). *Neuro-Fuzzy Adaptive Modeling and Control*. Upper Saddle River, NJ: Prentice Hall.

[26] Saridis, G. N., and Dao, T. K. (1972). A learning aproach to the parameter-adaptive self-organizing control problem. *Automatica* 8, 589–597.

[27] Saridis, G. N. (1989). Analytic formulation of the principle of increasing precision with decreasing intelligence for intelligent machines *Automatica* 25, 461–467.

[28] Rosenbrock, H. H., and Hayton, G. E. (1974). Dynamical indices of a transfer function matrix. *Int. J. Control* 20, 177–189.

[29] Rosenbrock, H. H. (1970). *State Space Multivariable Theory*. New York, NY: J. Wiley.

[30] Franklin, G. F., Powell, J. D., and Emani-Naeini, A. (1994). *Feedback Control of Dynamic Systems*. Reading, MA: Addison-Wesley.

[31] Jury, E. I. (1964). *Theory and Application of the Z-Transform, Method*. New York, NY: J. Wiley.

[32] Jury, E. I. (1958). *Sampled Data Control Systems*. New York, NY: J. Wiley.

[33] Jury, E. I. (1986). "Stability of multidimensional systems and related problems," in *Multidimensional Systems: Techniques Applications*, Chap. 3, ed. S. G. Tzafestas (New York, NY: Marcel Dekker).

[34] Horowitz, I. (1963). *Synthesis of Feedback Systems*. New York, NY: Academic Press.

[35] Horowitz, I., and Sidi, M. (1972). Synthesis of feedback systems with large plant ignorance for prescribed time domain tolerances. *Intl. J. Control* 16, 287–309.

[36] Houpis, C. H., and Lamont, G. B. (1992). *Digital Control System: Theory, Hardware, Software*. New York, NY: Mc Graw-Hill.

[37] Houpis, C. H., and Rasmussen, S. I. (1999). *Quantitative Feedback Theory: Fundamentals and Applications*. New York, NY: Marcel Dekker.

[38] MacFarlane, A. G. J. (1979). The development of frequency-response methods in automatic control. *IEEE Trans. Autom. Control* AC-24, 250–265.

[39] MacFarlane, A. G. J., and Postlethwaite, I. (1977). The generalized nyquist stability criterion and multivariable root loci. *Int. J. Control* 25, 81–127.

[40] Postlethwaite, I., and MacFarlane, A., G. J. (1979). *A Complex Variable Approach to the Analysis of Linear Multivariable Feedback Systems.* Berlin: Springer.

[41] Safonov, M. G. (1982). Stability margins of diagonally perturbed multivariable feedback systems. *IEE Proc.* 129-D, 251–256.

[42] Safonov, M. G., and Doyle, J. (1984). "Minimizing conservativeness of robust singular values," in *Multivariable Control*, ed. S. G. Tzafestas (Dordrecht: D. Reidel).

[43] Astrom, K. J., and Wittenmark, B. (1980). Self-tuning controllers based on pole-zero placement. *Proc. IEEE* 127, 120–130.

[44] Astrom, K. J., Hagander, P., and Sternby, J. (1984). Zeros of sampled systems. *Automatica* 20, 31–38.

[45] Astrom, K. J., and Wittennmark, B. (1989). *Adaptive Control.* Reading, MA: Addison-Wesley.

[46] Astrom, K. J., and Wittenmark, B. (1971). Problems of identification and control. *J. Math. Anal. Appl.* 34, 90–113.

[47] Kwakerrnaak, H. (1967). Optimal filtering in linear systems with time delays. *IEEE Trans. Autom. Control* AC-12, 169–173.

[48] Kwakernaak, H., (1972). Robust control and H_∞: optimization, *Automatica* 29, 255–273.

[49] Kwakernaak, H., and Sivan, R. (1972). *Linear Optimal Control Systems.* New York, NY: J. Wiley.

[50] Meditch, J. S. (1967). Orthogonal projection and discrete optimal linear smoothing, *SIAM J. Control* 5, 74–89.

[51] Meditch, J. S. (1967). On optimal linear smoothing theory. *Inform. Control* 10, 598–615.

[52] Meditch, J. S. (1969). *Stochastic Optimal Linear Estimation and Control.* New York: Mc Graw-Hill.

[53] Landau, I. D. (1979). *Adaptive Control: The Model Reference Approach.* New York: Marcel-Dekker.

[54] Landau, I. D. (1997). From robust control to adaptive control. *Control Eng. Pract.* 7, 1113–1124.

[55] Landau, I. D., Lozano, R., M'Saad, M. (1998). *Adaptive Control.* New York, NY: Springer.

[56] Narendra, K. S. and Parthasarathy, K. (1990). Identification and control of dynamical systems using neural networks. *IEEE Trans. Neural Netw.* 1, 4–27.

[57] Narendra, K. S. and Annaswamy, A. (1989). *Stable Addaptive Systems.* Upper Saddle River, NJ: Prentice-Hall.

[58] Slotine, J.-J. E., and Coetsee, J. A. (1986). Adaptive sliding controller synthesis for nonlinear systems. *Int. J. Control* 43, 1631–1651.

[59] Slotine, J.-J. E., and Li, W. (1991). *Applied Nonlinear Control.* Englewwod Cliffs, NJ: Prentice-Hall.

[60] Zames, G. (1996). "On the input-output stability of time-varying nonlinear feedback Systems, part I: conditions derived using concepts of loop gain, conicity, and positivity. *IEEE Trans. Automatic Control* 11, 228–238, 1966, Part II: Conditions Involving Circles in the Frequency Plane and Sector Nonlinearrities, *idid*, No. 3, pp. 465–476, 1966.

[61] Zames, G. (1981). Feedback and optimal sensitivity: model reference transfomations multiplicative semi-norms and approximate inverses, *IEEE Trans. Autom. Control*, 26, 301–320.

[62] Zames, G., and Shneydor, N. A. (1977). Structural stabilization and quenching by dither in nonlinear systems. *IEEE Trans. Autom. Control* 22, 352–361.

[63] Fradkov, A. L., Miroshnik, I. V., and Nikiforuk, V. O. (2007). *Nonlinear and Adaptive Control of Complex Systems.* Berlin: Springer.

[64] Astolfi, A. (2008). Towards applied nonlinear adaptive control. *Ann. Rev. Control* 32, 136–148.

[65] D'Azzo, J. J., and Houpis, C. H. (1966). *Feedback Control System Analysis and Synthesis*, 2nd Edn. New York, NY: Mc Graw-Hill.

[66] Sandberg, I. W. (1964). A frequency-domain condition for the stability of feedback systems containing a single time-varying nonlinear element, *Bell. Systems Tech. J.* 43, 1601–1608.

[67] Narendra, K. S., and Goldwyn, A. (1964). A Geometrical Criterion for the Stability of Certain Nonlinear Nonautonomous Systems. *IEEE Trans. Circuit Theory* 11, 406–407.

[68] Desoer, C. A. (1965). A generalization of the popov criterion, *IEEE Trans. Autom. Control* 10, 182–185.

[69] Draper Big for Space, C. S. (1961), *MIT and Project Apollo, MIT Institute Archives & Special Collections.* Available at: http://libraries.mit.edu/archives/exhibits/apollo/

[70] Houpis, C. H., and Lamont, G. B. (1992). *Digital Control Systems: Theory, Hardware, Software.* New York, NY: McGraw-Hill.

[71] Isermann, R. (1989). *Digital Control Systems, Fundamentals, Deterministic Control*, Vol. 1. Berlin: Springer.

[72] Katz, P. (1981). *Digital Control Using Microprocessors*. London: Prentice-Hall.

[73] Dorf, R. C., and Bishop, R. H. (1995). *Modern Control Systems*, 7th edn. Reading, MA: Addison-Wesley.

[74] DiStefano, J. J. III, Stubberud, A. R., and Williams, I. J. (1990). *Theory and Problems of Feedback Control Systems Design*. New York, NY: Mc Graw-Hill.

[75] Tzafestas, S. G. (ed.). (1982). *Distributed-Parameter Systems*. Oxford, UK: Pergamon Press.

[76] Tzafestas, S. G., and Watannabe, K. (1992). *Stochastic Large-Scale Engineering Systems*. New York, NY: Marcel-Dekker.

[77] Cassandras, C. G., and Lafortune, S. (1999). *Introduction to Discrete Event Systems*. Norwell, MA: Kluwer.

[78] Fofana, M. S., and Ryba, P. B. (2004). Parametric stability of non-linear time-delay equations. *Int. J. Nonlin. Mech.* 39, 79–91.

[79] Ray, A., Fu, J., and Lagoa, C. (2004). Optimal supervisory control of finite-state automata. *Int. J. Control* 77, 1083–1100.

[80] Capkovic, F. (1999). "Knowledge-based control synthesis of discrete event dynamic systems," in *Advances in Manufacturing: Decision, Control and Information Technology*, ed. S. G. Tzafestas (Berlin: Springer), 165–180.

[81] Zadeh, L. A. (1965). Fuzzy sets. *Inform. Control*, 8, 338–353.

[82] Tzafestas, S. G. (1994). Fuzzy systems and fuzzy expert control: An overview. *Knowl. Eng. Rev*. 9, 229–268.

[83] Tsoukalas, L. H., and Uhrig, R. E. (1997). *Fuzzy and Neural Approaches in Engineering*. New York, NY: John Wiley.

[84] Haykin, S. (1994). *Neural Networks: A Comprehensive Foundation*. New York, NY: MacMillan College Publishing

[85] Lin, C. T., and Lee, G. (1995). *Neural Fuzzy Control Systems*. Englewood Cliffs, NJ: Prentice Hall.

[86] Holland, J. H. (1975). *Adaptation in Natural and Artificial Systems*, Ann Arbor, MI: The University of Michigan Press.

[87] Goldberg, D. E. (1989). *Genetic Algorithms in Search, Optimization and Machine Learning*. Boston, MA: Addison-Wesley Publishing Co.

[88] C. Virgil Negoita, Expert Systems and Fuzzy Systems, Menlo Park, California, USA: The Benjamin/Cunnings Publ. Co., Inc., 1985.

[89] Forsyth, R. (1984). *Expert Systems*. Boca Raton, FL: Chapman and Hall.
[90] Boverman, R., and Glover, P. (1988). *Putting Expert Systems into Practice*. New York, NY: Van Nostrand Reinhold.
[91] Arkin, R. C. (1998). *Behavior-Based Robotics*. Cambridge, MA: The MIT Press.
[92] Tzafestas, S. G. (1997). *Soft Computing and Control, Technology*. London: World Scientific.
[93] Fernandez de Ganete, J. et al. (2012). *Neural and Genetic Control Approaches to Process Engineering*, Rijeka: Intech. Available at: http://www/intechopen.com/download/pdfs-id/37944
[94] Mester, G. (2014). Design of the fuzzy control systems based on genetic algorithms. *Interdiscipl. Descript. Compl. Syst.* 12, 245–254.
[95] Asif, S., and Webb, P. (2013). Networked control systems: An overview. *Int. J. Comput. Appl.* 115, 26–30.
[96] Tzafestas, S. G. (2009). *Web-Based Control and Robotics Education*. Berlin: Springer.
[97] Gupta, K. A., and Chow, M. Y. (2008). *Control Methodologies in Networked Control Systems*. Berlin: Springer.
[98] Wang, F.-Y., and Liu, D. (2008). *Networked Control Systems: Theory and Applications*. Berlin: Springer.
[99] Daoud, R. M., Amer, H. H., and ElSayed, H. M. (2010). "Performance and reliability of fault-tolerant Ethernet networked control systems," in *Industrial Engineering and Management: Factory Automation*, ed. J. Silvester-Blames (Rijeka: INTECH).

7

Complex and Nonlinear Systems

Chaos often breeds life, when order means habit.
Henry Adams

We grow in direct proportion to the amount of chaos
we can sustain and dissipate.
Ilya Prigogine

When a butterfly flutters its wings in one part of the world, it can eventually
cause a hurricane in another.
Edward Norton Lorenz

The theory of chaos and theory of fractals are separate,
but have very strong intersections.
That is one part of chaos theory is geometrically expressed by fractal shapes.
Benoit Mandelbrot

7.1 Introduction

The study of complex systems and complexity represents an important area of contemporary science. Complexity in a broad sense constitutes one of the most interdisciplinary issues that scientists and engineers have to face today in several theoretical and application domains. Complexity has crossed, besides the traditional disciplines of natural and physical sciences, the borders of areas like sociology, economics, psychology and others. The primary ancestor of complex systems and complexity is the theory of nonlinear dynamical systems (strange attractors, bifurcations, chaos) which originated from the early work of Poincare, and further developed by Lyapunov, Kolmogorov, Prigogine, and others. Prominent areas of study of complex systems as they have been progressed from nonlinear systems are chaos [1], fractals [2], self-organized criticality [3], cellular automata [4], network theory [5, 6], computational mechanics [7], and coupled map lattices [9].

The formal science of complexity is a relatively new branch of science (about 3 to 4 decades old). It is concerned with the issues of explaining some of the most difficult scientific questions of almost all branches of science and more. Stephen Hawking said: 'The present century will be the century of complexity', while Heinz Pagels stated that 'I am convinced that the nations and people who master the new sciences of complexity will become the economic, cultural, and political superpowers of the next century (meaning the present century)'.

The objective of the present chapter is to study the classes of complex and nonlinear systems. Specifically, the chapter:

- Gives the answer to the ontological questions 'what is complex system?' and 'what is complexity?', including a number of opinions about complex systems by eminent thinkers in the field..
- Discusses the issue of complexity measures, addressing the questions 'how hard is to describe?', 'how hard is to create?', and 'what is the degree of organization?', through an exposition of the complexity types and the corresponding measures.
- Provides a list of the classical and modern methods of analysis, stability and control of nonlinear systems.
- Discusses the ontological and epistemological issues of 'bifurcations'.
- Investigates the physical phenomena of chaos, strange attractors, and fractals.
- Summarizes basic ontological and epistemological issues of the 'emergence' phenomenon.
- Presents the definitions and properties of 'complex adaptive systems'.
- Provides a short discussion of the concepts of 'adaptation' and 'self-organization' focusing on their various definitions and mechanisms through which they can be achieved.

7.2 What is a Complex System?

A unique answer to this question does not exist. Typically, a complex system is made from a large number of simple agents, elements, and subsystems which interact with one another and with the surroundings, and which have the potential to generate *new* emergent collective behavior. The manifestation of this behavior is the spontaneous creation of new temporal, spatial, or functional structures. More specifically, a complex system is an *'open'* system incorporating *'nonlinear interactions'* among its subsystems that can exhibit, under certain conditions, a considerable degree of coherent or ordered behavior extending well beyond the range or scale of the individual subsystems or subunits. The term *'nonlinear'* in the above definition means that the *output* y of the system (interaction) for given input x has the form $y = \mathrm{A}(x)\,x$, i.e., the proportionality factor $\mathrm{A}(x)$ depends on x, which implies that the *superposition* (linearity) principle does not hold. As we will see later in the chapter this nonlinearity has far reaching consequences for the time evolution of the system. For example, the future progression of its events may become very sensitive to conditions at any given instant of time. A characteristic example of this is the *chaos* or *chaotic behavior.*

From a physical point of view, complex behavior arises in situations *far from thermal* equilibrium, although this is not always so. In such cases the entire framework of equilibrium thermodynamics may not be applied. In other words, one cannot speak of a complex system if its behavior can be described by the laws of linear thermodynamics. For complex behavior to emerge, the thermodynamic branch of a system needs to be already unstable. This implies that the concept of instability is an absolutely required tool for any reasonable understanding of complex systems [9].

Emergence is the most fundamental property of complex systems. It is the production of the system whole from parts and elements. The parts (subsystems) of a system are often complex systems themselves. Therefore, the primary question that can be asked here is how the complexity of the resulting whole is related to the complexity of the parts. Of course, a complex system can also be resulted from simple parts. This is an important possibility called *emergent complexity.* A full discussion of emergence and emergent complexity will be made later in the chapter (Section 7.7). We will now discuss the question *'what is complexity'* which puts more light to the issue of *'what is a 'complex system'*.

7.3 What is Complexity?

Complexity is not and cannot uniquely and precisely be defined. Its definition is also complex. Complexity is inherent in life, science, technology, society, and human performance in general. From the level of bimolecular interactions, to the global economy processes, complexity is the major aspect that has to be faced. The basic question here is whether there is a trend towards greater complexity over time among living beings, in nature, and in human activity. This can only be answered if we find a global measure of the world's complexity, something that seems to be very difficult, if not impossible. However, over the years many particular measures of complexity have been discovered, some of which will be discussed in Section 7.5.

Instead, of a unique formal definition of complexity, which seems to be impossible, we list here the following characteristics of complex systems that help us to understand what complex systems are and how they function:

- Complex systems are made from several simple or complex parts that interact nonlinearly.
- The parts of complex system are interdependent.
- Complexity systems involve the multi-cause paradox, i.e., there may exist more than one *'cause'*s' that lead to the same outcome.
- Complex systems are coherent, in the sense that there is *'an alignment of context,* viewpoint, purpose and action that enables further purposive action.
- Complexity exhibits a balance between *chaos* and *non-chaos* or as more often is called complex systems operate 'on the edge of chaos'.
- Complex systems exhibit *'emergence capability (holism)'*. Holism is distinguished in weaker or stronger. The stronger one postulates new system properties and relations among parts and subsystems that did not exist in the system components. This is what is called *emergence,* a creative principle. In the weaker interpretation, emergence simply means that the parts of a complex system have mutual relations that do not exist for the parts in isolation. By adopting the weak interpretation of holism we may be based on reductionism even though it is not easy to prove rigorously that the features of the whole can be obtained from features of parts. This is the standard concept of the science as building things from elementary parts.

A few examples of complex systems are the following:

- The *human body:* From a physiological perspective.
- *Family:* It is a set of individuals, each one having relationship with the other individuals, and involves an interplay between the relationships and qualities of the individual. The family has to interact with the outside world.
- *Government:* It has different functions namely taxation, immigration, income distribution, transportation, health, military, etc. There are many different levels and types of government: Local, state, federal, mayoral, council, etc.

Actually, the study of complex systems enables us to find the connections and reveal parallels among strongly distinct scientific subjects (e.g., biology, physics, medicine, behavioral sciences, humanities, etc.).

7.4 Opinions about Complex Systems

In this section we list a number of opinion statements of complexity scientists about what a complex system is:

- A complex system is literally one in which there are multiple elements adapting or reacting to the patterns these elements create *(W. Brian Arthur)* [10].
- 'Complexity theory indicates that large populations of units can self-organize into aggregations that generate pattern, store information, and engage in collective decision making' *(Julia K. Parrish and Leah Edelstein-Kesher)* [11].
- 'Common to all studies on complexity are systems with multiple elements adapting or reacting to the pattern these elements create' *(W. Brian Arthur)* [10].
- Complexity starts when causality breaks down [12].
- 'In general sense, the adjective 'complex' describes a system or component that by design or function or both is difficult to understand and verify [....] Complexity is determined by such factors as the number of components and the intricacy of the interface between them, the number and intricacy of conditional branches, the degree of nesting, and the types of data structures' *(Gezhi Weng, Upinder S. Bhalla and Ravi Iyengar)* [13, p. 92].
- 'You can only understand complex systems using computers, because they are highly nonlinear and are beyond standard mathematical analysis' *(Chris Langton).*
- The science of complexity has to do with structure and order, especially in living systems, social organizations, the development of embryo, patterns of evolution, ecosystems, business, and nonprofit organizations, and their interactions with the technological-economic environment.' *(Roger Lewin)* [16].
- 'The principal difference of chaos and complexity is their history, in the sense that chaotic systems don't rely on their history, whereas complex systems do'. *(M. Buchanan)* [17].

7.5 Measurement of Complexity

The measures of complexity are grouped according to three major questions, namely:

- How hard is to describe?
- How hard is to create?
- What is the degree of organization?

These measures are the following [19, 20].

1. **Difficulty of description** This is typically measured in binary digits (bits) using the following measures:
    - Information.
    - Entropy.
    - Algorithmic complexity or algorithmic information content.
    - Minimum description length.
    - Fisher information.
    - Renyi entropy.
    - Code length.
    - Chernoff information.
    - Fractal dimension.
    - Statistical complexity.
2. **Difficulty of creation** This is typically measured in time, energy, money, etc., using the following measures:
    - Computational complexity.
    - Time computational complexity.
    - Space computational complexity.
    - Information based complexity.
    - Logical depth.
    - Thermodynamic depth.
    - Cost.
    - Crypticist.
3. **Degree of Organization** Here use is made of two groups of measures:
    - *Effective complexity* (difficulty of describing organization structure, namely chemical, cellular, managerial, etc.).
    - *Mutual Information* (i.e., the information shared between the parts and the outcome of the organizational structure).

Measures for *effective complexity* are:

- Grammatical complexity.
- Hierarchical complexity.
- Conditional information.
- Schema length.
- Stochastic complexity.
- Excess entropy.
- Fractal dimension.
- Metric entropy, etc.

Measures for *mutual information* include:

- Organization.
- Correlation.
- Channel capacity.
- Stored information.
- Algorithmic mutual information.

A discussion of complexity measures in the context of dynamical systems can be found in [18], where a classification scheme of dynamical systems into four categories is given, along with a scheme of distinguishing between order and chaos using the concept of homogeneous partitions. In the following we discuss the computational and algorithmic complexities.

Computational Complexity Computational complexity is a measure of the shortest amount of computation needed (and so the minimum amount of computation time needed which is called 'time computational complexity') for the solution of a problem. Although this time will naturally depend on the kind of computer used and the efficiency of the computer program, we may still work with an asymptotic limit: For a given computer and a given computer program, how does the minimum necessary solution time increase as the size of the problem increases? The complexity of a system depends on the *level* of degree at which we want, or we can, describe it. Thus computational complexity is a measure of the effort needed to obtain an agreed amount of details of information. To the extent that the length of description of a problem or system is a measure of the degree of complexity, the complexity depends on the context of description, and on our previous knowledge about the regular features exhibited by the system.

Figure 7.1 gives the so-called *computational complexity chart* where the horizontal axis represents the number of elements n of the problem and the vertical axis gives the number of computational time (operations) required to solve the problem for the various degrees of complexity. The notation code of Figure 7.1 is:

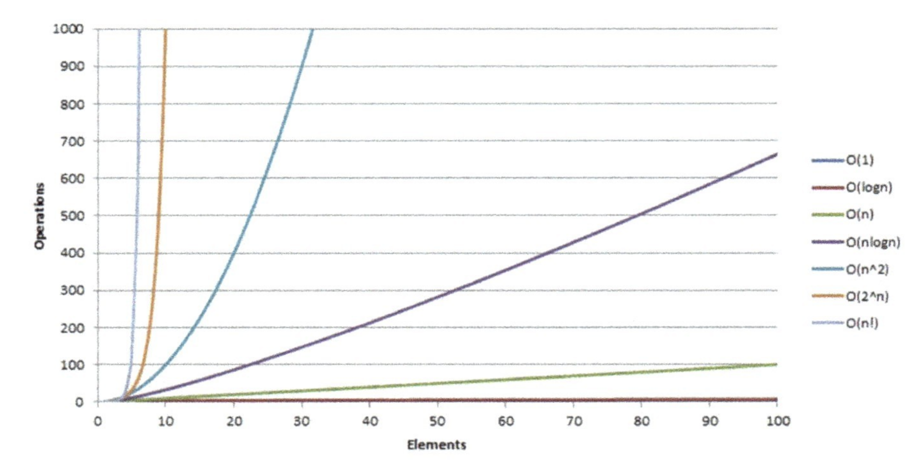

Figure 7.1 Computational complexity chart (the big O means order of magnitude of the computation time).

Source: http://arturmeyster.com; www.stackoverflow.com (polynomial time and exponential time, Haim Bendanan).

- O(1) Constant time (with an upper bound not depending on *n*).
- O(log *n*) Logarithmic time (usually base 2 logarithm, $\log_2 n$).
- O(*n*) Linear time (time proportional to *n*).
- O(*n*log n) Linearithmic (quasi-linear time/almost linear).
- O(n^A2) Polynomial time (*n* is in the exponent of the exponential).
- O($2^A n$) Exponential time (*n* is in the base of the exponential).
- O(*n*!) Factorial time.

Algorithmic information content (Algorithmic complexity) The *algorithmic information content* (**AIC**) of a system is defined as the minimum number of bits (strings of 0's and 1's) required to store the algorithm needed for computing the data/ information for describing the structure and function of the system (i.e., for 'summarizing' or 'explaining' the system). Therefore AIC is a measure of how 'difficult' is to represent a text or a bit-stream using a computer program' (i.e., it is the length in bits of the shortest computer program that can provide that text or bit-stream as output). The digital stream *has no redundancy* (i.e., it is *incompressible)* if it is such as the number of bits required to store and implement the algorithm required for generating or explaining it is no shorter than the data itself.

7.6 Nonlinear Systems: Bifurcations, Chaos, Strange Attractors, and Fractals

7.6.1 General Issues

As seen in Chapter 3 (Linear vs. Nonlinear Systems) the basic difference of the *nonlinear* systems from the *linear systems* is that in the former the superposition property which is possessed by the latter is not applied. The two classical methods for analysis of nonlinear systems are [21, 22]:

- The describing function method.
- The phase plane method.

The *describing function method* is an extension of the frequency-domain methods applied to linear systems. It can be used for systems of any degree (order) and gives information only for the stability of a system. The *phase plane method* can be applied only to 2nd degree systems, and gives detailed information (parameters) of the transient time response.

The principal modern methods of analysis and control of nonlinear systems are [23, 24]:

- Lyapunov stability analysis.
- Linearization around a stable nominal trajectory.
- Optimal control of the linearized system.
- Linear control methods applied to the linearized system.
- Trajectory tracking via the sliding mode control.
- Feedback stabilization via the Lyapunov stability.
- Computational optimal control using the differential dynamic programming method.
- Control via Pontryagin's maximum principle.
- Optimal Control via calculus of variations.

The theory of nonlinear dynamical systems finds applications to a wide repertory of areas (e.g., mathematics, physics, chemistry, biology, engineering, economics, medicine, and others). This theory actually brings scientists from many fields together with a common language. Dynamic systems are deterministic if there is a 'unique' consequent to every state, and 'stochastic' or 'random', if there is more than one consequent chosen from some probability distribution. Nonlinear dynamic systems have been shown to exhibit surprising and complex effects that would never be anticipated from a scientist trained only to linear theory. Dominant examples of these phenomena include bifurcation, strange attractors, chaos, and fractals.

7.6.2 Bifurcations

Very often, in practical applications which involve nonlinear differential equations, the differential equation contains parameters the values of which are known only approximately. In particular, they are generally determined by measurements that are not exact. In these cases it is important to study the behavior of solutions and examine their dependence on the parameters. This study leads to the area known as *bifurcation theory* [25]. It can happen that a slight change of the value of a parameter can have significant impact on the solution. Bifurcation theory is a deep and complicated field that still needs extensive research. Two fundamental concepts of bifurcation theory are the concepts of *fixed points and periodic trajectories.*

Consider a nonlinear dynamic system described by the state-space equation;

$$\dot{\mathbf{x}} = \frac{d\mathbf{x}}{dt} = \mathbf{f}\left(\mathbf{x}, \lambda\right) \tag{7.1}$$

where $\mathbf{x} = \mathbf{x(t)}$ is the n-dimensional state vector, $\mathbf{f}$ is an n-dimensional nonlinear function, and λ a parameter.

Fixed points A fixed point $\mathbf{x_f}$ satisfies the relation $\mathbf{f}(\boldsymbol{x}_\mathbf{f}, \lambda) = \mathbf{0}$. Thus if the system starts at $\mathbf{x_f}$, it will stay there forever.

Periodic Trajectories A periodic trajectory is a solution of (7.1) for which there is positive T $(0 < \mathrm{T} < \infty)$ such that $\mathbf{x}$(t)=$\mathbf{x}$(t +T) for all T .

Stable fixed point and periodic trajectory A fixed point or periodic trajectory is said to be *stable* if solutions starting close to it tends to it under the evolution of the flow. If this is not so, the solution is said to be *unstable.* We will now see how fixed points and periodic trajectories change as parameters are varied. For a given set of parameters a fixed point or periodic trajectory is stable, while for another set it is unstable. Such a change in stability under a change in parameters is a *bifurcation.* Another example of bifurcation is when as parameters vary new fixed points or periodic trajectories occur. The bifurcation diagrams are plotted in phase space, i.e., the space of the variable $\mathbf{x}$, and summarize the behavior near the bifurcation. An equilibrium point (node) that attracts nearby solutions at $t = \infty$ is called a *sink*, and an equilibrium point that repels nearby solutions is called a *source,* pictorially:

In the following we present the main types of bifurcations.

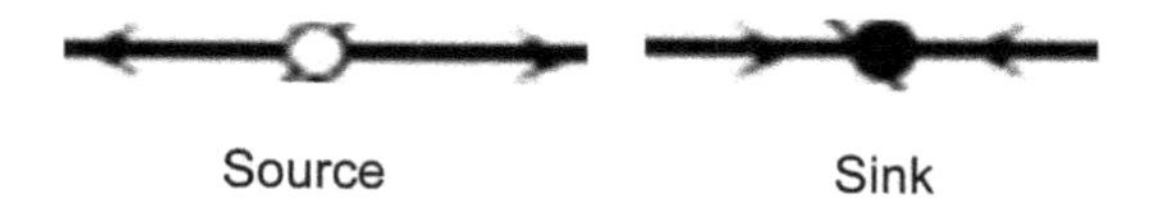

Type 1: *Saddle node Bifurcation*
The saddle node bifurcation corresponds to the creation and destruction of fixed points. This type of bifurcation is given by the system:

$$\dot{x} = r + x^2$$

The three cases of the parameter r, namely $r < 0, r = 0$ and $r > 0$ give different structure for the solutions as shown in Figure 7.2.

We observe that at $r = 0$ there is a *bifurcation.* For $r < 0$ there are two fixed points given by $x = \pm\sqrt{-r}$. The equilibrium $x = -\sqrt{-\tau}$ is stable, i.e., solutions starting near this equilibrium converge to it as time increases. Further initial conditions near $\sqrt{-r}$ diverge from it. The qualitative change in behavior at $r = 0$ is called a *saddle node bifurcation.*

Type 2: *Trans-critical Bifurcation*
The trans-critical bifurcation corresponds to the exchange of stability of fixed points. The normal form for this type of bifurcation is given by the system:

$$\dot{x} = rx - x^2$$

Here we have three fixed points one for $r = 0$, and two fixed points for $r \neq 0$. When $r = 0$ the only fixed point is $x = 0$ which is semi-stable (i.e., stable from the right and unstable from the left). For $r \neq 0$ we have two fixed points given by $x = 0$ and $x = r$. Therefore, in this case $x = 0$ is a fixed point for all r. For $r < 0$ the non-zero fixed point is unstable, but for $r > 0$ the non-zero fixed point becomes stable. Therefore, we see that the stability of the fixed point has changed from unstable to stable (as shown in Figure 7.3(a)). The diagram of the trans-critical bifurcation is shown in Figure 7.3(b).

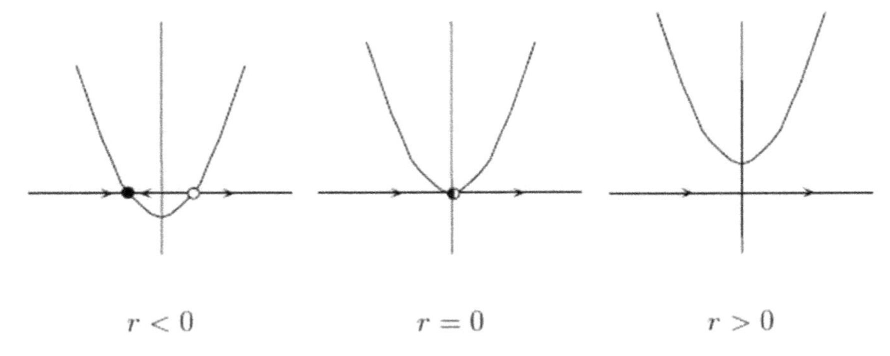

Figure 7.2 Saddle node bifurcation types for $\tau < 0, \tau = 0$ and $r > 0$.

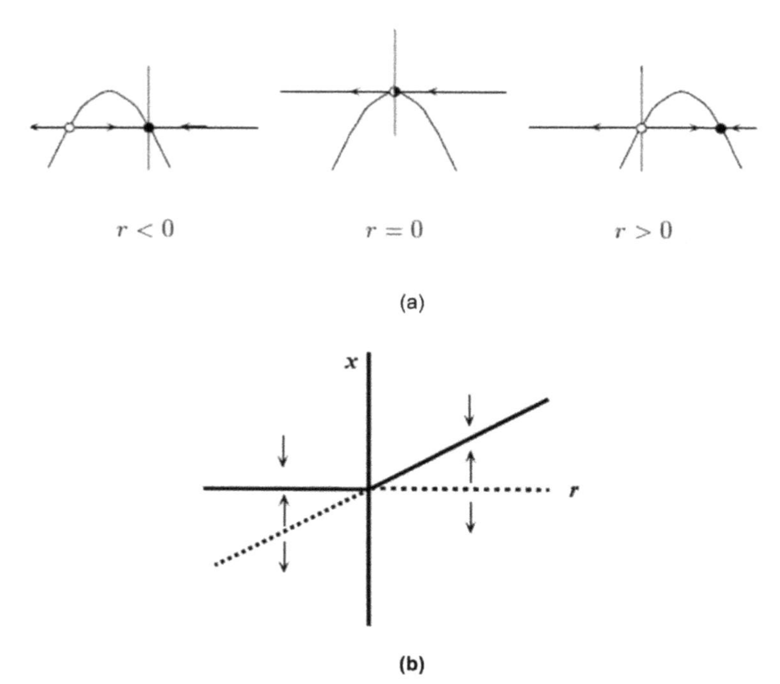

Figure 7.3 (a) The three parameter cases $r < 0, r = 0$ and $r > 0$, (b) Trans-critical bifurcation diagram.

Type 3: *Pitchfork Bifurcation*

The standard form of pitchfork bifurcation is given by the system:

$$\dot{x} = rx - x^3$$

Here again we have three fixed points. The cases of $r \leq 0$ and $r > 0$ once again give very different form for the solutions as shown in Figure 7.4.

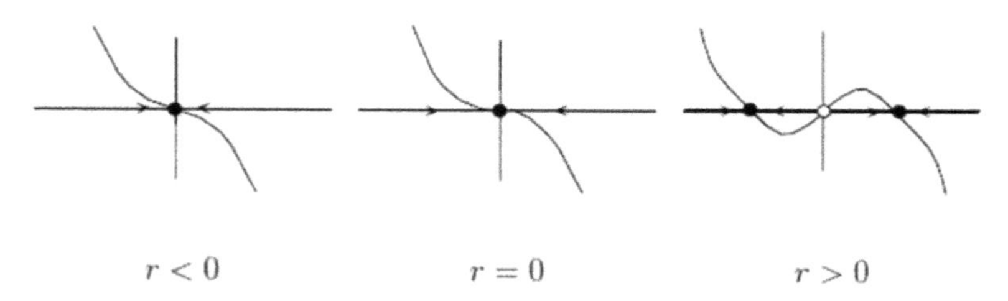

Figure 7.4 The three parameter cases $r < 0, r = 0$, and $r > 0$.

The associated pitchfork bifurcation diagram is *supercritical as* shown in Figure 7.5.

Now, consider the system:

$$\dot{x} = rx + x^3$$

For this system we get the so-called *sub-critical pitchfork bifurcation* (Figure 7.6).

Here we note that the solutions blow up in finite time, i.e., $x(t) \to \pm\infty$ as $t \to p < \infty$

Type 4: *Bifurcation of the Logistic Equation*
The logistic equation describes the evolution with time of the populations of various species. In discrete time ($n = 0, 1, 2, ...$) the logistic equation is:

$$x_{n+1} = rx_n (1 - x_n)$$

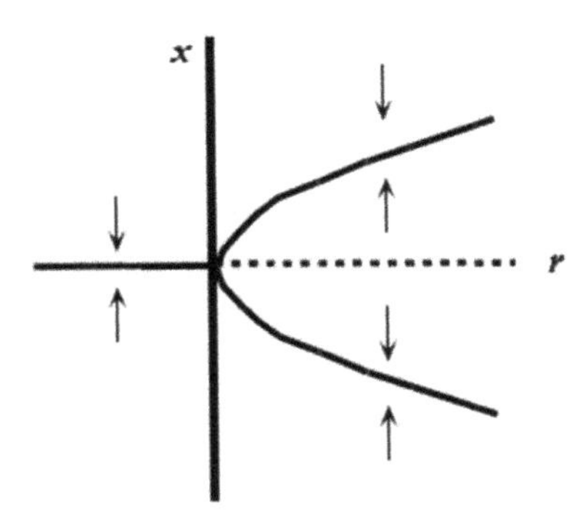

Figure 7.5 Supercritical pitchfork bifurcation diagram.

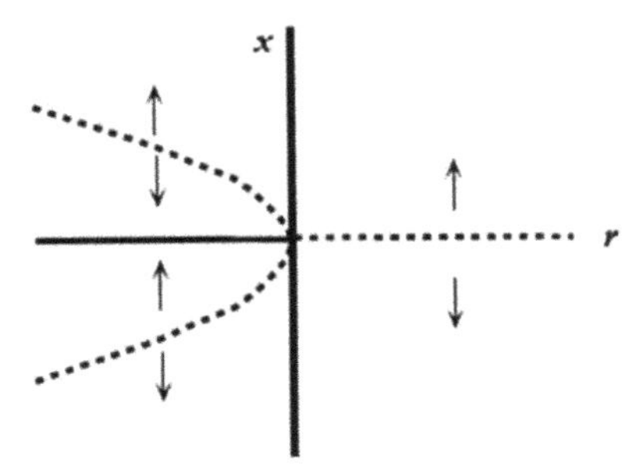

Figure 7.6 Sub-critical pitchfork bifurcation diagram.

where r is called the driving parameter. This equation is simulated in a computer with an initial value of the parameter r. For small values of r (up to $r = 3.0$), the sequence $x_0, x_1, x_2, ..., x_n$ as $n \to \infty$, converges to a single value x_∞. In biology this value x_∞ represents the population of the species.

When $r = 3.0$, x_n oscillates between two values, i.e., a *bifurcation* occurs. Increasing slightly the value of r further, x_n oscillates between not two, but four values. Further increase of r leads to oscillations (bifurcations) of period eight then sixteen and so on, until *chaos*. When r arrives at a value $r = 3.57$ x_n neither converges or oscillates and its value turns out to be completely random. For values $r > 3.57$ the behavior is *chaotic*. But there is a particular value of r where the sequence x_n again oscillates with period 3 and there appear bifurcations with period 6, 12, 24, then back to chaos. It was shown that any sequence with a period of three will exhibit regular cycles of every other period as well as exhibiting chaotic cycles. The overall bifurcation diagram of the logistic equation is shown in Figure 7.7(a). Figure 7.7(b) shows an example of 4-Dimensional bifurcation diagram [86].

7.6.3 Chaos and Strange Attractors

Chaos Theory is a mathematical sub-discipline of complex and nonlinear systems [26, 27]. Chaos is everywhere in nature and modern society from nature's deepest processes to economic systems. Chaos theory has helped the analysis of earth's weather system, migratory patterns of birds, the spread of vegetation across a continent, etc. In general, chaos theory has taught us that nature most often works in patterns, which are caused by the sum of many tiny pulses. Therefore nature is neither regular nor predictable.

Chaos in time appears due to the *'sensitivity of chaotic dynamic systems to initial conditions'*. The motion equations of dynamic systemshelp us to compute their state at the next instant of time, and in all successive time instants, thus providing the *trajectory (orbit)* of the system in the state space. Typically we start from the state of the system at some initial time, i.e., from a given set of *initial conditions.* Consider a *chaotic system* (or *time-chaos*) and two sets of initial conditions (i.e., two points in state space) arbitrarily close to each other. The two trajectories that correspond to these two sets of initial conditions are very close to each other at the beginning, but they don't stay close as time evolves. Actually, they diverge away from each other with time. This phenomenon, which as mentioned above is called *'sensitivity to initial conditions'*, was discovered by *Edward Lorenz,* who called it the

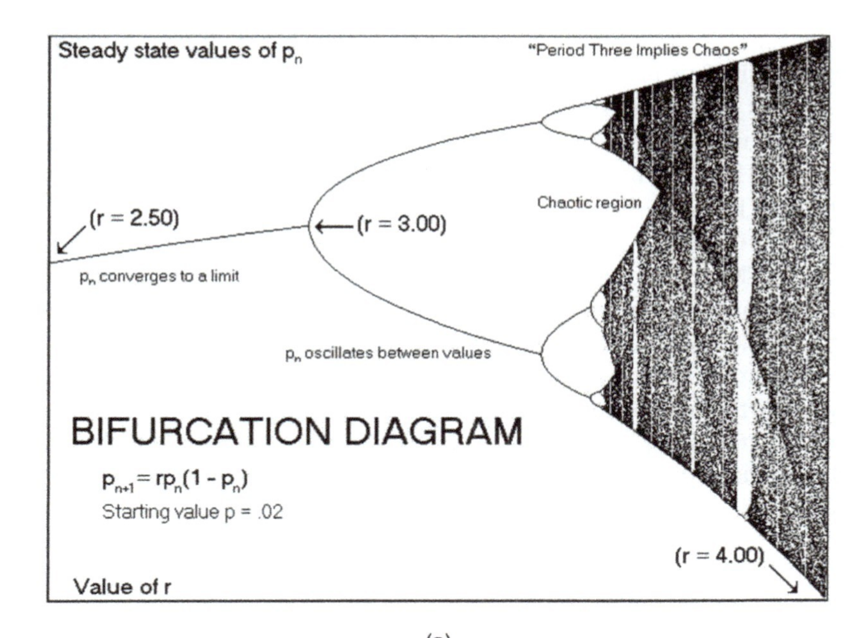

(a)

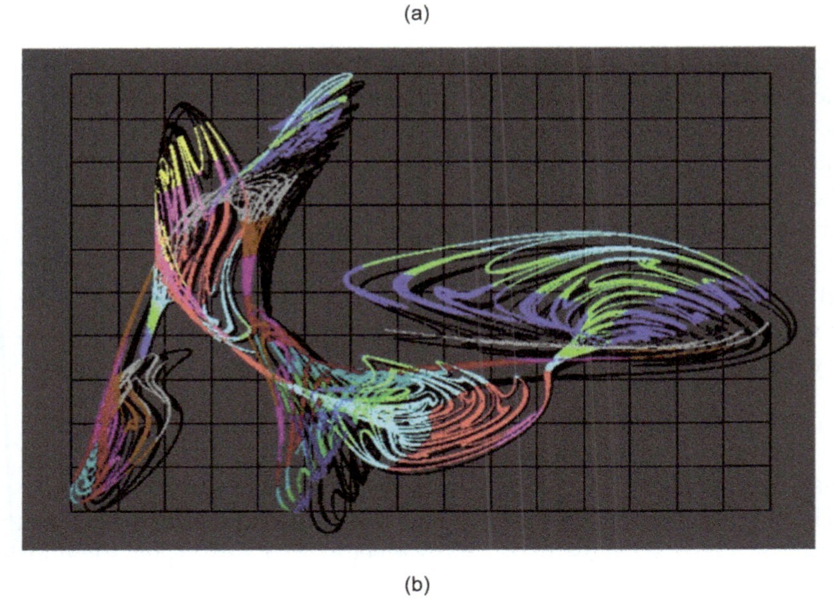

(b)

Figure 7.7 (a) Bifurcation diagram of the logistic equation, (b) 4-dimensional bifurcation.

Source: (a) https://www.zeuscat.com/andrew/chaos/bifurcation.html; (b) www.critcrim.org (readfeather/chaos/006challenges.html) [86].

'*butterfly effect*'. The probability of getting time-chaos in nonlinear systems is extremely high. This is the rule rather than exception (although generations and generations of scientists were taught that systems are *integrable* (multi-periodic), i.e., insensitive to initial conditions. Thus *chaos* is the *end of reductionism.* This feature has been hypothesized already from 1873 by *James Maxwell* [28].

Weather prediction, especially long term prediction, is an extremely difficult problem. Meteorologists can predict the weather for short periods of time, one or two days at most, but beyond that predictions are very poor. The long-term weather prediction has been extensively studied by the mathematician and meteorologist Edward Lorenz. To this end he developed a model of 12 nonlinear differential equations which exhibited chaotic behavior. He later simplified this model replacing it by a model of three equations, which was obtained via a simplification of the Navier–Stokes fluid dynamic equations.

These equations are the following:

$$\begin{aligned}\dot{x}_1 &= f_1(x_1, x_2, x_3) = -ax_1 + ax_2 \\ \dot{x}_2 &= f_2(x_1, x_2, x_3) = -x_2 + rx_1 - x_1x_3 \\ \dot{x}_3 &= f_3(x_1, x_2, x_3) = -bx_3 + x_1x_2\end{aligned}$$

An air layer is heated from bottom to top. The warm air that goes up interacts with the colder air that goes down and produces turbulent convection rolls. Here, x_1 is the rate of rotation of the convection rolls, x_2 is the temperature difference between the *upward* and the *downward* moving air masses, and x_3 is the deviation from linearity of the vertical temperature profile. We will study the performance of this system for the parameter values $a = 10$ and $b = 8/3$. The parameter r represents the temperature difference between the top and the base of the air layer. Increasing r we get more energy to the system which results in more vigorous dynamics. The equilibrium points are given by the solution of the algebraic system:

$$-ax_1 + ax_2 = 0, \quad -x_2 + rx_1 - x_1x_3 = 0, \quad -bx_3 + x_1x_2 = 0.$$

Obviously the point $(x_1, x_2, x_3) = (0, 0, 0)$ is an equilibrium point (a solution). The first equation gives $x_1 = x_2$ and so the second equation becomes $x_1(-1 + r - x_3) = 0$. Thus for $x_1 \neq 0$ we get $x_3 = r - 1$. Then, the third equation becomes $x_{l^2} = b(r - 1)$. For $0 < r < 1$ this equation has complex-valued roots, and so the only equilibrium point is $(x_1, x_2, x_3) = (0, 0, 0)$. For $r = 1$ the equilibrium point is again $(x_1, x_2, x_3) = (0, 0, 0)$. Finally for $r > 1$ we have three equilibrium points:

$$P_0 = (0,0,0)\,,\ P_1 = \left(\sqrt{b\,(r-1)}, \sqrt{b\,(r-1)}, r-1\right),$$
$$P_2 = \left(-\sqrt{b\,(r-1)}, -\sqrt{b\,(r-1)}, r-1\right)$$

To test the stability of these points we find the dynamic system's Jacobian matrix at the point $P_0 = (0, 0, 0)$:

$$\mathrm{A} = \frac{\partial \mathbf{f}}{\partial \mathbf{x}} = \begin{bmatrix} -a & a & 0 \\ r - x_3 & -1 & -x_1 \\ x_3 & x_1 & -b \end{bmatrix} = \begin{bmatrix} -10 & 10 & 0 \\ r & -1 & 0 \\ 0 & 0 & -8/3 \end{bmatrix}$$

where $\partial \mathbf{f}/\partial \mathbf{x}$ denotes the partial derivative of the vector-valued function $\mathbf{f} = [1, f_2, f_3]^{\mathrm{T}}$ with respect to the vector-valued variable $\mathbf{x} = [x_1, x_2, x_3]^{\mathrm{T}}$. For $r > 0$, the matrix A has the eigenvalues $\lambda_{1,2} = (1/2)\left(-11 \pm \sqrt{81 + 40r}\right)$ and $\lambda_3 = -8/3$. For $0 < r < 1$ all these eigenvalues are negative and so the equilibrium point $(0, 0, 0)$ is a stable equilibrium point. If $r > 1$, then $\lambda_1 > 0$ and the point $(0, 0, 0)$is an unstable equilibrium point. The eigenvalues of A in the cases of the two other equilibrium points P_1 and P_2 are of the same form (i.e., $\lambda_{1,2} = a \pm j\beta,\ \lambda_3 \leq 0$). The real part a is negative for $1 < r < r_0$ and positive for $r > r_0$ with $r_0 \approx 24.8$. Therefore, for $1 < r < r_0$, there are two stable equilibrium points, whereas for $r > r_0$ all equilibrium points are unstable. This means that for $1 < r < r_0$ two neighboring trajectories return in a spiral way to the nonzero equilibrium points, while for $r > r_0$ they spiral outward. The trajectories can be computed with the Runge–Kutta method (e.g., fourth-order R–K method) and are sensitive to the parameters of the method (e.g., the step size). Figure 7.8 shows a view of the Lorenz

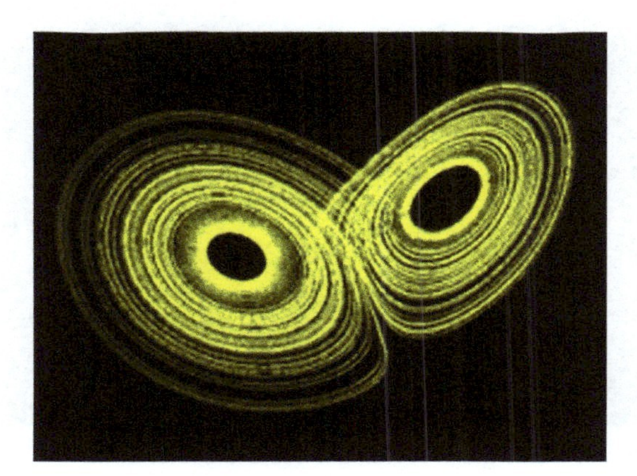

Figure 7.8 Lorenz strange attractor.

Source: www.ncku.edu.tw (cku.phys.ncku.edu.tw, strange attractors).

attractor's trajectories, for $r = 28$. Figures 7.9 and 7.10 show the trajectories of Rossler and Henon strange attractors.

The equations of the Henon attractor are:

$$\begin{aligned} X_{n+1} &= y_n + 1 - a{x_n}^2 \\ y_{n+1} &= bx_n \end{aligned}$$

Figure 7.10 shows the phase space trajectories for two sets of values.

7.6.4 Fractals

Chaos can appear in both space and time. In space, a chaotic system is known as *'fractal'*, which is a geometrical shape that does not change when it is analyzed in smaller parts continuously (i.e., fractals are not smooth). Similarly, simple fractals when emerged always reproduce the same shape (or structure), i.e., space fractals are *self-similar.* The term *'fractal'* was coined by *Benoit Mandelbrot* in 1977 [29] who showed via computer graphics that *strange attractors* have the *fractal* property. Every chaotic dynamical system is a fractal-generating mechanism, and conversely, every fractal can be regarded as the outcome of long-time acting time-chaos. Time-chaos and

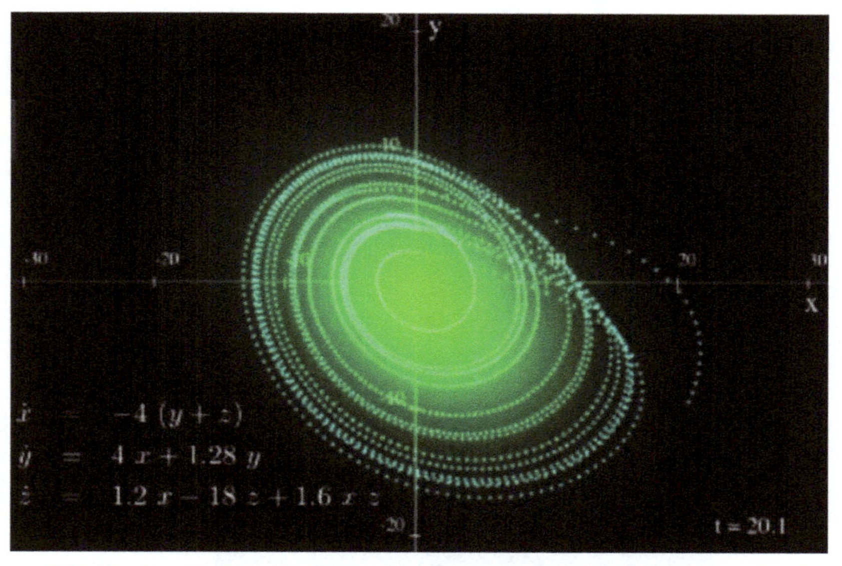

Figure 7.9 Rossler attractor (the green region is a representation of the probability density function of the attractor and the cyan dotted path is the actual phase space trajectory).

Source: www.theresilientearth.com (climate models irreducibly imprecise).
https://www.zeuscat.com/andrew/chaos/rossler.html

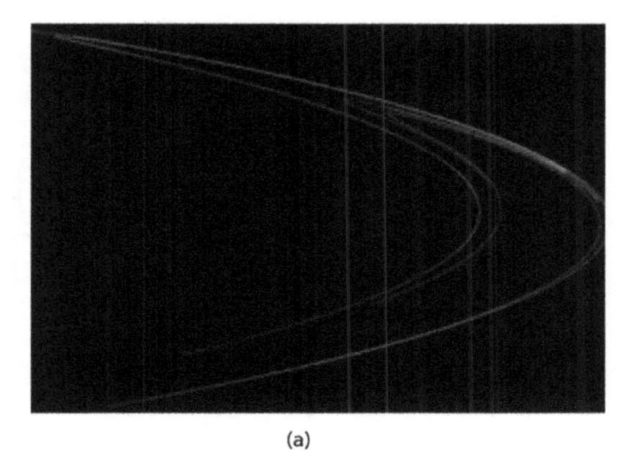

(a)

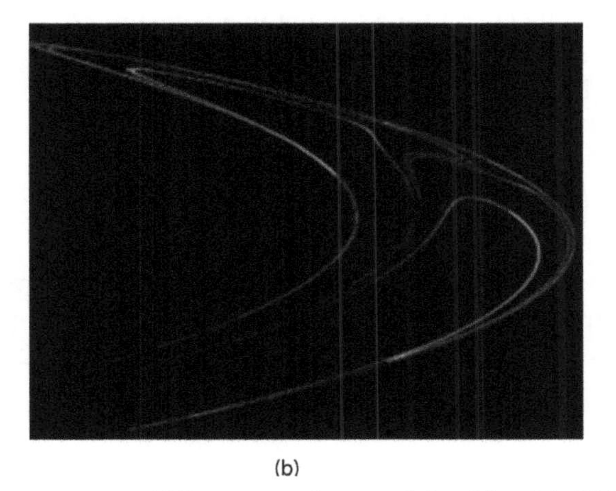

(b)

Figure 7.10 Henon attractor (a) Parameter values $a = 1.4$ and $b = 0.3$, (b) Parameter values $a = 1$ and $b = 0.642$.

Source: www.mathforum.org (The Math Forum, Images, Henon attractor). www.zeuscat.com/andrew/chaos/henon.html

space-chaos are closely related. For example, let us take a simple region (a cube or sphere, etc.) in the state space of a chaotic system. As each point of this region traverses its trajectory the region itself moves and changes shape. As time passes, the region will tend to a fractal which becomes perfect (complete) at infinite time. Nature is full of fractals. We actually live within fractals. The human body, a tree and its branches, the sky on a semi-cloudy day, the ocean surface waves, the avalanches, and so on, all are spatial fractals. Simple examples of fractals are given in Figure 7.11 where D is the spatial

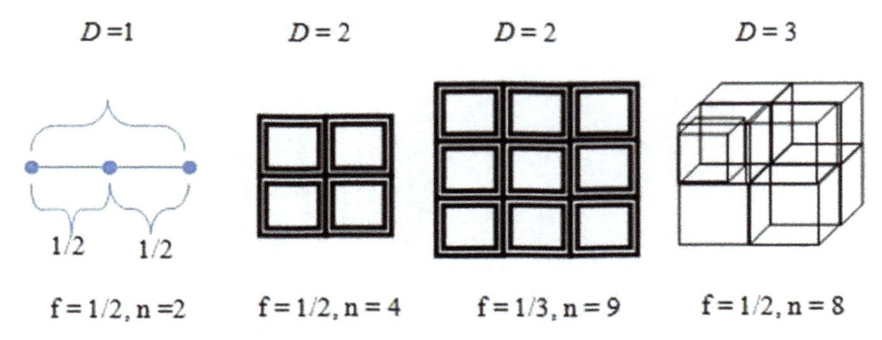

Figure 7.11 Spatial fractals (*f*, subdivision ratio and *n*, number of fractals).

dimension of the fractal ($D = 1$ for a straight line segment, $D = 2$ for a plane shape, $D = 3$ for a 3-D space shape, etc.).

The dimension D of a fractal is given by the relation:

$$n = \left(\frac{1}{f}\right)^D \quad \text{or} \quad D = \frac{\log n}{\log(1/f)} = \frac{-\log n}{\log f}$$

The *'snowflake'* fractal of Koch (Figure 7.12) is constructed by replacing, at every stage, each straight line segment by a broken line segment of length 4/3 as long. The results of stages 1, 2, and 5 are shown in Figure 7.12(a).

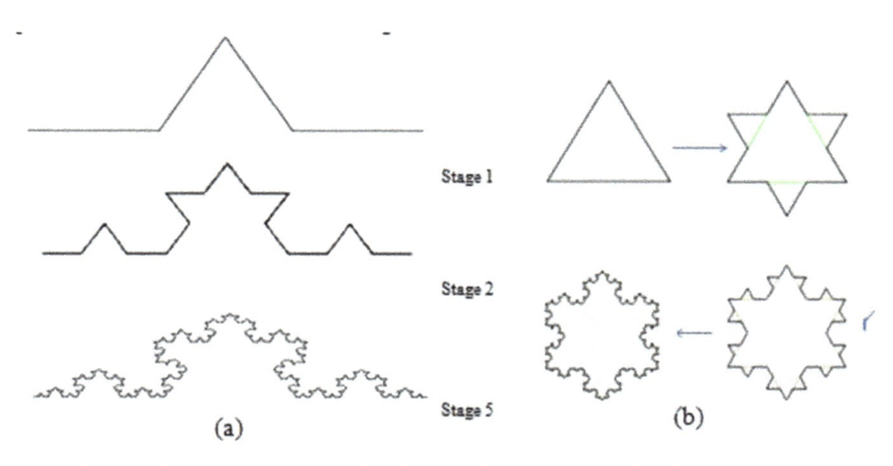

Figure 7.12 (a) Generation of Koch's 'snow-flake' fractal and (b) Application of Koch's fractal scaling to the sides of an equilateral triangle [51].

$$D = \frac{\log 4}{\log 3} = 1.26185 \quad (1 < D < 2)$$

If we apply this scaling to each one of the sides of an equilateral triangle we obtain, after 3 steps, the region shown in Figure 7.12(b). At each step, the perimeter of the region is increased by a factor 4/3. In the limit, the factor $(4/3)^N$ increases without bound, but the area enclosed by the curve is finite, equal to

$$\frac{\sqrt{3}}{4} + \frac{\sqrt{3}}{4}\left(\frac{1}{3} + \frac{4}{3^3} + \frac{4^2}{3^5} + \frac{4^3}{3^7} + ...\right) = \frac{2\sqrt{3}}{5}$$

i.e., equal to 8/5 the area of the initial triangle.

Similarly the tree fractal has the form in Figures 7.13(a,b) show a real-plant-looking fractal generated via a Lindenmayer Grammar (LG) when adding morphological perturbations.

In Figure 7.13(a) for the braches to not overlap we must have in the limit:

$$f \cos 30^o = f^3 \cos 30^o + f^4 \cos 30^o + f^5 \cos 30^o + ...$$

i.e.,

$$f = f^3 + f^4 + f^5 + ... = f^3\left(1 + f + f^2 + ...\right) = \frac{f^3}{(1-f)}$$

$$f^2 = 1 - f \quad \text{or} \quad f = \frac{\sqrt{5}-1}{2} = \frac{1}{\phi} = 0.618,$$

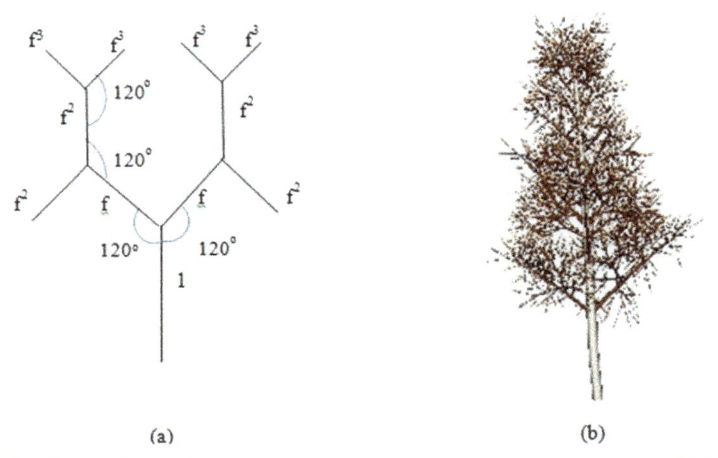

Figure 7.13 The tree branches fractal (a) Golden division-based tree fractal (b) Realistic tree produced by LG.

where ϕ is the '*golden division ratio*' (well-known from Geometry) which is defined as follows. Let a straight line segment (Figure.7.14).

We say that the point C is a golden division point of (AB) , if:

$$\frac{\mathbf{1}}{x} = \frac{x}{x+\mathbf{1}} \quad \text{or} \quad x^2 = x+1$$

which has the positive solution $x = \left(1+\sqrt{5}\right)/2 = 1.618 = \phi$, hence $1/\phi = 0.618$. Some other well-known expressions of ϕ are:

$$\phi = \sqrt{1+\sqrt{1+\sqrt{1+\ldots}}}, \phi^2 = 1+\sqrt{1+\sqrt{1+\sqrt{1+\ldots}}} = 1+\phi$$

$$\frac{1}{\phi} = \frac{1}{1+\frac{1}{1+\ldots}} = \frac{1}{1+\frac{1}{\phi}}, \text{ i.e., } \phi^2 = 1+\phi.$$

Actually, in addition to the natural fractals, there exist a large number of man-made fractals designed by mathematicians, physicist and computer scientists (e.g., those seen on animated movies, computational games, etc.). Three artificial fractals (cardial fractal, seahorse fractal, and tetrahedron fractal) are shown in Figure 7.15(a–c), and three natural fractals (butterfly fractal, cell fractal, and rain forest fractal) are shown in Figures. 7.16(a,b) and Figure 7.17.

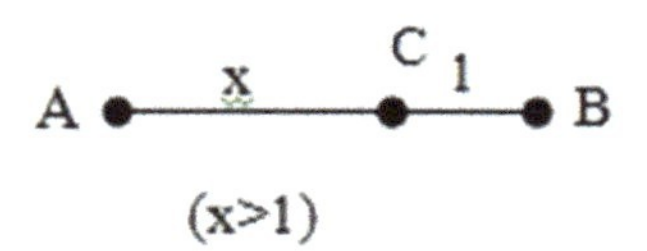

Figure 7.14 Definition of the golden division.

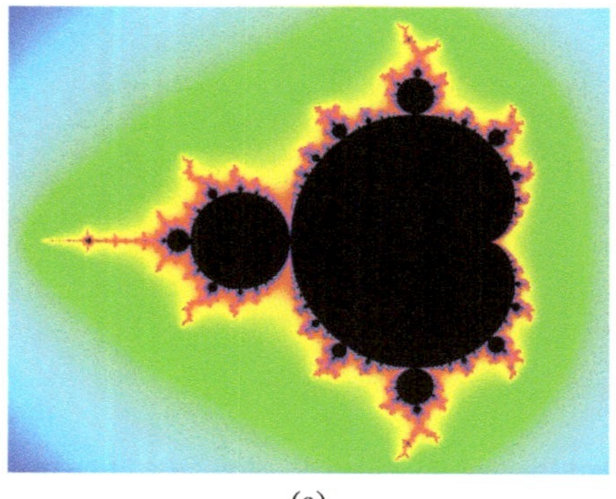

(a)

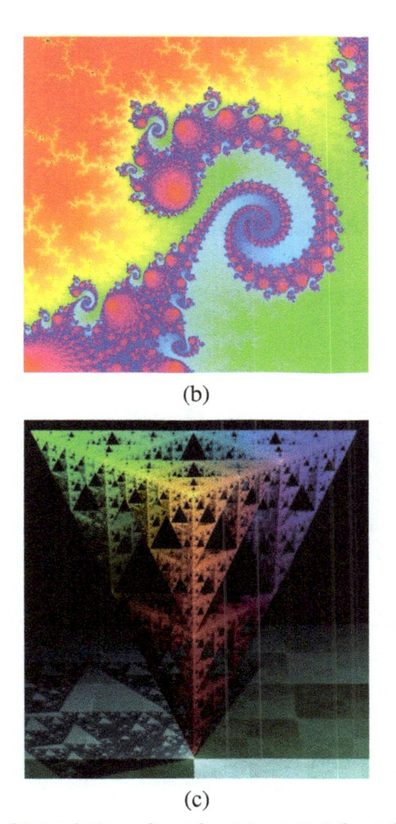

(b)

(c)

Figure 7.15 Examples of Mandelbrot fractals: (a) cardial fractal, (b) seahorse fractal, (c) tetrahedron fractal.

Source: (a) https://www.wired.com/2013/01/mandelbrot-images
(b) https://gr.pinterest.com/jimwrenholt/tetrahedron;
(c) www.superliminal.com/fractals/mbrot/mbrot.htm

7.7 Emergence

The concept of emergence has a long history in which many philosophers and scientists have attempted to define it in several different ways. The debate about what, actually, is emergence, has started in the early 1900s and has strongly revived during the last three decades. This re-appearance of the emergence debate is mainly due to the development of *complexity science*, in general, and *complex adaptive systems,* in particular, which have led to the death of *positivistic reductionism* and the resurgence of '*emergentism*' (the non-reductive materialism). We start our discussion with a few of the opinions and definitions of emergence given over the years.

(a)

(b)

Figure 7.16 (a) Butterfly fractal and (b) Cell fractal.

Source: (a) http://www.mirrorofnature.org/BBookButterfllyFritillary.jpg; (b) http://www.mpg.de (19191/Selforganization_in_biology).

Figure 7.17 Rain forest fractal.

Source: https://news.mongabay.com (2015/09/u-n-data-suggests-slowdown-in-forest-loss).

Samuel Alexander (1920) [30] has viewed the concept of emergence within the framework of metaphysics, according to which the activity of a human is the outcome of a unique type of process of a physico-chemical nature. In volume II of this work [30] he states: 'We are forced, therefore, to go beyond the mere correlation of the mental with these neural processes and to identify them. There is but one process, which, being of a specific complexity, has the quality of consciousness [...]'. This means that there appear new qualities and corresponding patterns of high-level causality that cannot be expressed directly via the more fundamental principles and entities. Actually they are macroscopic patterns taking place within microscopic level interactions. Emergent qualities are something new on Earth, but the world's fundamental dynamics are kept invariant.

Lloyd Morgan (1923) [31, 32] describes emergent evolution as follows: 'Evolution, in the broad sense of the word, is the name we give to the comprehensive plan of sequence in all natural events. But the orderly sequence, historically viewed, appears to present from time to time something genuinely new. Under what I call here *emergent evolution* stress is laid on this incoming of the new'. According to Morgan 'the emergent step...is best regarded as a qualitative change of direction, or critical turn point, in the course of events, and emergent events are related to the expression of some new kind of relatedness among pre-existent events'. More specifically, Morgan's view of emergence gives us the basic features that must be possessed by a phenomenon, process, property, event, etc. to be characterized as emergent. These features are [33]:

- It must be genuinely new, i.e., something that never happened before in the course of evolution.
- It is something tightly connected with the appearance of a new kind of relatedness among pre-existent events or entities.
- It changes the evolution mode, as the way pre-existent events run their course is altered in the context of new kind of relatedness.

C. D. Broad (1925) [34] was concerned with the so-called *'mechanistic-vital'* debate on the question: 'Are the apparently different kinds of material objects irreducibly different?' and with the broader question: 'Are the special sciences reducible to more general sciences (e.g., biology to chemistry) and finally to physics, the base-level science?'. He concluded that *emergence* is the result of primitive high-level causal interactions which are beyond those of the more fundamental levels.

Stephen Pepper (1926) [35] was concerned with the question: 'Are there emergents?' His first argument was that 'indeterminism is neither essential to, nor characteristic of, theories of emergent evolution'. His second conclusion was that 'a theory of emergent qualities is palpably a theory of epiphenomena'. Finally, he classified emergence theories in theories of *emergent qualities* and theories of *emergent laws.* This classification yields actually the following three categories of emergent theories.

- Theories of emergent qualities without emergent laws.
- Theories of emergent qualities with emergent laws.
- Theories of emergent laws without emergent qualities.

Paul Meehl and Wilfrid Sellars [36] remark that only *class 1* is committed to epiphenomenalism, and that to make it consistent with determinism one must refuse to call *'laws'* the regularities between emergent qualities and the contexts in which they emerge, or, calling them *'laws,* refuse that they are emergent. After a thorough reasoning, Meehl and Sellars conclude that *'Peppers'* demonstration of the impossibility of non-epiphenomenal emergent is invalid'.

Cruchtfield (1994) [37] introduced the concept of *intrinsic emergence* which involves behaviors that possess the following properties:

- They are compatible with the model used by the observer concerned.
- Their occurrence cannot be foreseen in advance on the basis of the adopted model only.
- They are macroscopic, i.e., they occur persistently despite any changes in the observational scale.

Obviously, the results of the observer depend on the means he/she has available to observe the behavior at hand, or measure operations, and on his/her mental schemas. Thus, emergence is a concept that can be defined only relative to the observer. The question here is of course whether there really exist mathematical models that allow intrinsic emergence. The answer is *affirmative.*

We now proceed to a short general discussion of some general issues on emergence [38–40].

Emergence is actually a *philosophical concept of art* that applies on substances or properties:

- that 'rise' from more fundamental substances or properties.
- that 'are new' or 'irreducible', i.e., in some sense they do not depend on more fundamental substances or properties.

This means that we have to face the phenomenon of two sets of properties/substances that are distinct but yet closely related. The second feature above is what it makes emergent properties really interesting, and singles them out of any odd distinct properties. Emergence is generally distinguished in:

- Epistemological emergence.
- Ontological emergence.

Epistemology is concerned with the questions: 'what do we know or can we know, and how do we come to know certain things?

Ontology is concerned with the question 'what kind of entities, substances, properties etc., exist? Therefore, emergent properties are either new properties from an *epistemological* point of view or new properties from an *ontological* point of view. In any case they add something to our knowledge of the world. Of course, their common aspect is their dependence (in some way) on the more fundamental properties. Epistemological emergence deals with what we can know about the behavior of complex systems and not what comes into existence, and so epistemological emergent properties would not be known on the basis only of our knowledge about the parts of the system, and their interactions Epistemological emergence is either *strong* (C.D. Broad) or *weak* (S. Alexander). Strong epistemological emergence declares that not even an ideal cognizer could be able to know which properties will emerge from a complex system given its parts, their internal interactions, and their interaction with the environment. The two most common versions of epistemological emergence are:

- *Predictive* Emergent properties are features of complex systems which could not be predicted from the stand-point of a pre-emergent stage, despite a thorough knowledge of the features of, and laws governing, their parts.
- *Irreducible* Emergent properties and laws are systematic features of complex systems governed by true, law-like generalizations within a particular science such as these generalizations cannot be reduced to (captured in) the concepts of physics.

Epistemological emergence is *weaker* than ontological emergence. In ontological emergence the physical world is seen as completely consisting of physical structures (simple or composite). It is noted that composite structures are not (always) mere sums or aggregates of the simples, because there are levels, or layered strata, or objects as complexity increases. Each other higher layer is the outcome of the emergence of an *'interacting gamma of new-qualities'*. As J. Kim pointed out [41] in order to have *robust emergence* (natural relation) something more than *supervenience* and *functional irreducibility* is needed, because supervenience allows for different background/base relations and so it is not a robust concept.

Weak emergence includes properties or states P that are characterized by one or more features from the following list:

- The outcome of many lower-level states (microstates) non-linearly interacting.
- Unanticipated/unpredicted.
- Unintuitive or counter intuitive.
- Qualitativelly different to any of the lower-level states that lead to P.
- Nomologically different to any of the lower-level states that lead to P.

Weak emergent properties are encountered in networks of biological signaling pathways [42], in microtubules [43], in conspecific attraction in fragmented landscapes [44], in functional structure tree growth models [45], etc. A short discussion on these properties, of course taking into account the various definitions of emergence, is collectively provided in [49]. Among the noticeable references on the concept of emergence in complexity science is [56]. Figure 7.18 gives a schematic of the emergence of global properties/patterns and complex adaptive behavior from local interactions/relationships between system agents.

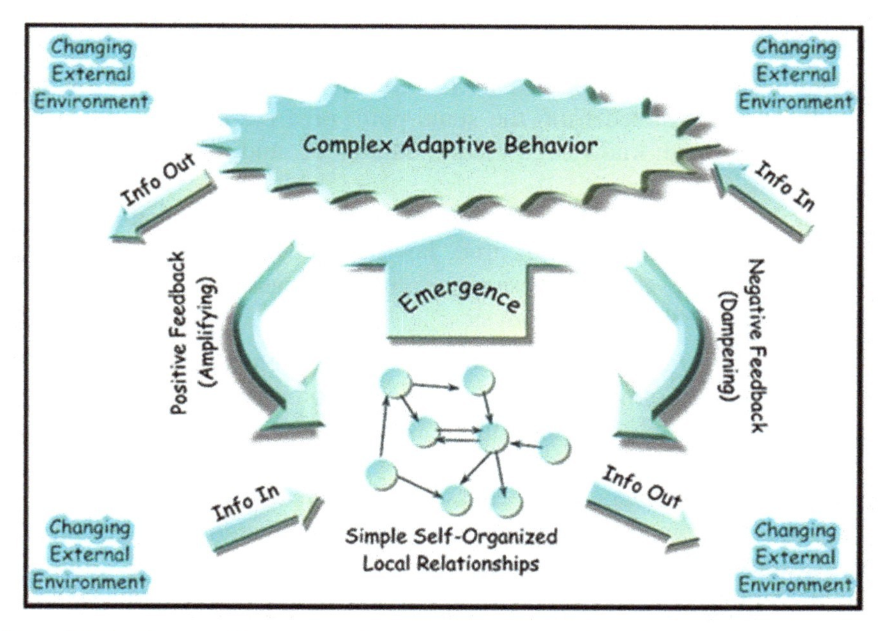

Figure 7.18 Pictorial illustration of the emergence of complex adaptive behavior (Information flows back to the environment allowing both negative and positive feedback to contribute to the next cycle of response).

Source: https://www.learner.org (Complex Adaptive Behavior).

7.8 Complex Adaptive Systems

The field of *Complex Adaptive Systems* **(CAS)** is concerned with the study of high-level instantiations of natural and man-made systems that cannot be studied by traditional analysis techniques. A CAS is a complex, self-similar set of interacting agents and its study requires a multi-disciplinary methodology. Examples of complex adaptive systems are rain forest, ant colonies, the ecosystem, the biosphere, the immune system, the brain, the stock market, the global economy, large computer networks, a human society or community, a manufacturing enterprise, etc. *John Holland,* one of the inventors of evolutionary computation and genetic algorithms, defines a *Complex Adaptive System* as many agents that act in parallel, act persistently, and react to the actions of other agents. Speaking about internal agent models he says: 'This use of building blocks to generate internal models is a pervasive feature of complex adaptive systems'. A CAS is controlled by decentralized and distributed controllers. The performance of a CAS is the result of the competition and cooperation between the agents, which involves a large number of decisions made all the time by the individual agents [46–48]. Four basic goals of CASs are the following:

- To understand the ways in which complex adaptive systems work and the unified underlying principle of complex adaptive behavior in life and society.
- To compare natural and man-made manifestations of CAS and reveal their possible differences.
- To study and understand the interplay of behavior at different scales, and reveal what is universal or not, and when averaging is applicable or not.
- To study and evaluate computer simulations of simplified models of natural systems and their actual importance for human-life applications.

The term *'complex adaptive systems'* involves three words complex, adaptive, and systems:

- *Complex* means composed of many parts which are joined together.
- *Adaptive* refers to the fact that all living systems adapt (dynamically) to the changing environments in their effort to survive and thrive.
- *System* is the concept which implies that everything is interconnected and interdependent.

Contrary to non-adaptive complex systems (e.g., the weather system), CAS have the capability to internalize information, to learn and to evolve (modify their performance) as they adapt to changes in their environments. A CAS produces macroscopic global patterns that emerge from the agents' nonlinear and dynamic interactions at the microscopic (low) hierarchical level. These emergent patterns are more than the sum or aggregation of their parts. Here is exactly the reason why the conventional reductionist approach cannot describe how the macroscopic patterns emerge.

According to Holland the following three reasons explain why a CAS is difficult to be studied by the conventional system's theory approach [49]:

- A CAS loses the majority of its features when the parts are isolated.
- A CAS is highly dependent on its history and so the comparison of instances and identification trends is difficult.
- A CAS operates far from global optimum, and equilibrium points are difficult to be secured that concerned with system 'end points', by conventional approaches.

Fundamental higher-level features that discriminate a CAS from a pure *multi-agent system* (**MAS**) are:

- Complexity.
- Self-similarity.
- Emergence.
- Self-organization.

Complexity is characterized as the '*edge of chaos*' since it lies at the middle area between two extremes, viz. static order and chaos [46, 50, 51], and a discussion of complexity was provided in Section 7.3. The self-similarity property was discussed with relation to fractals in Section 7.6.4. The emergence concept was outlined in Section 7.7, and self-organization will be discussed in Section 7.10.

A *multi agent system* (**MAS**) is simply composed of multiple interacting agents, whereas in a CAS the agents and the overall system are adaptive, and the system is self-similar.

According to Holland, a unified theory of CAS must incorporate at least three mechanisms [49]

- Parallelism.
- Competition.
- Recombination.

On the basis of the above, the detailed structure of a CAS is as shown in Figure 7.19.

Also, in order for a system to be characterized as a CAS it must have the following seven characteristic (properties and mechanisms):

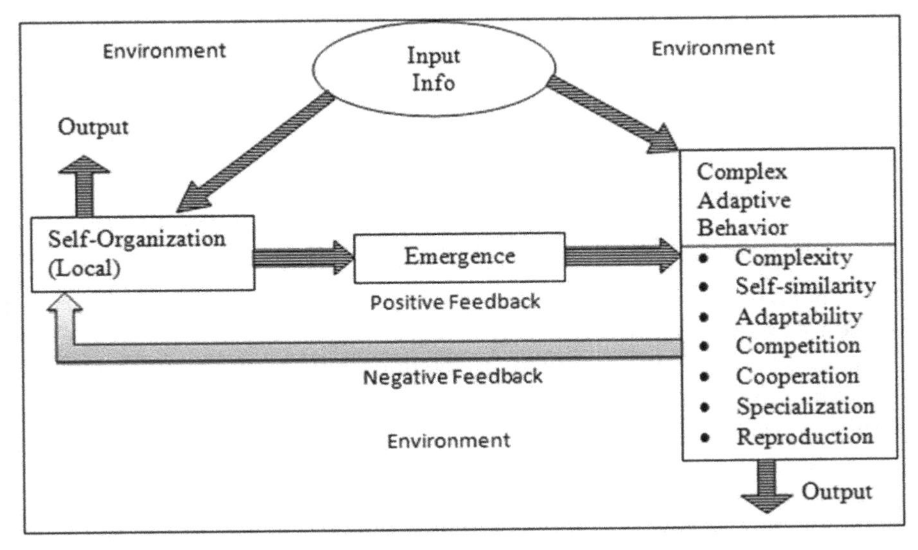

Figure 7.19 Basic structure of a complex adaptive system.

Properties

- *Aggregation* (complexity emerges from interaction of smaller parts).
- *Nonlinearity* (agents interact dynamically and non-linearly).
- *Flows* (network of agent interactions).
- *Diversity* (agent evolution goes toward filling diverse niches, whereby niche evolution has larger impact on the system than the evolution of agents).

Mechanisms

- *Tagging* (Agents have ways to discriminate agents with particular properties).
- *Internal models* (Agents change via their interactions, and the changes specify future action, i.e., agents adapt).
- *Building blocks* (Reuse of components for multiple purposes).

A Concise Definition of CAS

The Definition of a CAS given by Holland is just one of the many alternative definitions available in the literature. Our purpose here is to provide a concise (nominal) definition of a CAS that attempts to capture all the features involved in the definitions given by *Prigogine* and *Stengers* [52]. Jantsch [53], Maturana and Varela [54], and Holland [46]. This definition was composed by *Kevin Dooley* in 1996 [55] and merges into one master list all conceptual lists (and principles) involved in the above works.

Definition of CAS *(Dooley)*
This definition involves the following four aspects:

1. A CAS is composed by *agents* (as basic elements) which are semi-autonomous units that try to maximize their fitness by evolving over time. Agents scan their environment and develop '*schemas*' (schemata), i.e., mental templates that interpret reality and define the proper responses to given stimuli. These schemas are subject to change and evolution via a selection–enactment–retention process in their effort to survive in their competition environment.
2. When a mismatch between observation and expected behavior is detected an agent can take action in order to adapt the observation to fit an existing schema. An agent can also alter schemas as it likes to better fit the observation. Shemas can change via random or purposeful mutation and/or combination with other schemas. Changes of schemas lead to agents that are more robust to increasing variety or variation, adaptable to a broader range of conditions, and more predictable in performance.
3. Schemas define the way an agent interacts with its neighboring agents, whereby actions among agents involve a nonlinear exchange of information and/or or resources. Agent tags helps to identify what other agents are capable of transaction with a given agent, and also facilitate the formation of meta-agents that distribute and decentralize functionality so as to allow diversity to thrive and specialization to occur. Agents can also reside outside the bounds of a CAS, and schemas associated to them can also define the rules of interaction and resource/information flow externally.
4. The agents' fitness involves in a complex way several local and global factors. Unfit agents are likely to initiate changes of schemas. The optimization of local fitness allows differentiation and novelty/diversity. Global fitness optimization leads to improved CAS coherence (as a system) and induces long-term memory.

Figure 7.20 shows pictorially a generalized CAS model which includes its position in the environment and how the feedback is realized.

A useful approach to CAS model construction is to use the so-called '**8-R** *scheme*' which is based on the answers to the following questions:

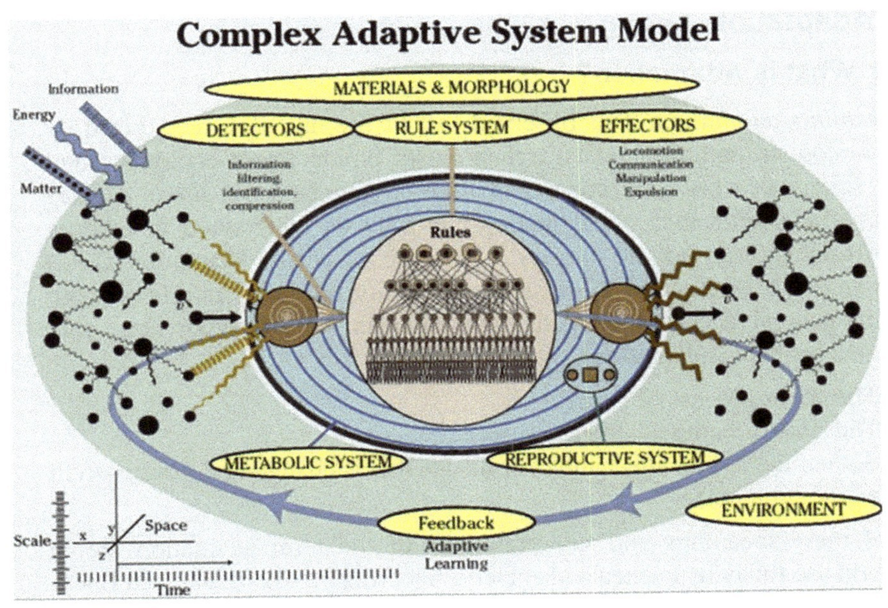

Figure 7.20 The position and functional interaction of a CAS with its environment.

Source: http://necsi.edu/procts/mclemens/casmodel.gif

- **Right view** What is the type of information that flows from the world to the agents and between them? Is it filtered or processed by other agents?
- **Right intention** What are the goals of the agents? Are the goals common to all agents?
- **Right speech** What is the information communicated by the agents and to whom?
- **Right action** What are the actions that the agents can take? Are these actions the right ones? When are they taken?
- **Right livelihood** What are the payoffs (utility) that the agents receive in return to their decisions under different conditions? How do payoffs influence agents?
- **Right effort** What principles, rules or policies do the agents follow to make their decisions? How much they learn from their experience, properly changing policies?
- **Right mindfulness** How intelligent are the agents and how sophisticated are their internal models?
- **Right concentration** In what aspects should the models be accurate and at what level of detail?

7.9 Adaptation

7.9.1 What is Adaptation?

Life exhibits adaptation. This observed fact of life has been recognized by most biologists, anthropologists, archeologists, philosophers, ecologists, and other scientists who were concerned with 'living beings'. *Julian Huxley* notes: 'The significance of adaptation can only be understood in relation to the total biology of the species' [57]. The object of adaptation varies according to the nature of the system or function to which is referred, but it is in general accepted that adaptation means the following (see, e.g., answers.com/topic/adaptation):

- The act or process of adapting.
- The state of being adapted.
- Something (e.g., a mechanism or device that is changed or changes to fit a new situation).

All of these meanings and interpretations of the term adaptation are in common use today in science and society. According to the Britannica Encyclopedia (Britannica.com): 'Adaptation in biology is the process by which an animal or plant becomes fitted to its environment. It is the result of natural selection acting on inherited variation'. According to Sci-Tech Encyclopedia: 'Adaptation is a characteristic of an organism which makes it fit for its environment or for its particular way of life'.

The research on *adaptation* has revealed several forms of adaptation: These include, but are not restricted to, the following:

- Structure adaptation.
- Physiological adaptation.
- Function adaptation.
- Evolutionary adaptation.
- Generic adaptation.

Structure adaptation includes, among others, the anatomical adaptation of animals. For example, the shape of the body of the fish is adaptive to life in water, the body of the bird id adapted for flight, horse legs are adapted for running on the grass, and the long ears of the rabbits living in desert-like environments allow them to radiate heat more efficiently and so survive under harsh conditions.

Physiological adaptation includes the responsive adaptation of sense and action organs. For example, the eye has the ability to adjust to changing conditions of light intensity and the heart functioning can be adapted to higher work load, etc. Clearly, physiological adaptations span over an organism's lifetime.

Function adaptation includes the historical adjustments of the way in which the organs of an organism work to fit the changing long-term conditions of the organism's habitat. It is again achieved through natural selection that forms and maintains the function concerned. For example, the function of the heart is pumping blood. The resulting sound is not a function but a side-effect of pumping. Every organ of an organism has a functional history, which has undergone *selection* for its *survival* in its environment.

Evolutionary adaptation refers to adjustment of living matter to environmental conditions and to other living things over very long periods (e.g., over many generations of a given population). The adaptation property is a property of life which is not possessed by nonliving matter.

Genetic adaptation takes place in a population via *natural selection* which affects its genetic variability (i.e., the population undergoes adaptive genetic adjustments to cope with the circumstances). Genetic adaptation (also called genetic improvisation) occurs at the **DNA** (**d**eoxyribo **n**ucleic **a**cid) level and is performed by *mutation and selection.* Genetic adaptation includes adaptations that may result in visible structures or changes to physiological activity so as to fit the environment's changes.

According to *Darwin* [58], the life processes are distinguished in:

- **Internal processes**, i.e., the processes which generate the organism.
- **External processes**, i.e., the processes that take place in the environment where the organism must live.

Before Darwin there was not a clear distinction between internal and external processes. Adaptations of internal processes have to do with the totally coordinated adjustments in the organism's (or system's) body, e.g., temperature control, blood pressure control, suitable changes in the immune system, etc. The adaptation which comes from the interaction of the organism or population with the nature (habitat) is performed through a proper feedback process. This means that nature *'selects'* the *'design'* which best solves the adaptation problem. Prior to Darwin the only systematic explanation of design in nature was the existence of a designer and a purpose in life's diversity. That is, adaptation was thought to be a sign of design and purpose. With his theory of evolution by natural selection, Darwin has changed this line of thought underwriting a fully naturalistic explanation of function in the biological world. However, the point of view that adaptation provides evidence that the World is governed by Design is still adopted and promoted by a large number of thinkers. Adaptations observed in the 'Nature' which reveal the sensitive fit between live beings and their habitat are considered by them as a proof of *'Creator's Design'*.

There is still a continuing debate regarding adaptations as the act, state, or mechanism of change to fit new circumstances: and adaptations as an indicator of design and purpose (i.e., between *evolutionists and creationists*) [59].

Adaptation is a must for the sustainable development of all sectors of modern society. It involves the actions that must be made so as to assure the adjustment to new conditions, and thus reduce harm or take advantage of new opportunities. Major adaptation sectors of society are shown in Figure 7.21.

7.9.2 Historical Note

Over the years biologists and other scientists have tackled the issue of adaptation and developed many different theories. Here we will present only a few of them. In ancient times *Aristotle* regarded life adaptation as a process towards a purpose, while *Empedocles* had exactly the opposite opinion, i.e.,

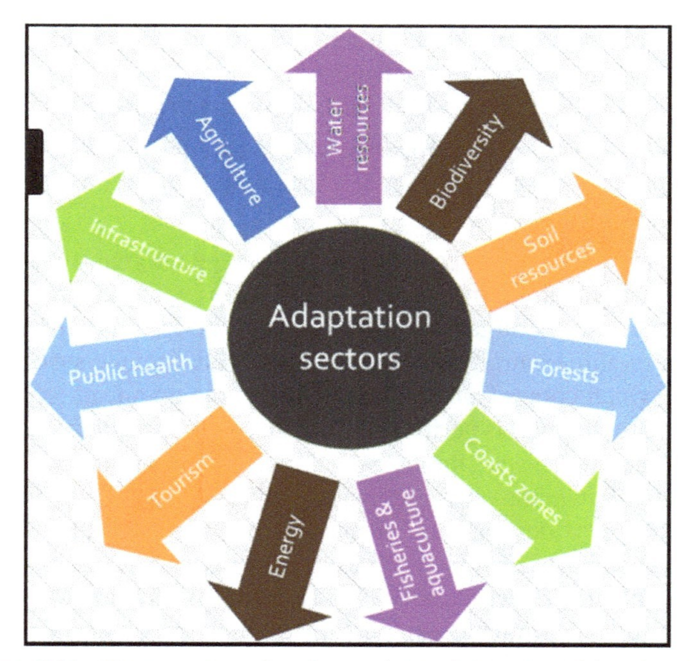

Figure 7.21 Eleven sectors of modern society activity potential for adaptation.

Source: www.globalchange.gov (report/response-strategies/adaptation).

adaptation had not required a purpose, but came about naturally, since such things survived [60].

Descartes viewed the world as consisting of two separate parts, viz. an active, striving purposeful psychological part (thinking, perception) and a 'dead' physical part (matter, body) defined fully by its extension to space and time without any inherent *arrow of time.* According to Descartes the active part of the world is limited to human minds [61].

Kant has presented another fundamental duality in the world, namely the dualism between the active striving of living beings and their dead environments [61].

Boltzmann *(1886),* promoted the view that the physical world is not actually dead or passive, but persistently acting towards increasing disorder (contrary to Newtonian physics).

Darwin *(1872)* studied the imperfections and limitations of animal and plant life that have existed over the time, and concluded that as the *habitat* changed so did the *biota* and that their habitats were subject to changes in their biota (e.g., by invasions of species from other regions) [58]. Darwin has divided life processes in *internal* and *external* and developed his theory of *evolution by natural selection.* Building on the Cartesian–Kantian dualistic point of view, most ascendants of Darwin *(neo Darwinists)* promoted the theory of complete separation between living-beings and environments or, equivalently, the full independence of biology from physics [61, 62].

Donald Fisher *(1930)*, a neo Darwinist, examined deeply the apparent incommensurability between biology (evolution) and physics (thermodynamics), whereby evolution leads towards increase of order (organization) and entropy change leads towards decrease of order (disorganization). He has put forward the idea that the two opposite directions of evolution and thermodynamics could be unified under a more general principle [61].

Lotka *(1922)* conclude that the natural selection tends to make energy flux a maximum, so far as compatible with the constraints to which the system is subject, and so the law of natural selection becomes also the law of evolution (maximum energy flux principle).

Odum *(1963)* and his colleagues have developed further Lotka's 'maximum energy flux principle' providing a corollary of it called 'maximum empower principle' and pointing out that 'in surviving designs a matching of high-quality energy with larger amounts of low-quality energy is likely to occur'.

Johannes von Uexkull did not accept the Darwinian separation of organism and habitat, promoting the thesis that: 'Every animal carries its environment about with it, line an impenetrable shell, all the days of its life. Every animal is bound to its environment by the circle of functions. Uexkull developed the *'pure natural science'* approach of *living beings,* and investigated the *'structural plan' (blueprint)* of them, their origin and their outcomes. According to this theory the stimuli of the environment, which an animal is able to receive via its blueprint are the only reality present for it. Due to this physical limit, the organism closes itself to all other spheres of existence [62].

***Ehrlich and Raven** (1964),* introduced the *co-evolution* concept to describe the two way dynamic interaction between organisms and their environment. Living beings adapt to their environment (living and nonliving), and in this adaptation process environments are many times modified, probably in such a way that influences the living-beings [63].

***John Maynard Smith** (1975)* explains that the *a-priori* reasoning of *evolutionists* (according to which there must be an adaptive/functional explanation for any trait, and conversely that natural selection provides an explanation for every biological phenomenon), is not necessarily wrong and may be the best way to proceed. Most traits of living-beings probably evolved towards survival and earning a living, despite the fact that they may not currently be optimal or any way adaptive [64, 65].

***Theodosious Dobzhansky** (1956–1968),* defined the terms adaptation, adaptedness and adaptive trait as follows [60]: (i) *Adaptation* is the evolutionary process whereby an organism becomes better able to live in its habitat or habitats, (ii) *Adaptedness* is the state of being adapted, the degree to which an organism is able to live and reproduce in a given set of habitats, and (iii) *An adaptive trait* is an aspect of development of pattern of the organism which enables or enhances the probability of that organism surviving and reproducing.

***Donald Hardesty** (1977),* an ecological anthropologist regarded the adaptation process 'as any beneficial response to environment' meaning 'biologically beneficial' [64, 66].

***Richard Lewontin** (1978, 1979)* adopted the constructionism view, according to which the world is changing because the living beings are changing. According to the evolutionary biologists the relation between adaptation and ongoing changes in the environment is analogous to what happened in the Queen *Through the Looking Glass,* who realized that she had to continue running just to keep in place. On this basis they have formulated the so-called *'Red Queen's Running Hypothesis'* which makes the constructionist view even stronger [62]. Lewontin reexamined the relationship between outside (environment) and inside (organism) and argued that the evolution process can be better explained by the construction metaphor [67, 68].

Bronislaw Malinowski *(1944)*, an anthropologist adopted the *basic needs* approach to adaptation in a society. For him a society is actually, a self-organized system of cooperatively pursued activities. It has a purposive behavior, whereby its purpose is primarily the *'satisfaction of basic needs'*, i.e., the conditions in the human organism within the cultural and natural environment that are necessary and sufficient for the survival of the organism and the society or group.

Peter Corning [64] has developed further this 'basic needs approach' to societal adaptation on the grounds of the biological need of survival and reproduction, i.e., in connection with the biological fundamentals.

7.9.3 Adaptation Mechanisms

The basic mechanisms of *adaptation* (or *adaptability)* are the following:

- Feedback mechanism
- Feedforward mechanism

and have a behavioral structure. This means that in the interaction of an organism or population with the nature, adaptation is performed through feedback, i.e., the nature selects the design which best solves the adaptation problem applying *positive feedback* to the change that enhances the capability to cause the selected design's own reproductions. If a change leads to fewer (or no) offsprings, a *negative feedback* is applied and the design will become extinct. Thus evolution by *natural selection* always tends to enhance *fitness* to the environment, making organisms to *better adapt* to their habitat and way of life. Of course, this adaptation which proceeds in small steps may not be ultimately perfect in all cases. The feedforward and feedback mechanisms are embedded in the so-called *program of a living being* [69] i.e., in the plan that specifies the organisms body ingredients and the interaction as the organism persists overtime. The feedforward and feedback processes are responses at the molecular level, the organ and body levels of the organism, and at the population level which secure *'survival'* in quickly varying habitats. When the environment (habitat) changes, the resident population (e.g., flying insects, oceanic organisms, etc.), goes to another more suitable place. This feedback response is the so-called *'habitat tracking'*, which however does not result in adaptation. At the *program level* adaptation is particularly called *improvisation* which is implemented by the DNA and performed by *mutation* and *selection.* Program improvisation is actually an optimization of the program such that to face new coming environment conditions. According to Koshland [69] adaptation and improvisation are handled on Earth by different mechanisms, and so they must be considered as different concepts and treated by different mechanisms in any newly devised or newly discovered system. The relative types of species in a given environment are always subject to change, so in nature *'change is the rule'*. According to *Darwin* the following three conditions are necessary for evolution of natural selection [67]:

- **Variation** This refers to the variation in phenotype characteristics among population members.
- **Inheritance** This refers to the fact that these characteristics are heritable to some extend (i.e., offsprings have the characteristics of their parents more likely than the average characteristics of the population).
- **Differential reproductive success** This refers to the fact that different variants leave different numbers of off springs in succeeding generations.

As *Robert Brandon* points out [70] these conditions are not sufficient. He explains 'insufficiency' by the following example. 'If two physically identical coins (as much as can be) are tossed 100 times it is highly likely that one of them will yield more heads than the other. Similarly, in the absence of selection, *drift* will occur. Therefore change (in gene frequencies or in phenotype distribution) is by no means indicative of selection. What is needed more than change to invoke selection is *differential adaptedness* to a *common selective environment'*.

The Darwinian explanation of *'differential reproductive success'* is the *Principle of Natural Selection* (**PNS**). Brandon has formulated the PNS principle as follows: 'If A better adapted than B in environment E then, (probably) A will have greater reproductive success than B in E'. He has argued that the propensity interpretation of adaptedness (or propensity interpretation of fitness) renders the PNS explanatory [71, 72].
In [70] *Brandon* discussed the following three issues of adaptation:

- Adaptation and environment.
- Adaptation and function.
- Hierarchical selection and adaptation.

Basic questions that must be addressed before adaptation planning is performed in real-life applications are:

- What do we need to adapt or change?
- What are the alternatives to changing/innovation?

- What roles are to be assigned and to whom?
- Who pays for what?
- How and what is it to be evaluated?

Figure 7.22 depicts the cycle that is behind the principles of adaptive behavior. The meaning of each arc in the diagram is:

Given **x** *then via* **P** *emerges* **y**,

where **x** is any of the four box descriptions, **y** is the next box in the direction of the arrow, and **P** is the process associated with the arc. This model makes a compound of the continuous and discrete, the words and the reality, the system and the physical, etc. and provides much prospect in the analysis and application of adaptive behaviors.

A scheme for implementing an adaptation process in organizations/enterprises with stakeholder participation is shown in Figure 7.23. This scheme has the following steps:

- Identify risks and vulnerabilities.
- Plan, assess, and select options.
- Implement the selected strategy option.
- Monitor and evaluate the outcome.
- Revise the applied strategy and share with others the lessons learned.
- Repeat the algorithm.

Clearly, this scheme is general and can also be appropriately applied to most engineering and social situations.

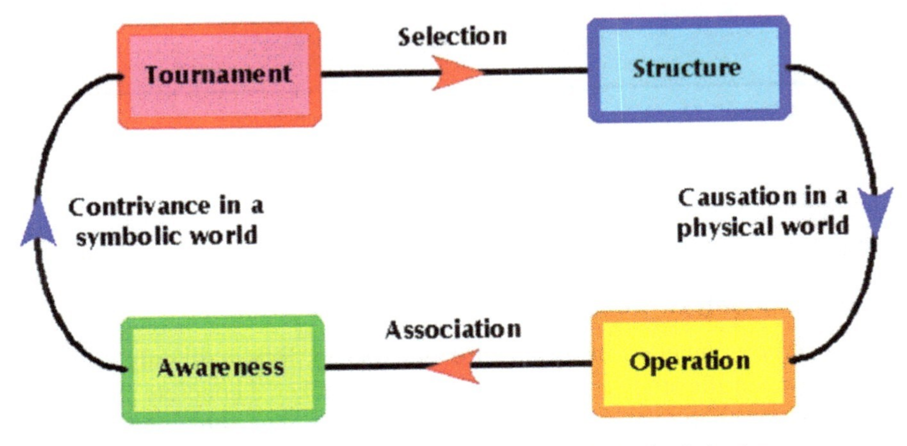

Figure 7.22 The cycle underying the laws of adaptive behavior.

Source: www.abooth.co.uk (/sand/sand.html#Old).

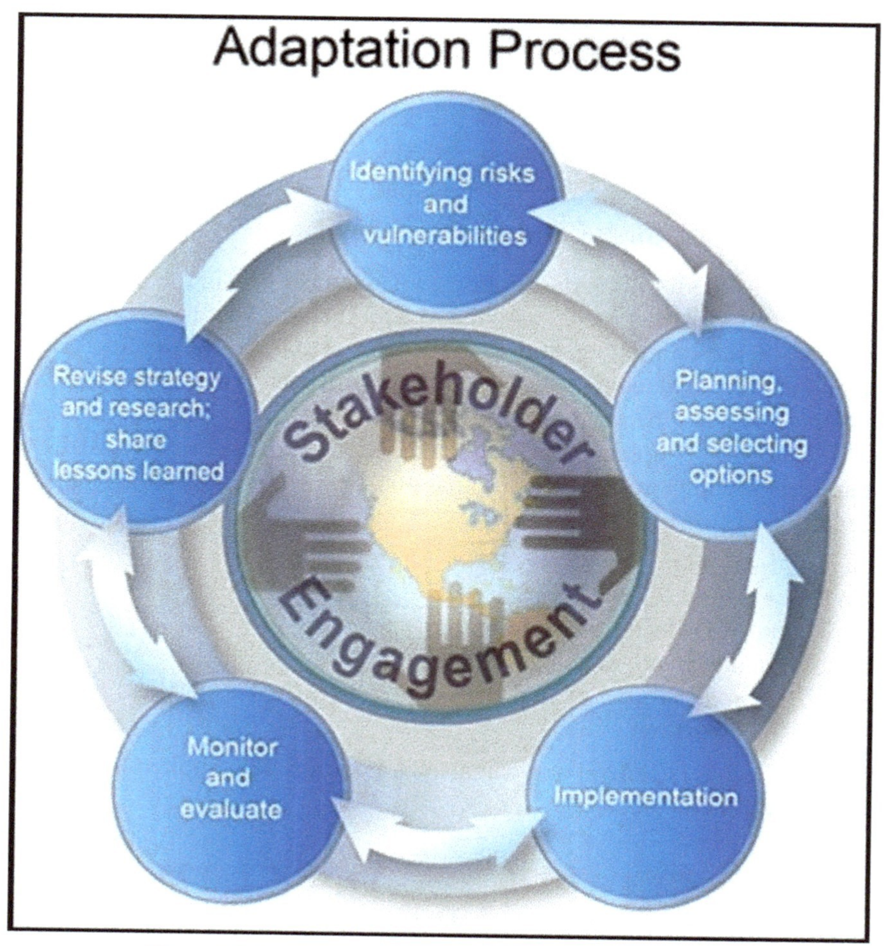

Figure 7.23 Adaptation scheme for business organizations.

Source: https://www.weadapt.org/knowledge-base/adaptation-decision-making/adaptation-planning-process

7.10 Self-Organization

7.10.1 What is self-organization?

Self-organization is inherent in life, nature and society. However, only after the 1950s the scientific study of self-organization has taken a concrete shape. According to Longman Dictionary, the word organization has three linguistic meanings [73]:

- The way in which different parts of a system are arranged and work together.
- Planning and arranging something so that it is successful or effective.
- A group such as a club or business that has formed for a particular purpose.

These meanings are used in our present scientific, information, technological, cultural and economic society, and cover both cases: external and internal organization of a system. In general, all these definitions imply that organization is some kind of order and excludes randomness produced by any cause at any level. The alternative definitions presented here are the following.

Francis Heylinghen Definition According to him: 'Self-organization is the spontaneous emergence of global structure out of local interactions [74]. *'Spontaneous'* has the meanings that no internal or external agent is in control of the process; for a sufficiently large system, any individual agent can be removed or replace without any effect on the resulting structure. The self-organization process is fully parallel and distributed over all the agents, i.e., it is truly collective. This implies that the organization which is achieved is inherently robust to faults and perturbations.

Chris Lucas Definition He states that: 'Self-organization is the evolution of a system into an organized form in the absence of external constraints. It is a move from a large region of state space to a persistent smaller one, under the control of the system itself [75]. Here, the term 'organized form' is taken in the sense described before (i.e., non-random form).

Scott Camazine Definition According to him 'Self-organization in biological systems is a process in which pattern at the global level of a system emerges solely from numerous interactions among the lower-level components of the system, and the rules that specify interactions, among system components, are executed using local information, without reference to the global pattern' [76]. This definition implies that the pattern is an emergent property of the system and not a property imposed on the system by an external ordering influence.

A. N. Whitehead Definition He states that, 'Self-organization of society depends on commonly diffused symbols evoking commonly diffused ideas, and at the same time indicating commonly understood action' [77]. He argues that the human mind is functioning symbolically when some components of its experience elicit consciousness, beliefs, emotions, and usage, respecting other components of its experience. The former set of components involves the 'symbols', and the latter set constitutes the 'meaning' of the symbols. He remarks that 'symbolism plays a dominant part in the way in which all higher organisms conduct their lives. It is the cause of progress and the cause of error'.

M. B. L Dempster Definition Dempster worked on the distinction between *autopoietic* (self-producing) and *sympoietic* (collectively-producing) systems. These two contrasting lenses offer alternatives views of the world, forcing recognition of system properties frequently neglected. Taking into account Andrew's remark that it is difficult, probably impossible, to find a precise definition of what is understood by a self-organizing system, he is not attempting to give such a precise definition by stating that: 'On an intuitive level, self-organization refers to exactly what is suggested: systems that appear to organize themselves without external direction, manipulation, or control [78]. Self-organization in human society occurs at different levels (*vertical self-organization*) and different activities or processes (*horizontal self-organization*). From top level to bottom level vertical self-organization involves [79]:

- Human-non human environment.
- Society establishment.
- Groups and communities.
- Individuals.

In the horizontal dimension we have:

- Culture
- Ideology
- Politics
- Religion
- Economy
- Industry
- Agriculture
- Education, etc.

All processes are interdependent and influence each other. This implies that, co-evolution occurs within and between vertical and horizontal processes.

Another more concrete way of representing self-organization in society is to split the self-organizing processes involved in four dimensions, namely:

- Social dimension.
- Institutional dimension.
- Economic dimension.
- Environmental dimension.

Self-organization along these dimensions is fully interrelated and interdependent as shown in Figure 7.24:

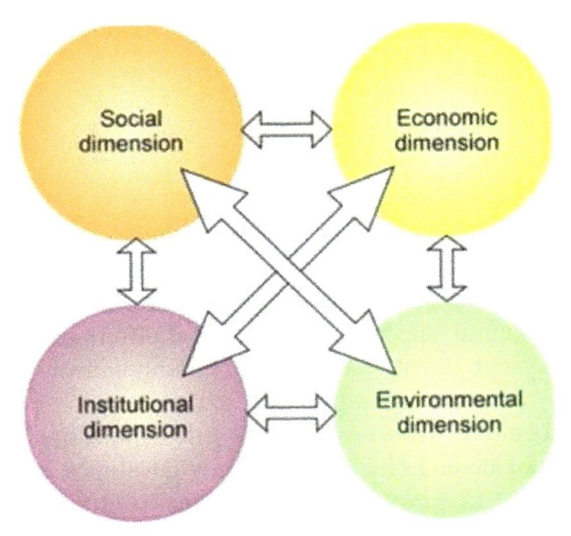

Figure 7.24 The four dimensions of society self-organization.

Source: www.eols.net/CF03-1.jpg

7.10.2 Mechanisms of Self-Organization

The fundamental natural mechanisms by which self-organization is achieved are the following:

- Synergetics.
- Export of entropy.
- Positive/negative feedback interplay.
- Selective retention.

The *synergetics* mechanism (from the Greek $\sigma\upsilon\nu\rho\gamma\varepsilon\iota\alpha$=synergia=act together) has been discovered by the German physicist *Hermann Haken* [84] who studied lasers and other similar phenomena and was surprised by the apparent cooperation (synergy) between the interacting components. The elements (agents, components) of a complex system at the beginning interact only locally (i.e., with their close neighbors), but due to the direct or indirect connection and interaction of the agents the changes are gradually propagating to far-away regions leading finally to an obvious synergy at the system level. Examples of such collective patterns resulting from many interacting components include (besides the lasers), chemical reactions, molecular self-assembly, crystal formations, spontaneous magnetization, etc. This synergy in the laser light production is explained as follows. When atoms or molecules receive an energy input, they emit the surplus energy as 'photons' at random times and directions. This leads to *diffuse light.* But under certain conditions the atoms can be synchronized and emit the photons at the same time in the same direction, with the outcome of a highly coherent and *focused beam of laser light* [79].

The achievement of a *synergetic state* is in general a 'trial-and-error' or 'mutual adaptation' process. System's components (agents, etc.) handle permissible or plausible actions (or sometimes select them randomly), and maintain or repeat those actions that bring them nearer to their goals. This process is actually a natural selection process, but it differs from Darwinian evolution since the system agents are functioning simultaneously until they mutually fit, i.e., they *co-evolve* (mutually adapted) so as to minimize friction and maximize synergy. The *export of entropy* self-organization mechanism was revealed by Prigozine and Nicolis [80]. They developed and promoted the theory of *dissipative structures* (i.e., systems that continuously decrease their entropy). Dissipation (i.e., entropy export) is the mechanism that leads to self-organization. This means that a self-organizing system imports high-quality (usable) energy from the environment and exports entropy back to it. Prigozine formulated a new world view. He sees the world as an irreversible 'Becoming', which produces novelty without an end. This is the opposite to the Newtonian reduction to a static framework, i.e., to the 'Being' view. This point of view is compactly expressed by *Prigozine's* quote: 'The irreversibility of time is the mechanism that brings order out of chaos'. Speaking about chaos *James Gleick* states that 'Where chaos begins, classical science stops'. In other words this means that chaos is our third great revolution in physical sciences after relativity and quantum mechanisms.

Prigozine and Stengers [81] state that 'order creation' at the macro-level is a way of dissipating (exporting) entropy caused by energy flux at the micro-level. For example, a whirlpool is formed spontaneously in a draining bathtub because in this way the potential energy of the standing water is dissipated better than a laminar (smooth) or turbulent (chaotic) flow [82].

A nonlinear system has in general a multiplicity of attractors. Each one of these attractors corresponds to a self-organized configuration. Therefore, the study of self-organization is equivalent to the study of the system attractors' properties and dynamics. If the system starts out in a basin state, it will settle down to the corresponding attractor, but if it starts between different '*basins',* it has the freedom to choose the basin and the attractor to which will end up. This depends on the unpredictable fluctuations that may exist. The self-organized configuration is of course more stable than the configuration from which the system has started. We call this phenomenon *'order from noise'* [83], but thermodynamicists [80] call it *'order through fluctuations'* or *'order out of chaos'*.

The processes and fields that contribute to the achievement of self-organization are the following:

Processes

- Feedback (positive, negative).
- Homeostasis.
- Emergence.
- Internal pattern.

Fields

- Physical sciences.
- Natural sciences.
- Biological sciences.
- Computer science.
- Open systems (far from equilibrium).
- Networks.

The three most important processes taking place in self-organization are: feedback, homeostasis, and emergence. Feedback is illustrated in Chapter 6 (Figures 6.1 and 6.2), and the emergence process in Figure 7.18. Figure 7.25 illustrates the concept of homeostasis of living systems (animals) which involves the *control center* (the brain), the *sensor* (receptor), and the *effector* (a gland). The nervous and endocrine systems cooperate to complete the stages of homeostasis.

Self-organization can be modeled with the aid of:

- Multi-agent systems (MAS)
- Cellular automata
- Other modeling techniques

7.10.3 Examples of self-organization

Here we will give five examples of self-organization.

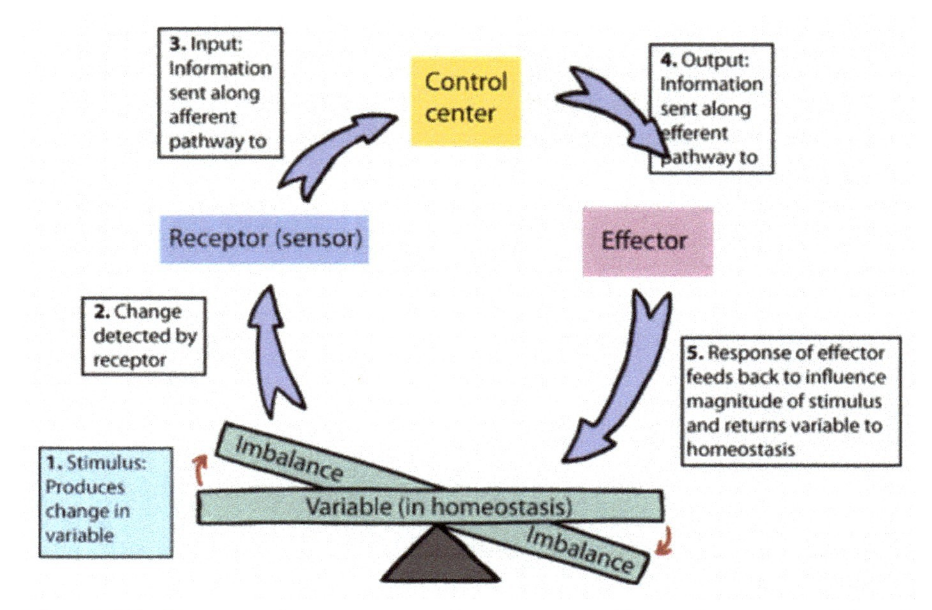

Figure 7.25 The cycle of homeostasis involves negative feedback and is performed through the stages of stimulus, change, input, output, and effectors response. The path that leads towards the brain is called '*afferent path*', whereas the path leading away from the brain towards the body is called '*efferent path*'.

Source: http://www.shmoop.com/animal-movement/homeostasis.html

Example 1: Self-organization of heat convection (heated liquid)

A liquid contained in an open container is heated only from below (via hot plate) [74, 84]. Hot liquid is lighter than cold liquid, and so it tends to move upwards. In the same way the cold liquid tries to sink to the bottom (convective instability). These two opposite movements take place in a self-organized way in the form of parallel 'rolls', with an upward flow of the roll on one side of the roll and downward flow on the other side. Originally, the molecules of the liquid have random movement but finally all 'hot' molecules are moving upward on the one side of the roll and 'cool' molecules are moving downward on the other side (Figure 7.26).

Example 2: Self-organization in a liquid flow

By simulating the rotation of disks that create vortices in liquids it was found that the mechanism of self-organization of moving particles in liquid flow is realized entirely from the flow alone and does not require any form of static interaction (Figure 7.27) [85].

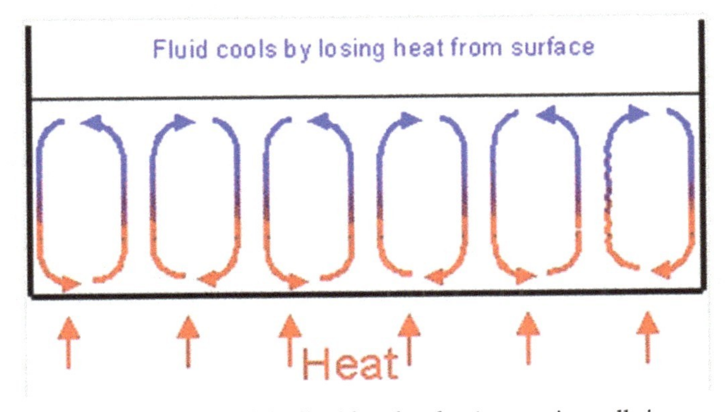

Figure 7.26 Self-organization of the liquid molecules (convection cells in a gravity field).
Source: www.mwthermalenergy.com/convection.html and www.ghaley.com/Images/convection_cells.gif

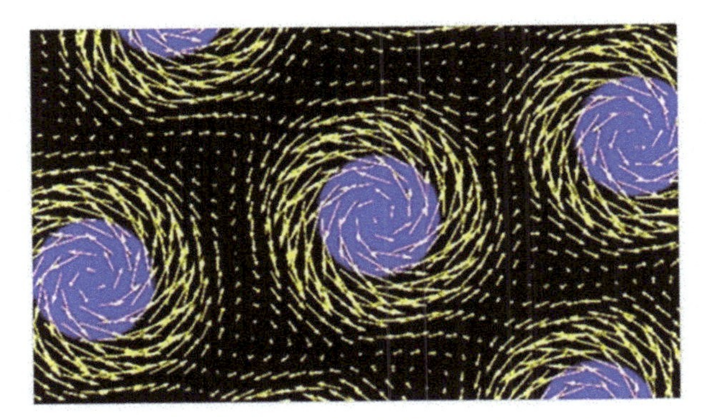

Figure 7.27 Self-organization of liquid flow without static interactions.
Source: https://www.asianscientist.com

Example 3: Self-organization in magnetization

Consider a piece of magnetic material (e.g., iron) which involves a huge number of microscopic magnetic *'dipoles'* (known as *'spins'*) [74]. At high temperatures these dipoles move quite randomly (i.e., they are disordered) with the orientation of their magnetic fields quite random and canceling each other, which results in non-magnetized overall configuration (state) of the material (Figure 7.28(a)). But if the temperature is lowered the spins 'are' spontaneously aligned and point in the same direction (Figure 7.28(b)). The outcome of this alignment is that now the magnetic fields are added-up

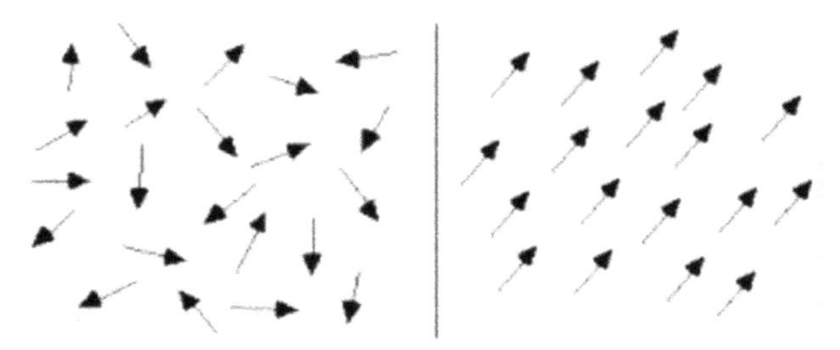

Figure 7.28 Self-organization leading to magnetization: (a) disordered spins and (b) ordered (aligned) spin [74].

giving a strong overall magnetic field. This performance is due to that dipoles pointing in the same direction attract each other (the North pole of a magnetic dipole attracts the South pole of another dipole), while dipoles with opposite direction repel each other. This spontaneous alignment (magnetization) process shows that 'self-organization' is occurring.

Example 4: Self-organization in bird flocking
Bird flocking is a good example of self-organization in animal societies. Each bird gets local information about the position/orientation of neighboring birds and adapts its own position/orientation such as a global moving pattern organization emerges. (Figure 7.29).
Other examples of self-organization of physical, biological, and natural processes include:

- Spontaneous crystallization.
- Laser superconductivity and Bose–Einstein condensation, in the quantum domain (leading to macroscopic manifestations).
- Critical opalescence of fluids at the critical points.
- Self-assembled materials.
- Percolation in random models.
- Structure formation in thermodynamic systems away from equilibrium.
- Ant colony self-organization.
- Fish shoal organization.
- Ecological systems self-organization.

Example 5: Self-organization of miniature robots
A group of miniature robots initially randomly oriented is self-organized getting the same final orientation, with the aid of suitable sensors and actuators. This collective self-organizing behavior of intelligent robots is known

Figure 7.29 Two instances of bird flocking.

Source: www.complexityacademy.io (self-organization overview), https://www.linkedin.com (topic/implementation-engineer).

as 'swarm intelligence' or 'swarm robotics'. Figure 7.30 gives two examples of swarm intelligence.

Some other examples of self-organization in man-made systems are:

- Self-organizing control systems.
- Self-organizing system fault diagnosis.
- Self-organizing economic systems.
- Self-organizing decision systems.
- Self-organizing linguistic systems.
- Self-organizing maps.
- Self-organizing knowledge networks.
- Self-organizing traffic systems.

Figure 7.30 Self-organization of miniature robots (swarm robots).

Source: www.rahalco.co.uk (kilobot, Rahal Technology Ltd.), www.obodarshan.in

7.11 Conclusions

In this chapter, we have presented ontological and epistemological issues of complex and nonlinear systems. We started with a discussion on the fundamental questions 'what is a complex system?' and 'what is complexity?' including a number of quotations of complexity scientists on what is a complex system, and we continued with the issue of complexity measurement. Next, we turned our attention to the nonlinear systems. First, we discussed the stability and control issues listing the available classical and modern methodologies. Then the concept of *bifurcation* was examined: what it is, and how it is generated, including a number of examples of different bifurcation types. The next topics discussed are 'chaos and strange' attractors' (definition, *Lorenz strange attractor)* and 'fractals (ontology, epistemology). The chapter continued with the concepts of 'emergence' and 'complex adaptive systems'. Several opinions on the nature and application of emergence and a discussion of some epistemological issues on emergence were included. The complex adaptive system (CAS) covers high-level instantiations of natural and man-systems that cannot be analyzed by standard analysis methods. The definition, properties and *mechanisms of (CAS)* firstly presented by *John Holland* (the originator of CAS studies) were outlined, and a concise definition of CAS composed by Kevin Dooly, merging most previous definitions, was given.

Regarding system patterns, we recall here that the concepts of *complex* and *complicated* patterns are quite different because the nature of their ambiguities is different. A *complicated* pattern is one that is intricate in the number of parts and their hidden relationships to each other. Such a system seems to be 'folded' such that parts are hidden from view. To understand a complicated system, the parts must be separated and the relationships clearly defined. A complex pattern on the other hand includes the bonding together of parts into an intricate whole. Each part is massively connected with others, and the emergent complex pattern cannot be determined from its components. The whole emerges from the interaction of the parts. In summary, if a system can be understood in terms of its parts it is a *complicated system.* If the whole of a system is different from the sum of its parts, then it is a *complex system.* For this reason, the initial situation of a complex system cannot be accurately determined, and, therefore, its evolution in time cannot be accurately predicted. An extremely small change in initial parameters may result in completely different behavior of a complex system, which is the dominant characteristic of chaos. The chapter ends with a brief discussion of the concepts of *adaptation* (what is adaptation? historical note, adaptation mechanisms) and *self-organization* (what is self-organization? mechanisms of self-organization, including representative examples).

References

[1] Stewart, I. (1990). *Does God Play Dice?* New York, NY: Penguin.

[2] Mandelbrot, B. B. (1977). *The Fractal Geometry of Nature*, 3rd Edn. San Francisco, CA: Freeman.

[3] Bak, P. (1966). *How Nature Works: The Science of Self-Organized Criticality*. New York, NY: Copernicus.

[4] Wolfram, S. (1986). *Theory and Applications of Cellular Automata*. Singapore: World Scientific.

[5] Albert, R., and Barabasi, A. L. (2002). Statistical mechanics of complex networks. *Rev. Mod. Phys.* 74, 47–97.

[6] Newman, M., Barabasi, A., and placeWatts, D. (2006). *The Structure and Dynamics of Networks*. Princeton, NJ: Princeton University Press.

[7] Shalizi, C. R., and Crutchfield, J. P. (2002). Computational mechanics: pattern and prediction structure and simpli. *J. Stat. Phys.* 104, 817–879.

[8] Kaneko, K. (ed.) (1993). *Theory and Applications of Coupled Map Lattices*. New York, NY: John Wiley.

[9] Atmanspacher, H., and Demmel, G. (2016). "Methodological issues in the study of complex systems," in *Reproducibilty Principles, Problems, Practices, and Prospects*, eds H. Atmanspacher and S. Maassen (New York, NY: Wiley), 233–250.

[10] Brian Arthur, W. (1999). Complexity and the economy. *Science* 284, 107–109.

[11] Parrish, J. K., and Edelstein-Kesher, L. (1999). Complexity, pattern, and evolutionary trade-offs in animal aggregation. *Science* 284, 99–101.

[12] Editorial (2009). No man is an island. *Nat. Phys.* 5: 1.

[13] Weng, G., Bhalla, U. S., and Iyengar, R. (1999). Complexity in biological signaling systems. *Science* 284, 92–96.

[14] Bar-Yam, Y. (ed.) (2000). "Unifying themes in complex systems I," in *Proceedings of the 1st International Conference on Complex Systems* (Reading, MA: Perseus Press).

[15] Baranger, M. (2002). *Chaos, Complexity, and Entropy: A Physics Talk for Non-Physicists, Center for Theoretical Physics, Department of Physics, MIT, Cambridge, MA, USA*. Available at: http://necsi.org/projects/bouranger/cce.pdf

[16] Lewin, R. (1992). *Complexity: Life at the Edge of Chaos*. New York, NY: MacMillan.

[17] Buchanan, M. (2000). *Ubiquity: Why Catastrophes Happen?* New York, NY: Three TypeRiver Press.

[18] Goldenfeld, N., and Kadanoff, L. P. (1999). Simple Lessons from Complexity. *Science* 284, 87–89.

[19] Ladyman, J., Lambert, J., and Wiesmer, K. (2013). What is a complex system? *Eur. J. Philos. Sci.* 3, 33–67.

[20] Wadhaman; Lorenz Attractor; Nino El-Hami and Pihlstrom; Matthies et al. *Explained: 5. Defining Different Types of Complexity*. Available at: www.nirmukta.com/2009/09/14/complexity-explained-5-defining-diffe- rent-types-of-complexity
[21] Raven, F. (1961). *Control Engineering*. New York, NY: McGraw-Hill.
[22] Gibson, J. (1963). *Nonlinear Automatic Control*. New York, NY: McGraw-Hill.
[23] Letov, A. M. (1961). *Stability in Nonlinear Control Systems*. Princeton, NJ: Princeton University Press.
[24] Vincent, T., and Gratham, W. (1997). *Nonlinear and Optimal Control Systems*. New York, NY: John Wiley & Sons, Inc.
[25] Kuznetsov, Y. (2004). *Elements of Applied Bifurcation Theory*. Berlin: Springer.
[26] Mandet, P., and Erneux, T. (1984). Laser Lorenz Equations with a Time-Dependent Parameter. *Am. Phys. Soc.* 53, 1818–1820.
[27] Lorenz Attractor, *Wolfram Mathworld*. Available at: http://mathworld.wolfram.com/LorenzAttractor.html
[28] Maxwell, J. C. (1873). *Teaching Nonlinear Phenomena*. London: King's College.
[29] Mandelbrot, B. (1977). *The Fractal Geometry of Nature*. New York, NY: Freeman.
[30] Dobzhansky, T. (1970). *Genetics of the Evolutionary Process*. New York, NY: University of Columbia Press, 1–6, 79–82, 84–87.
[31] Morgan, C. L. (1929). The case of emergent evolution. *J. Philos. Stud.* 4, 431–432.
[32] Morgan, C. L. (1923). *Emergent Evolution*. London: Wlliams and Norgate.
[33] Nino El-Hami, C., and Pihlstrom, S. *Emergence Theories and Pragmatic Realism*. Available at: http://www.helsinki.fi/science/commens/papers/emergentism.pdf
[34] Broad, C. D. (1925). *The Mind and Its to Nature*. London: Routledge and Kegem Paul.
[35] Peper, S. C. (1926). Emergence. *J. Philos.* 241–245.
[36] Meehl, P. E., and Sellars, W. (1956). "The concept of emergence," in *Minnesota Studies in the Philosophy of Science: The Foundations of Science and the Concepts of Psychology and Psychoanalysis*, Vol. 1, eds H. Fregl and M. Soriven (Minneapolis, MN: University of Minnesota Press), 239–252.

[37] Cruchtfield, J. P. (1994). The calculi of emergence: computation, dynamics, and induction. *Phys. D* 75, 11–54.

[38] O'Connor, T., and Wong, H. Y. (2006). *Emergent Properties, Stanford Encyclopedia of Philosophy, October 23, 2006*. Available at: http://plato.stanford.edu/entries/properties-emergent/

[39] O'Connor, T. (2000). Causality, mind and free will. *Philos. Perspect.* 14, 105–117.

[40] Matthies, A., Stephenson, A., and Tasker, N. *The Concept of Emergence in Systems Biology, A Project Report*. http://www.stats.ox.ac.uk/_data/assets/pdf_file/0018/3906/Concept_of_Emergence.pdf

[41] Kim, J. (2006). *Being Realistic About Emergence: The Re-Emergence of Emergence: The Emergentist Hypothesis from Science to Religion* (eds P. Clayton and P. Davies), Oxford: Oxford University Press, 189–202.

[42] Bhalla, U. S., and Lyengar, R. (1999). Emergent properties of networks of biological signaling pathways. *Science* 283, 381–387.

[43] Tabony, J. (2006). Self-organization and other emergent properties in a simple biological system of microtubules. *ComPlexUs* 3, 200–210.

[44] Flecher, R. J. (2006). Emergent properties of conspecific attraction in fragmented landscapes. *Am. Nat.* 168, 207–219.

[45] Eschenbach, C. (2005). Emergent properties modeled with the functional structural tree growth model ALMIS: computer experiments on resource gain and use. *Ecol. Model.* 186, 470–488.

[46] Holland, J. H. (1995). *Hidden Order: How Adaptation Builts Complexity*. Reading, MA: Addison Wesley.

[47] Holland, J. H. (1992). Complex adaptive systems. *Daedalus* 121, 17–30.

[48] Brownlee, J. (2007). *Complex Adaptive Systems. CIS Technical Report 070302A*. Melbourne, VIC: Complex Intelligent Systems Laboratory, 1–6.

[49] Holland, J. H. (1975). *Adaptation in Natural and Artificial Systems: An Introductory Analysis with Applications to Biology, Control and Artificial Intelligence*. Cambridge, MA: MIT Press.

[50] Holland, J. H. (1998). *Emergence: From Chaos to Order*. Redwood, CA: Addison-Wesley.

[51] Waldrop, M. M. (1992). *Complexity: The Emerging Science at the Edge of Order and Chaos*. New York, NY: Simon and Schuster.

[52] Prigogine, I. and Stengers, I. (1984). *Order Out of Chaos*. New York, NY: Bantam Books.

[53] Jantsch, E. (1980). *The Self-Organizing Universe*. Oxford: Pergamon Press.

[54] Maturana, H. and Varela, F. (1992). *The Tree of Knowledge*. Boston, MA: Shambhala.

[55] Dooley, K. (1996). A nominal definition of complex adaptive systems. *Chaos Netw.* 8, 2–3.

[56] Clayton, P. (2000). "Conceptual foundations of emergence theory," in *The-Reemergence of Emergence: The Emergentist Hypothesis from Science to Religion,* eds P. Clayton and P. Davies (Oxford: Oxford University Press).

[57] Huxley, J. (1942). *Evolution the Modern Synthesis*. London: Allen and Unwin.

[58] Darwin, C. (1872). *The Origin of Species*. London: John Murray.

[59] Wikipedia (2017). *Reaction to Darwin's Theory*. Available at: http://en.wikipedia.org/wiki/Reaction_to_Darwin's_theory

[60] Wikipedia (2017). *Adaptation*. Available at: http://en.wikipedia.org/wiki/Adaptation

[61] Swenson, R. (1997). "Thermodynamics, evolution and behavior," in *The Encyclopedia of Comparative Psychology,* eds G. Greenberg and M. Haraway (New York, NY: Garland Publishers).

[62] Hubert, C. (2017). *Adaptation*. http://christianhubert.com/writings/adaptation.html#16

[63] Ehrlich, P. R., and Raven, P. H. (1964). Butterflies and plants: a study in coevolution. *Evolution* 18, 586–608.

[64] Corning, P. A. (2000). Biological adaptation in human societies: a basic needs approach. *J. Bioecon.* 2, 41–86.

[65] Maynard Smith, J. (1975). *The Theory of Evolution*. New York, NY: Penguin.

[66] Hardesty, D. L. (1977). *Ecological Athropology*. New York, NY: J. Wiley.

[67] Lewontin, R. C. (1978). Adaptation. *Sci. Am.* 239, 213–230.

[68] Lewontin, R. C. (1984). "Adaptation," in *Conceptual Issues in Evolutionary Biology*, ed. E. Sober (Cambridge, MA: Harvard University Press).

[69] Koshland, D. E. Jr. (2002). The seven pillars of life. *Science* 295, 2215–2216.

[70] Brandon, R. (2017). *Adaptation and Representation: The Theory of Biological Adaptation and Function, Interdisciplines*. Available at: http://www.interdisciplines.org/adaptation/papers/10

[71] Brandon, R. (1990). *Adaptation and Environment*. Princeton, NJ: Princeton University Press.

[72] Brandon, R. (1996). The principle of drift: biology's first law. *J. Philos.* 7, 319–335.
[73] Organization (2017). *Longman Dictionary of Cotemporary English.* http://www.Idoceonline.com/Organizations-Topic/organization
[74] Heylighen, F. (2008). "Complexity and self-organization," in *Encyclopedia of Library and Information Sciences,* eds M. J. Bates and M. N. Mack (London: Taylor and Francis).
[75] Lukas, C. (1997). *Self-Organization FAQ.* Available at: http://psoup.math.wisc.edu/archive/sosfaq.html
[76] Camazine, S., Deneubourg, J. L. N., Franks, R., Sneyd, J., Theraulaz, G., and Bonablau, E. (2001). *Selforganization in Biological Systems.* Princeton, NJ: Princeton University Press.
[77] Whitehead, A. N. (1927). *Symbolism: Its Meaning and Effect.* London: MacMillan.
[78] Dempster, M. B. L. (1998). *A Self-Organizing Systems Perspective on Planning for Sustainability.* Master thesis, University of Waterloo, Waterloo, ON.
[79] Zeiger, H. J., and Kelley, P. L. (1991). "Lasers," in *The Encyclopedia of Physics,* eds R. Lerner and G. Trigg (Chichester: VCH Publishers), 614–619.
[80] Prigozine, I., and Nicolis, G. (1997). *Self-Organization in Non-Equilibrium Systems.* New York, NY: Wiley.
[81] Prigozine, I., and Stengers, I. (1984). *Order out of Chaos.* New York, NY: Bantam Books.
[82] Deker, E. H. (2001). *Self-Organizing Systems: A Tutorial in Complexity.* Albuquerque, NM: University of New Mexico.
[83] Von Foerster, H. (1960). "On self-organizing systems and their environments," in *Self-Organizing Systems,* eds M. C. Yovits and S. Cameron (London: Pergamon Press), 31–50.
[84] Haken, H. (2000). *Information and Self-organization: A Macroscopic Approach to Complex Systems.* Berlin: Springer.
[85] Goto, Y. and Tanaka, H. (2015). Purely hydrodynamic ordering of rotating discs at a finite Reynolds number. *Nat. Commun.* 2015:5994.
[86] Geyer, F. (1994). "The challenge of sociocybernetics," in *Proceedings of the 13th World Congress of Sociology (Symposium VI Challenges to Sociological Knowledge: Session 04),* Bielefeld, 18–24.

8

Automation

Design should begin by identifying a human or societal need – a problem worth of solving – and then fulfill that need by tailoring technology to the specific, relevant human factors.
Kimon Vicente

With automation, jobs are physically easier, but the worker takes home worries instead of an aching back.
Homer Bigart

Machines will be capable of doing any work Man can do.
Herbert Simon

8.1 Introduction

The term '*automation*' is somehow misunderstood in that it is mostly taken as referring to 'automatic configuration', where software scripts and programs replace operator functions and actions to configure a system. But, automation is a broader term referring to the 'automatization' of many functions in the operation of man-made systems such nuclear reactor plants, commercial and military airplanes, process and electric power generation plants, sea transportation systems, office automation, etc. Automation systems have to be able to adapt to fast internal and external changes something that can be accomplished by a variety of successful models, control, and supervision techniques. An issue of primary concern and importance in modern automation systems is the study and understanding of the physical, mental, and psychological features of the human at work. This has produced the field of '*human-centered (human-minding) automation*' which is studying the human–machine interfaces and human factors in automation and is trying to design automation systems that assure 'high degrees of *job satisfaction*', and 'minimize the *human job stress*' during his work. This chapter provides an outline of the field of automation, at a level of detail compatible with the purpose of this book. In particular the chapter:

- Presents several existing answers to the ontological question 'what is automation?
- Provides a review of a few selected representative works drawn from the literature.
- Discusses epistemologically two classes of automation (industrial automation, and office automation), which differ in their nature and operational requirements.
- Outlines four classes of automation systems, namely: robotic automation, aircraft automation, air traffic control, and automated driving.
- Discusses fundamental issues of human–machine interfaces (HMIs) used in human-automation, communication, and interaction.
- Presents the definitions of virtual, augmented, and mixed reality, along with VR devices and VR application examples with emphasis on medical systems.
- Investigates three key human factors (function allocation, stimulus response compatibility, and internal model of the operator) that must be considered in the design and implementation of flight decks and other systems, with a view to achieve more human-friendly operational characteristics.

8.2 What is Automation?

Modern systems are large and complex and so for their analysis and design one needs to study not only the characteristics of their subsystems, but also the interactions of them (as a 'Holon'). The term *automation* (automatic organization) was coined in 1952 by *D.S. Ha*rder of Ford Company to involve the methodology which analyzes, organizes, and controls the production means such that all material, machine and human resources are used in the best way [1, 2]. Actually, many different definitions of automation have been proposed. Some of them are the following:

- *Dunlop* [3]: Automation is defined as a mechanical or chemical process directed, controlled, and corrected within the limits such that no human intervention is needed once a system is established.
- *Thomas* [4]: Level of automation is defined as the extent to which decision making functions associated with control of a man/machine system are performed by machines.

- *Dieterly* [5]: Automation is the operation of a system or production process by mechanical or electronic devices that takes the worker's place in terms of effort, observation, and decision-making.

The last definition is an integration of various other definitions of automation. As pointed out by many authors, this definition does not differentiate between mechanization and automation. These authors point out the fact that mechanization includes only the machine performing tasks which do not require decision making, whereas the concept of automation includes the machine performing tasks that need decision-making.

The principal goal of automation is the *optimal allocation of human effort* (muscular, mental) so as to maximize the *productivity*, i.e., the ratio of the product obtained over the human effort needed to do this. Today, the term automation is used in all situations where the system operation is automated to various degrees. The operation of the system is usually performed in a sequential or parallel multi-stage manner.

Computerized systems, with terminals, displays, sensors, and other human-computer interfaces are now considered as part of automation (even if they do only processing, and not control or supervision). The system in which the operation is to be automated can be any man-made systems (physical, chemical, electrical/electronic, managerial, etc.). An overall self-explained pictorial representation of an automation system which shows how the above concepts and ideas can be integrated is given in Figure 8.1 [6].

All automation systems contain, besides the information processing, and control elements that implement the required feedback loops, appropriate communication links (channels)—analog and/or digital—through which the various parts of them are communicating and exchanging the proper signals and messages.

We note here that Figure 8.1 contains two additional boxes, namely:

- *Values, goals and specifications*: This box refers to the human values and goals that must be respected by the automation system, and to the technical specifications of the operation of the system which assure the desired quality of the product or service delivered to human user/customer.
- *Output to nature:* This box refers to the effects of the automation system to the nature (environment, ecosystem) in which the humans live.

Reviews of automation literature are presented in [6, 7]. A few selected references from them are:

Mueller [8]: He provides a cross sectional survey on changes brought about by automation, including attitudes, and offers a good method for defining

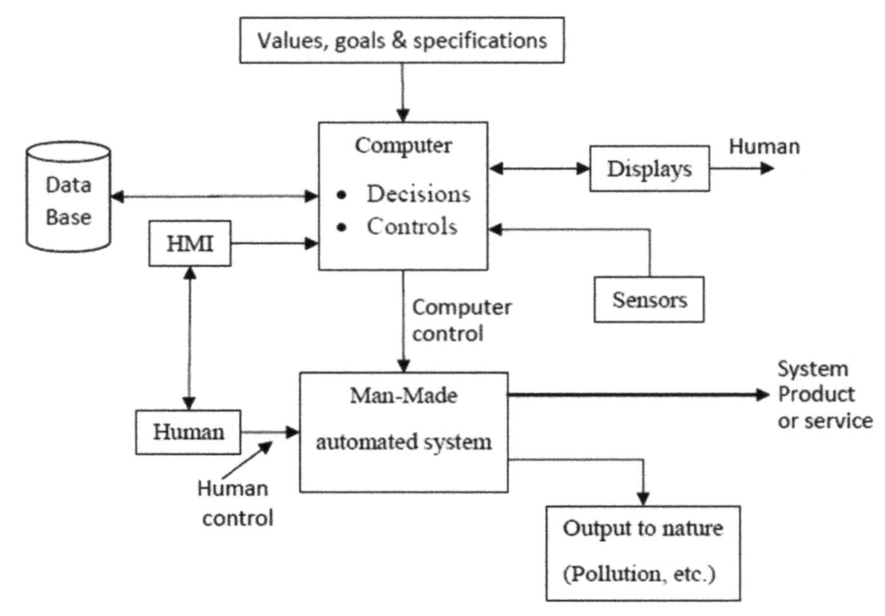

Figure 8.1 Block diagram of a generic automation system.

and measuring different levels of automation (although these levels were not quantified).

Ratner and Williams [9]: These authors compared four levels of automation for the air traffic controller's contribution to ATC system capacity.

Noll, Evans, and Simpson [10]: These authors compared the impact on controllers of two ATC systems that operated on two different levels of automation by different types of ATC systems (a typical Air Traffic Control system and an advanced system of the future).

Sheridan [11]: He has offered many contributions on automation with emphasis on the role of supervisory control (architectures, human–machine interfaces, interacting computers in automation, coupling of human with automation, etc.). Especially in **Sheridan** [11], the author reviews the development and evolution of automation and computerized methodologies and systems over the years, emphasizing on the human-automation interaction, and the humanization of automation.

Tzafestas [6]: This book provides an overview of the fundamental principles, problems and issues in the field of 'human and nature minding automation and industrial systems'. Regarding the human–automation symbiosis (living together), the book studies the concepts and techniques of the human factors and economics, supplemented by the study of human–computer (or human–machine) interfaces, standard, intelligent, multi-modal, etc. For automation-nature sustainable symbiosis more complex and difficult tools and decisions are required. The partners here are not only the machines and the scientists or engineers, but also the politicians, governors, and decision makers, world-wide. It is argued that techniques and equipment for nature minding (clean or green) automation industry should be designed so as to protect the environment and ecosystem including pollution, waste generation, biodiversity, decrease, and consumption of the earth resources.

8.3 Industrial versus Office Automation

Two large classes of automation are:

- Industrial automation.
- Office automation.

8.3.1 Industrial Automation

Industrial automation is one of the areas of engineering which over the years has undergone continuous, strong and fast developments since its inception [12]. Although great achievements have been realized in the last decades in this field, the technology is still under improvement. According to the control variables involved, industrial automation is distinguished in two categories, namely [13]: (i) *Process Automation* and (ii) *Power System Automation.* The difference of nature of control variables between process and power systems have led to difference in the methodology and philosophy of control, fault detection, general monitoring and system protection. A *process* is defined as an operation sequence which manipulates physical quantities, i.e., transforms inputs into outputs, monitoring and controlling them such as to lead to the desired end product. All process control systems involve three main elements, namely: (i) the inputs (manipulated inputs, disturbances), (ii) the outputs (measured, non-measured), and (iii) the controlled variables (Figure 8.2). Typical manipulated inputs include motor speed, damper position, or blade pitch.

The controlled variables include temperature, pressure, pH, level position, moisture content, etc. Disturbances enter or affect the process and tend to drive the controlled variables away from their desired value or set-point

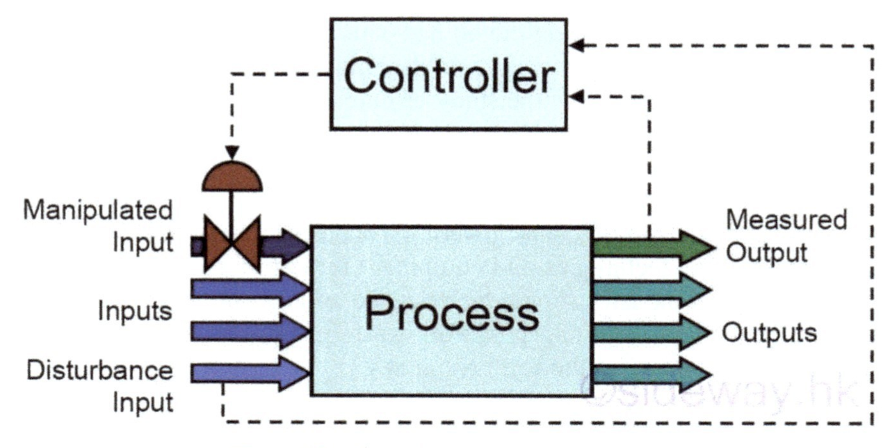

Figure 8.2 General process control diagram.

Source: http://z-diagram.com/block-diagram/process-control-block-diagram-in-operating-system.html (Process control block diagram).

position. The controller's purpose is to reject or eliminate the disturbances and thus maintain the controlled variables to the desired values regardless of the disturbances. The controller inputs are the measured output(s) and the disturbance input(s).

Process automation technology is the means that enables manufacturers to keep their operations running automatically and unperturbed within the set limits of the process, also assuring safety and quality. The most popular type of control for process variables is the **PID** *control* typically designed empirically using variants of the *Ziegler–Nichols* method as discussed in Section 6.4.4.

Today, **PLC**s (Figure 8.3) are used combined with **SCADA** facilities (Figure 8.4(a)). The use of SCADA technology is not restricted to industrial automation systems but finds many other applications as shown in Figure 8.4(c).

More advanced controllers used in process control include: **DCSs**, ***STCs***, and **ACs**, which are implemented by mini and microcomputers [14].

Control loops in the process industry need three tasks namely: *measurement, evaluation*, and *adjustment.* Instruments and devices used in control loops include sensors, transmitters, and actuators. The *measurement* is carried-out by the sensors and transmitters, and the *adjustment* is performed by the actuators. The *evaluation* is performed by controllers such as those mentioned above.

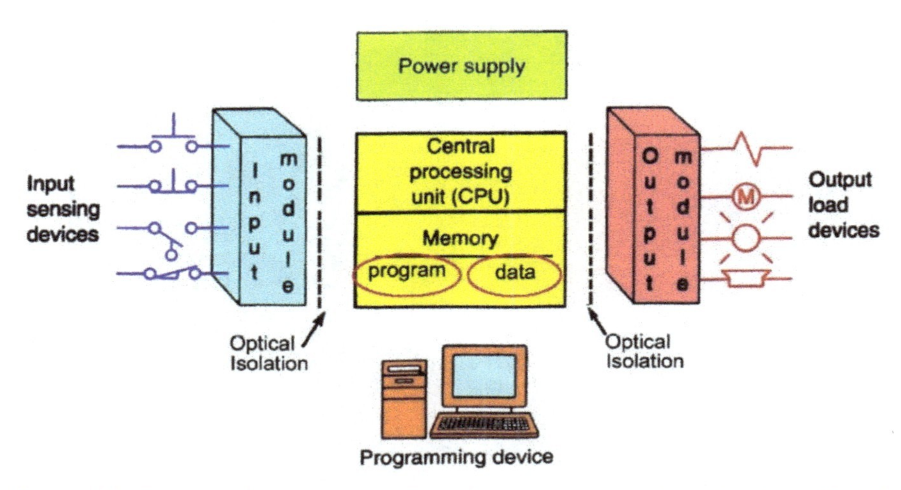

Figure 8.3 Programmable logic controller (APLC contains the following: Input Module, CPU, Memory, Output module, Programming Unit).

Source: https://www.elprocus.com/understanding-a-programming-logic-controller

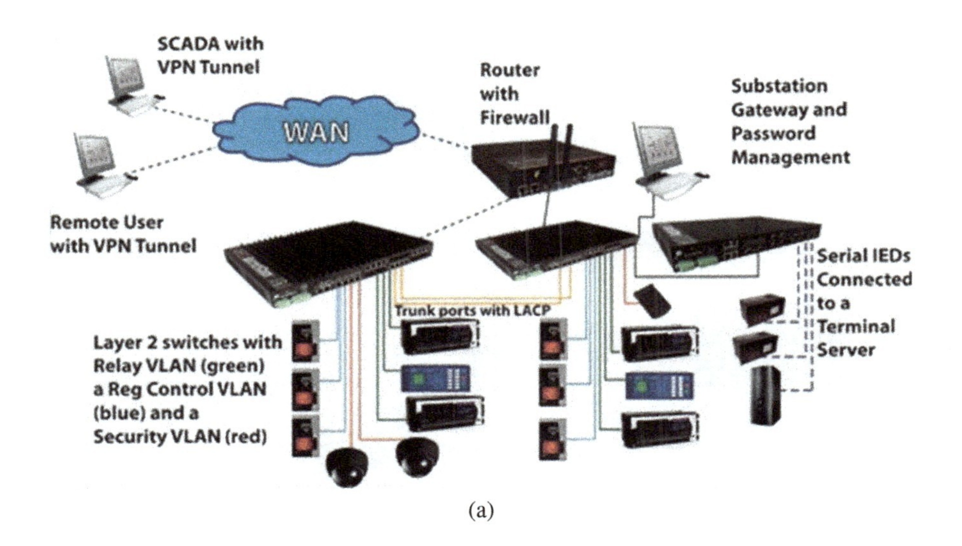

(a)

(b)

SCADA

Industrial site

Storage site

Domestic site

Reservoir site

Farming site

Pumping station site

(c)

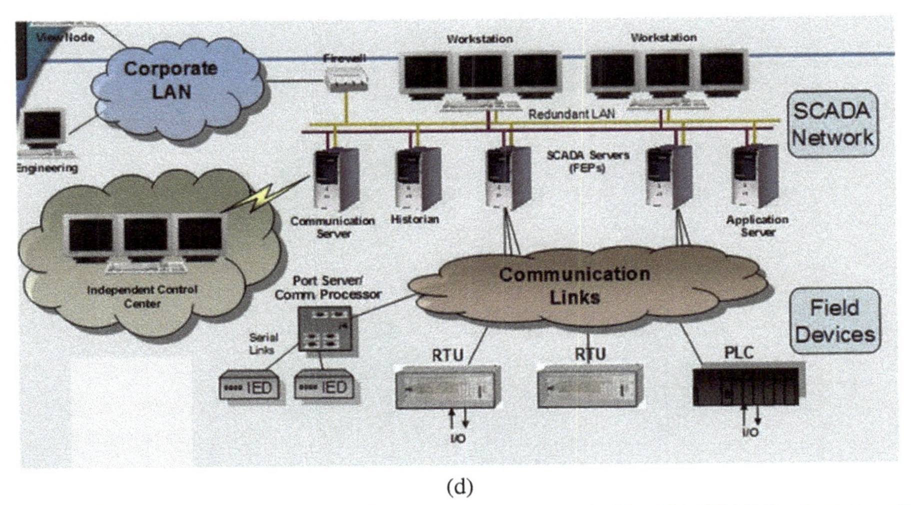

(d)

Figure 8.4 (a) Supervisory Control and data acquisition facility, (b) SCADA screen panel room, (c) SCADA applications (cloud computing), and (d) SCADA networked control architecture example.

Source: www.electronicshub.org/scada-system; http://www.iethanoi/2015/06/12/scada-va-dcs

Besides the process control function, process/factory automation performs the following functions (Figure 8.5(a)):

- Power/process monitoring.
- Energy management and saving.
- Preventive maintenance.
- Fault diagnosis.
- Entry and exit management.

Figure 8.5(b) shows a typical overall process control and management system that involves three hierarchical levels of control/management, namely: (i) *Operating level* which consists of several operator stations connected via *Ethernet*, (ii) *Local control cabinets* which contain the control elements (PLC, etc.) and form an industrial network, and (iii) *Field equipment level* (see Section 6.6).

Power system automation refers as a whole to the *generation*, *transmission*, and *distribution* of electrical energy, and performs automatic control and monitoring of the electrical parameters such as power flow voltage, current, etc. The four basic elements of power systems are:

(a)

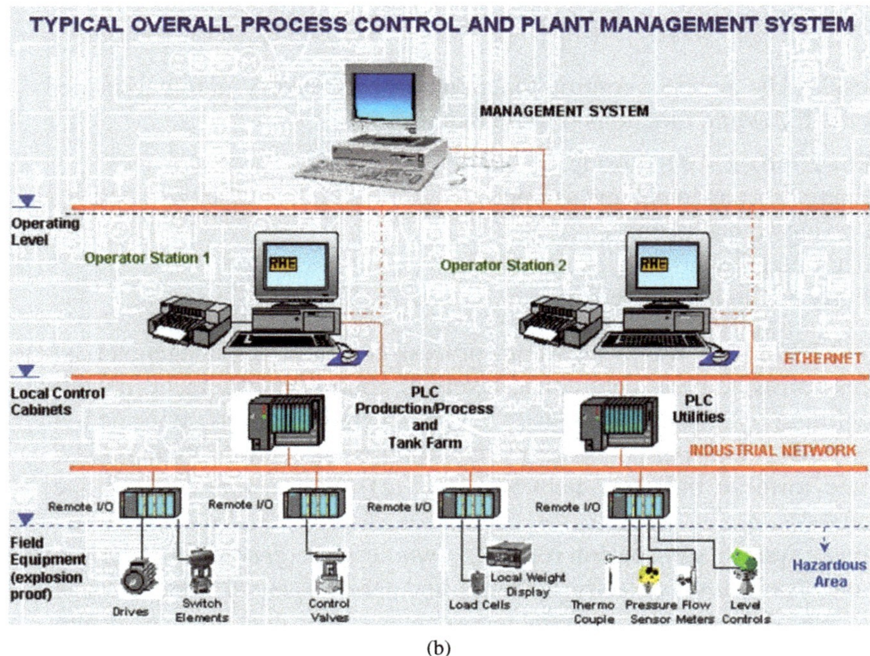

(b)

Figure 8.5 (a) Factory automation functions, (b) Overall multi-level structure of process control.

Source: http://www.rhe-america.com/control.html

- *Generating units* (which produce electricity).
- *High voltage transmission lines* (which transport electricity over long distances).
- *Distribution lines* (which connect the generation, transmission, and distribution with each other).
- *Energy control centers* (which coordinate the operation of the components).

The electrical network automation system involves the following subsystems:

- ***Power generation control/Power management system (PMS)*** The main functions of this subsystem are the active and reactive power control, and the optimization of the power generation (with minimal fuel cost).
- ***Electrical substation automation* system** Substations are the key of the power grid that facilitate the efficient transmission and distribution of electricity and help in the monitoring and control of power flows.
- **Communication networks system (CNS)** This subsystem allows the interconnection and coordination of the generating facilities, transmission and distribution networks, and end customers.

A typical architecture of electric power utility communication network is shown in Figure 8.6, and an example of power plant control room is shown in Figure 8.7.

A process automation system example is shown in Figure 8.8, and a power plant example is shown in Figure 8.9.

8.3.2 Office Automation Systems

Office automation systems are information systems that are able to handle, process, transfer, and distribute all the data/information involved in the operation of an enterprise or organization [15]. In handling of information, the following functions are included: *acquisition, registration, storage*, and *search.* The type of information treated should be recognizable by the human, and so an office automation system must be capable to deal with oral, written, numerical, graphic, and pictorial information. Other names of office automation systems are: 'office information systems, 'bureautique', etc.

Modern office automation involves the following components:

- Computer systems (hardware/software).
- Communication network(s).
- Procedures.
- Human decisions and actions.

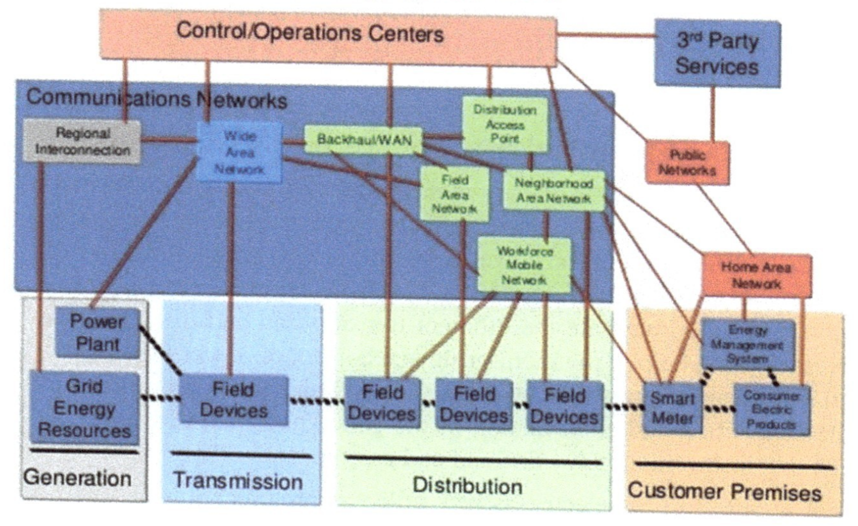

Figure 8.6 Electric utility communications architecture.

Source: www.slideshare.net (SandeepSharma65/smart-meter-3).

Figure 8.7 Power grid control room.

Source: https://linkedin.com/pulse/south-australia-powering-up-grid-takes-time-justin-wearne (Justin Wearne, Sept. 30, 2016).

Figure 8.8 Process industrial plant.

Source: https://www.gore.com/products/categories/filtration

Figure 8.9 A power plant example.

Source: https://www.rfa.org/commentaries/energy_watch/china-coal-fired-power-glut-grows-08012016105217.html

It uses in an integrated and synergetic way the following technologies:

- Data processing.
- Word and text processing.
- Image processing.
- Voice processing.
- Communications processing.
- Office technology.

Office automation is one of the newest achievements of information and computer science and has started being applied immediately after the appearance of microcomputers and word processors. The purpose of the office automation systems is to make the office more 'humanized', where the machines perform all the routine-like and non-mental tasks leaving for the humans the tasks that need thinking and responsibility. Figure 8.10(a) shows a typical set-up of an office automation system, and Figure 8.10(b) shows examples of office automation systems equipment.

According to Zisman (MIT) [16], the evolution of office automation from the mid-seventies and end of the seventies up to now is as shown in Figure 8.11.

The goals of each stage in Figure 8.11 are shown in Table 8.1.

Decision-Support Systems are very important components of office automation. A DSS is a computer-based interactive system that helps decision makers utilize data and models to solve unstructured problems (Scott Morton). Typically a DSS focuses on the principal (key) decisions made by a manager or professional. Its components may vary, and very often they include a regression-based forecasting tool for evaluating alternative decisions, and direct access to organizational and outside data bases for input to the analysis. Decision-support systems emphasize computer support for a *few key managers* in an organization and help them to make decisions as *individuals*. Decisions, most scholars now agree, are the outcome of a complex social and political process involving many people and a variety of special interests [1]. A DSS has the ability to try different strategies under different configurations, providing fast, consistent, and objective decisions. For group decision making settings there are available *Group DSSs* which, in addition to standard DSS features, contain special software modules that effectively support collective decision making by groups of experts. According to the methodology used, DSS are distinguished in:

- Statistical DSS.
- Optimization DSS.
- Forecasting DSS.

(a)

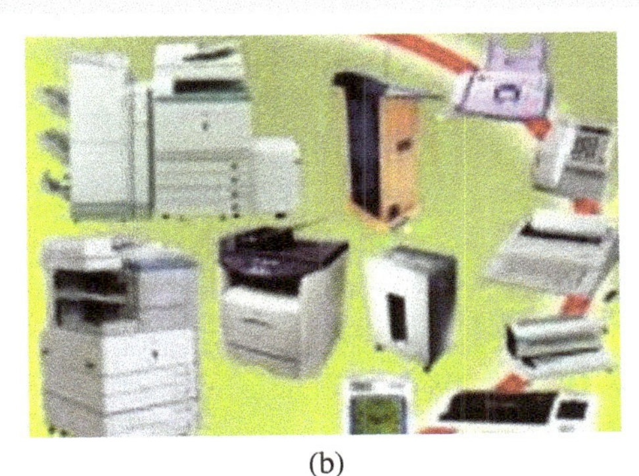

(b)

Figure 8.10 The synergy of automation technologies/equipment in the modern office.

Source: https://trade.indiamart.com/search.mp?search=office-automation+equipment

Analytical modeling of DSS can be performed by:

- What-if analysis.
- Sensitivity analysis.
- Goal seeking analysis.
- Optimization analysis.

Figure 8.12 shows the architecture of a generic DSS application, namely **TPS** which includes payroll, sales, ordering, inventory, purchasing, receiving, shipping, etc.). The central unit of the architecture is the DSS software program which is built following one of several available DSS models. The DSS program receives and processes external data and data from a **MIS**. The output of the system is given in the form of reports and graphs. This architecture is applicable to all decision-making applications (managerial, economic, office automation, manufacturing, medical, environmental, etc.).

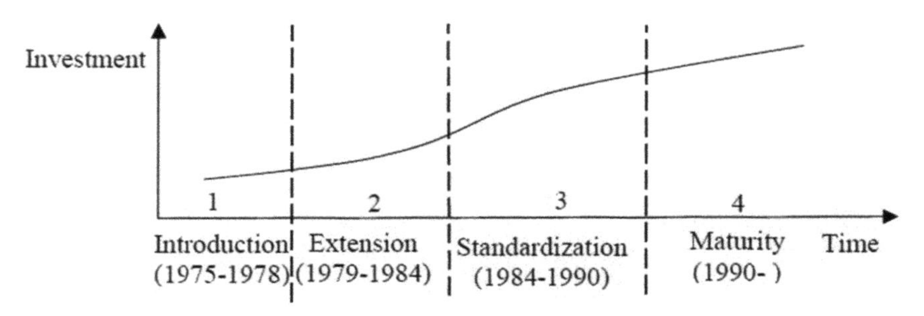

Figure 8.11 The evolution stages of office automation.

Table 8.1 Description of evolution stages of office automation

Stage 1: Introduction	Stage 2: Extension	Stage 3: Standardization	Stage 4: Maturity
Technological progress	Replacement of paper	Integration	Decision support
Productivity/cost improvement	Automated means (mail, telephone, diary)	Automated processes	Operational stability
Text processing	Organization/system approach	Compatibility	Integration
Use of paper	Experimental future offices	Active systems with memory	Technology assimilation
		Abolition of routine tasks	Work methods modifications

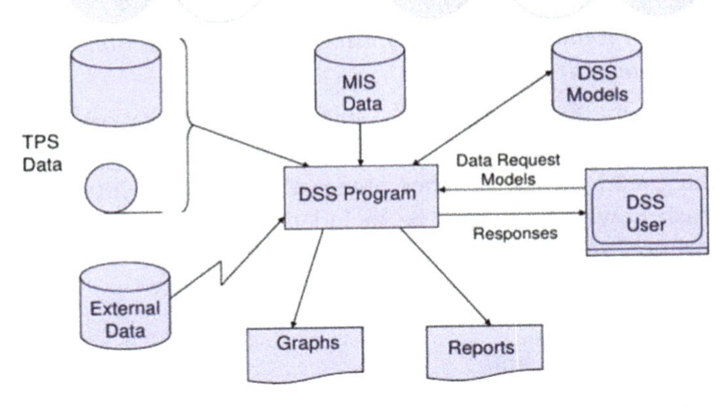

Figure 8.12 General architecture of DSS for TPS and other applications.

Source: www.slideshare.net (PP Presentation: Kinds of Information System – Module 2).

Types of TPS include: *Online TPS systems* (direct connection between the operator and the TPS), *Batch TPS systems* (grouping transactions in a file), and *Real-time TPS systems*. Transaction processing systems process business exchanges, maintain records about them, and collect, store, modify and retrieve them.

The operations of an office are related to the following:

- The type of information (texts, pictures, graphs, etc.).
- The modification of the above types of information.
- The reproduction and distribution of the information to people needing it.
- The transmission of the information to remote agent via computer networks (LANs, WANs, and Internet).
- The retrieval of documents.

Today concepts such as e-banking, e-commerce, e-conferencing, e-advertising, e-training, etc. are used in the processes of automation office.

The primary advantages of office automation include:

- *Improved productivity* (many tasks get accomplished faster).
- *Keep tab on inventor* (thus leading to *fast progress).*
- *Optimum usage of power* (big savings on electricity bills every month).
- *Big savings* (on overall operational costs).

- *Monitoring and control* (of complete space-lights, HVAC, security, inventory from a smart user interface, etc.).
- *Keep check on disguised physical threats and cyber-attacks.*
- *Reduce carbon footprint* and so contribute towards *creation of a green (pollution protected) environment.*

Most enterprises and office automation systems use a proper workflow management system (software engine) to scale up their productivity. Workflows work in an integrated way with other processes such as documents, sales reports, and cash flows which must be included as a part of the work flow itself. Any workflow management model predefines operational aspects like the following:

- A set of business rules.
- Authorization requirements.
- Roles and tasks for stakeholders.
- Routine options.
- Task sequences.
- Data bases used.

Kissflow Company proposes that the ten top features every workflow should have are those shown in Figures 8.13(a,b) show a representative organization architecture of office workflow automation.

A brief description of the components of the workflow system shown in Figure 8.13(b) is as follows:

- **Workflow designer**: An environment for the enterprise analyst to design and model unique business workflows.
- **Workflow engine**: Modules in this component interface with human operators and external systems for executing and monitoring the execution of the defined workflows.
- **User administration**: Modules in this component manage the users by name, department, roles, and hierarchy and authority level.
- **Reporting**: Modules in this component help the users to produce analysis reports based on the automated workflows.
- **System integration**: Integration with the 'human resource management' and 'accounting system' using interfaces that connect the company with external enterprise systems.
- **Workflow portal**: This component enables employees and business partners to execute the business processes set-up in workflow designer.

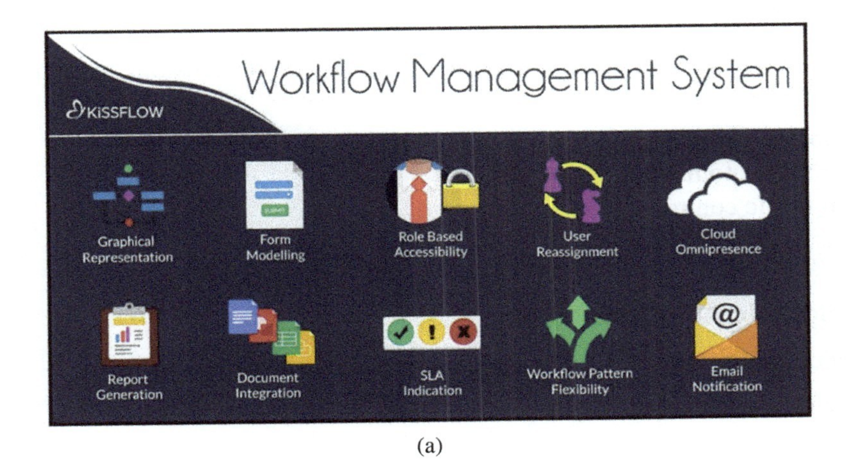

(a)

(b)

Figure 8.13 (a) Ten top features of workflow management systems and (b) Typical office work flow automation architecture (KBQuest).

Source: (a) https://kissflow.com and (b) www.kbquest.com/index.php?page=content/solutions/products/details&pid=29

8.4 Further Examples of Industrial Automation

Here we will outline four further representative classes of industrial automation systems, namely [19, 20]:

- Robotic automation.
- Aircraft automation.
- Air traffic control.
- Car driving automation.

8.4.1 Robotic Automation

The major applications of robotic automation include the following:

- Welding and assembly.
- Material handling Machine loading and unloading.
- Medical and assistive robotics.

Welding and Assembly *Welding* is distinguished in *spot welding* and *arc welding*. Spot welding automotive bodies represents the largest single application. Spot welding is typically performed by point-to-point servo-robots holding a welding gun. Arc welding is performed by robots using non-contact seam trackers. Currently, robotic arc welders are low-cost easily programmable, and durable. Figure 8.14 shows a welding robot cell at work. The design of a *robotic assembly system* needs a combination of economic and technical considerations. In our days, the complexity of products, the requirement to manufacture the products in many models that change design rapidly, and the need the manufacturers to be more responsive to changing demand and just-in-time manufacturing, enforce to design flexible assembly systems. The robotic assembly must compete against manual assembly, rigid automation, or some combination of them. Robotic assembly provides an alternative with some of the flexibility of humans and the uniform performance of fixed automation. Robotic assembly includes two phases: the *planning phase* where a review is made of the design of the product at a variety of levels, and the *assembly phase* where the product first comes to life and can be tested for proper functioning. Assembly is also the phase where the production directly interfaces with customer orders and warranty repairs. Thus assembly is not just putting parts together, but includes the task of determining how to meet a variety of business requirements, ranging from quality to logistics. A programmable robotic assembly system typically consists of one or more robot workstations and their associated grippers and parts presentation equipment (Figure 8.15).

Figure 8.14 A robotic welding cell.

Source: https://www.thekharkivtimes.com/2015/12/29/an-industrial-robot-appeared-at-the-kharkiv-plant/

Figure 8.15 Robotic car assembly shop-floor.

Source: https://thinking.com/scene/489433944765235201

Material Handling and Machine Loading and Unloading The robots used in purely material handling operations are usually *'pick-and-place'* robots. These applications make use of the basic capability of robots to transport objects (the robot's manipulative skills are of less importance here). The main benefits of using robots for material handling are reduction of direct labor costs and removal of humans from tasks that may be hazardous, tedious, or fatiguing. Also, robots typically result in less damage to parts during handling, a major reason for using robots for moving fragile objects. Robots are used in many machine loading and unloading applications, such as die casting, etc. Loading and unloading is actually a more sophisticated robot application than simple material handling. Such applications include grasping a work piece from a supply point, transporting it to a machine, orienting it correctly, and then inserting it into the work holder on the machine. After processing, the robot unloads the work piece and transfers it to another machine or conveyor. Figure 8.16 shows a typically machine loading and unloading, and loading and loading.

Medical and Assistive Robotic Services The potential use (and market) for robotic automation in service is expected to be much larger than that of manufacturing, but service robots should have more capabilities than industrial robots, such as intelligence, user-friendliness, higher manipulability and dexterity. Advanced sensing capabilities (visual, tactile, sonar, and speech),

Figure 8.16 A machine loading and unloading robot.

Source: www.roboticsbible.com/machine-loading-and-unloading-by-industrial-robots.html

and so on. Robots are used for hospital material transport, security and surveillance, floor cleaning, inspection in the nuclear field, explosives handling, pharmacy automation systems, integrated surgical systems, and entertainment. In the following, we discuss a little more medical robotics and computer-integrated surgery. Medical robots are programmable manipulation systems used in the execution of intervention medical procedures, mainly surgery. For centuries surgery (now called classical surgery) has been practiced in essentially the same way. The surgeon formulates a general diagnosis and surgical plan, makes an incision to get access to the target anatomy, performs the procedure using hand tools with visual or tactile feedback, and closes the opening. Modern anesthesia, sterility methods, and antibiotics have made classical surgery extremely successful. However, the human has several limitations that brought this classical approach to a point of diminishing returns. These limitations can be overcome by *automated surgery* via computer-integrated robotic surgical systems (Figure 8.17(a,b)). These systems exploit a variety of modern automation technologies such as robots, smart sensors, and human-machine interfaces, to connect the 'virtual reality' of computer models of the patient to the 'actual reality' of the operating room. The *da Vinci* surgical system (Figure 8.17(a)) includes:

- A surgeon's console.
- A patient-side robotic cart with four arms controlled by the surgeon (one to control the camera and three to manipulate instruments).
- A high definition 3D vision system.

Articulating surgical instruments are placed on the robotic arms which are introduced into the body through cannulas. The robotic system senses the surgeon's hand movements and translates them electronically into scale-down micro-movements to manipulate tiny proprietary instruments. The camera of the system provides a true stereoscopic picture transmitted to a surgeon's console. The da Vinci system also detects and filters any tremors in the surgeon's hand movements, so that are not duplicated electronically. Details on computer-aided surgery are provided in Taylor [21].

Assistive Robotics: **AR** is a branch of **AT**, which develops adaptive and intelligent systems capable to serve **PwSN** in several environments (home, professional, etc.). These systems have been classified according to several criteria. The dominant categorization has been produced by the **WHO**, and internationally standardized (ISO999, see *Wikipedia*). Assistive robotics encompasses all robotic systems that are developed for PwSN and attempt to enable disabled people to reach and maintain their best physical and/or social

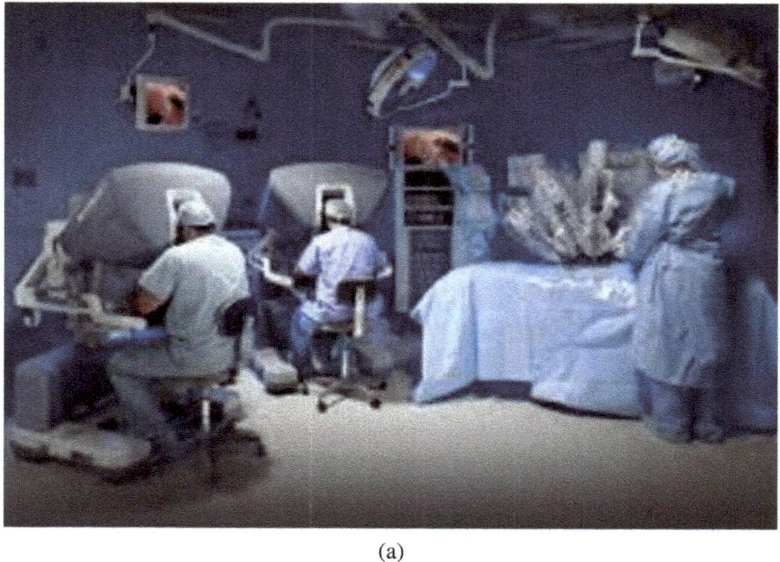

(a)

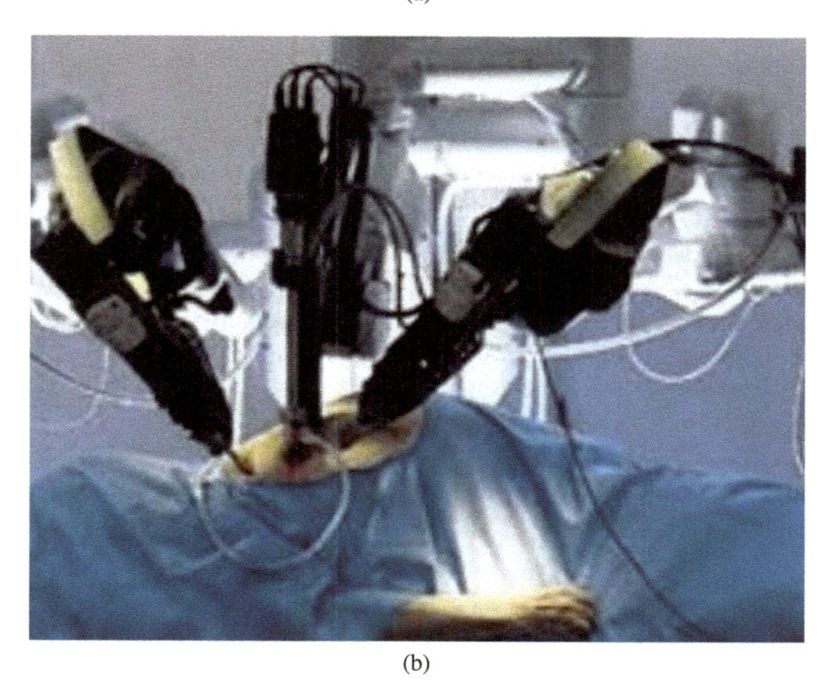

(b)

Figure 8.17 (a) Typical computer-integrated robotic surgical system (DaVinci Robotic system) and (b) Robotic surgery in action.

Source: www.medexpertchile.com/surgeries/surgical-service (b) htm.wikia.com/wiki/Robotic_surgery

functional level, improving their quality of life and work productivity. The main categories of PwSN are [22, 23]:

- PwSN with loss of lower limb control (paraplegic patients, spinal cord injury, tumor, degenerative disease).
- PwSN with loss of upper limb control (and associated locomotors disorders).
- PwSN with loss of spatiotemporal orientation (mental, neuropsychological impairments, brain injuries, stroke, ageing, etc.).

The field of AR was initiated in North America and Europe in the 1960s. A landmark assistive robot is the so-called Golden Armo developed in 1969, a 7 degrees-of-freedom orthosis moving the arm in space (*Rancho Los Amigos Hospital*, California). In 1970 the first robotic arm mounted on a wheelchair was designed. Today many smart AR systems are available, including:

i. *Smart-intelligent wheelchairs* that can eliminate the user's task to drive the wheelchair and can detect and avoid obstacles and other risks.
ii. *Wheelchair mounted robots* which offer the best solution for people with motor disabilities increasing the user's mobility and the ability to handle objects. Today WMRs can be operated in all alternative ways (manual, semiautomatic, automatic) through the use of proper interfaces,
iii. *Mobile autonomous manipulators*, i.e., robotic arms mounted on mobile platforms, that can follow the user's (PwSNs) wheelchair in the environment, can perform tasks in open environments, and can be shared between several users.

Figure 8.18 gives a general illustration of assistive technology that enables impaired persons to communicate, learn, work, and be entertained with the aid of several means including the Internet.

Figures 8.19–8.25 show a few examples of assistive robots, namely: rehabilitation robot, smart intelligent wheelchair, wheel chair mounted robot, mobile autonomous manipulator, and three examples of socialized/ entertainment robots, namely:

- **Cosmobot** humanoid robot.
- **Aibo** robot dog.
- **Paro** robot baby seal.

A discussion of a wide gamma of socialized robots (sociorobots) is presented in Tzafestas [24].

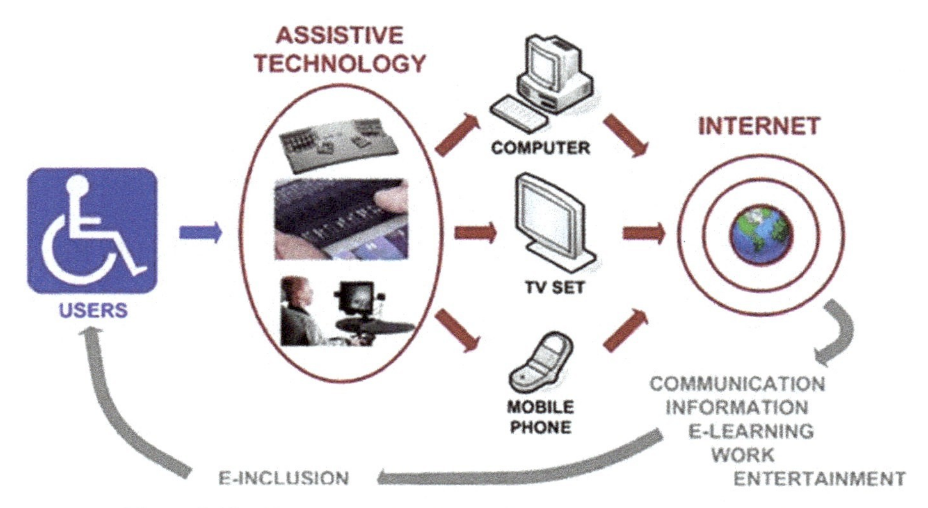

Figure 8.18 General pictorial illustration of assistive technology aids.

Source: http://dom-iris.si/images/prilagoditve-komunikacija1/en.gif; http://parentedge.in/assistive-technology-for-the-differently-abled/

Sociorobots are characterized by human-like cognition, intelligence and autonomy, and in some cases partial emotion consciousness and conscience. Sociorobots are distinguished in *anthropomorphic (humanoid), zoomorphic* (animaloid), and *wheeled* mobile sociorobots. *Humanoid* robot examples are *ClRIO, NAO* and *Cosmobot, Zoomorphic* robot examples include *i-Cat, Aibo, Paro, Teddy Bear, NeCoro*robo-cat, etc. Wheeled sociorobots include, *Maggie* sociorobot, *Robovie, PEARL, Bandit*, etc. [24]. The primary function of sociorobots is the human-robot interaction, which is needed for a smooth and profitable symbiosis (co-habitation) of human and robots. This function is achieved by using advanced intelligent human robot interfaces. These interfaces are designed by including the human's perspective through the entire design and development process.

8.4.2 Aircraft Automation

Automation in aviation systems are undoubtedly among the higher degree of automation in man-made systems (e.g., ship automation, railway, etc.). In this section we provide a brief description of aircraft automation [25].

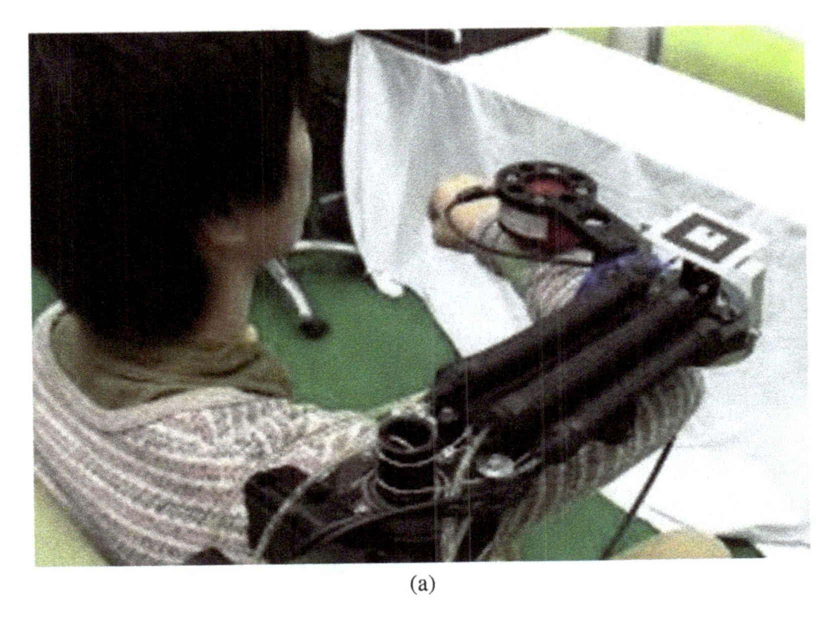

(a)

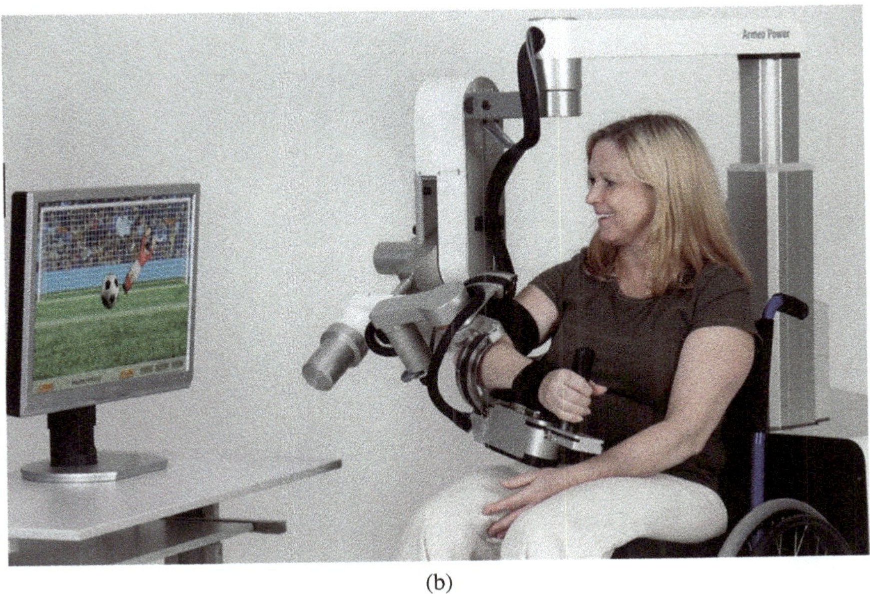

(b)

Figure 8.19 Two examples of arm rehabilitation robots.

Source: https://gr.pinterest.com/pin/382665299574362315

ultrasonic
sensors
embedded
joystick
LineDrivers
Encoders

Local USER

embedded Wireless Internet
Camera Server (TV-IP110W)
embedded hardware and
software architectures
AX2550 controller
Battery

(a)

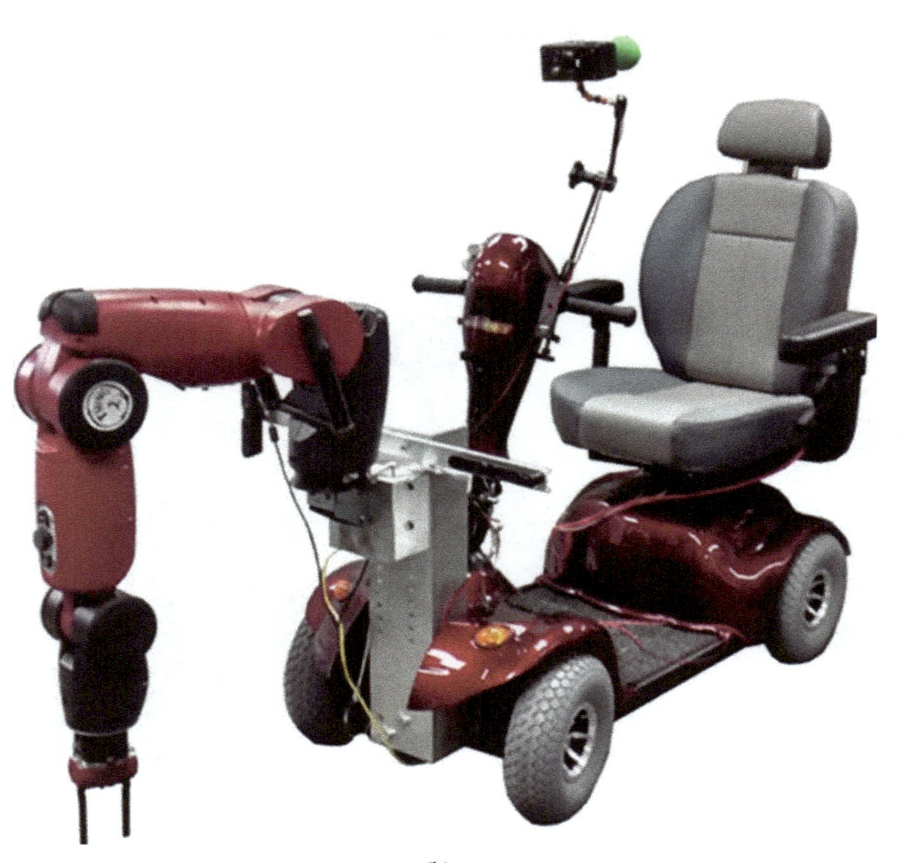

(b)

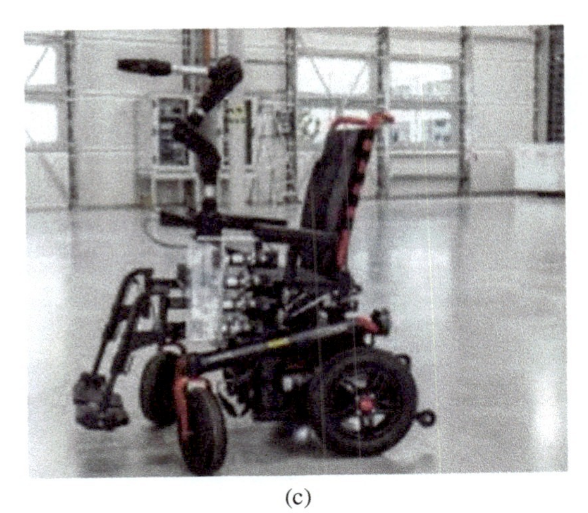

(c)

Figure 8.20 (a) Smart/intelligent wheelchair example with its components (joystick, ultrasonic sensors, wireless camera server, and line driver encoders). (b,c) Wheel-chair-mounted service robots.

Source: (a) http://article.sapub.org/10.5923.j.se.20120203.02.html (Touati and Ali-Cherif, Scientific and Academic Publishing, 2012), (b) robotics.cs.uml.edu/robots.php (UMass Lowel Robotics Lab), (c) www.roboticsupdate.com/2015/02/auxiliary-robot-arm-for-wheelchair-users

Figure 8.21 Wheel-chair mounted service robot.

Source: www.neobotix-robots.com/applications-research.html

Figure 8.22 Mobile autonomous manipulator.

Source: www.neobotix-robots.com/applications-research.html

(a)

(b)

Figure 8.23 (a) The Cosmobot socialized robot and (b) Emotional interaction of Cosmobot with children.

Source: (a) www.51voa.com/VOA_Special_English/Education_Report_34960.html;
(b) www.voanews.com/a/sociallyassistiverobots08dec2009-78761362/416256.html

(a)

(b)

Figure 8.24 (a) AIBO socialized robot dog (two different models) and (b) AIBO-child interaction.

Source: (a) www.laht.com/article.asp? Article id=2360317&Categoryid=13136 and
(b) www.nytimes.com/learning/teachers/featured_articles/20010802thursday.htm

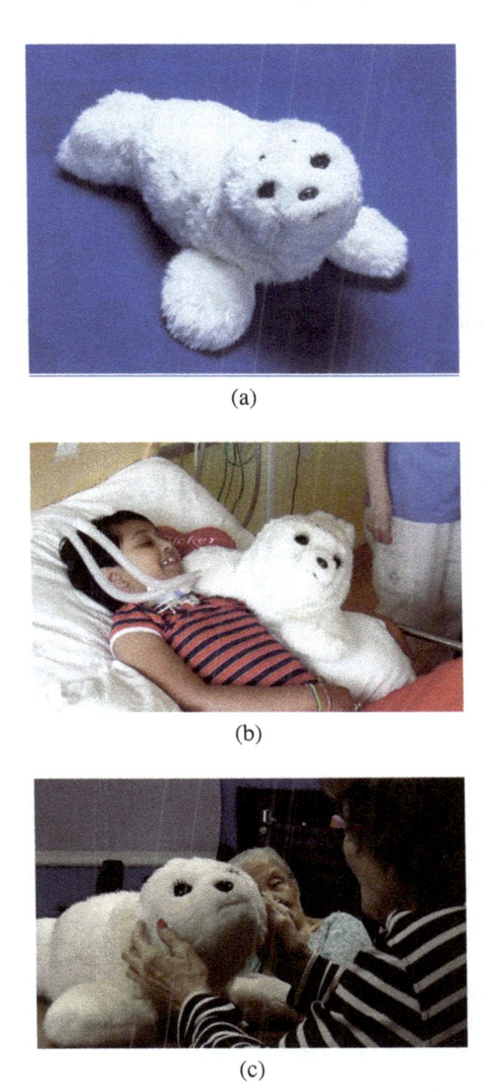

(a)

(b)

(c)

Figure 8.25 (a) Paro baby seal therapeutic robot (it can be used in place of animal therapy with the same documented benefits in patients treated in hospitals and extended care facilities) (b) Paro emotional company to a child, and (c) The dementia lady started talking and asking questions about Paro.

Source: (a) www.roboticstoday.com/robots/parowww.parorobots.com,
(b) www.web.cecs.pdx.edu/~gerry/class/ME370/notes/cases/Case_study_Paro_robot.html and
(c) www.robotcenter.co.ukhttps://youtube.com/watch?v=OPIDdwBw1NNc

Despite the advanced control and automation devices and techniques used in commercial aircrafts, the accidents continue to occur and are mostly attributed to the human pilots. Cockpit automation is a mixed blessing. Warnings of possible problems with cockpit automation were raised as early as the late 1970s, [26], and since then the concerns have become more severe because of the incidents and accidents involving automated aircraft. Actual experiences with advanced cockpit technology verify that automation did have a positive effect regarding the pilot's workload, the operational cost, the precision, and the human errors. But, the impact of automation turned to be different and much more complex than predicted. Work-load and pilot faults were not simply reduced and advanced automation has led to new requirements which were qualitative and context dependent rather than quantitative and uniform in nature. In the 1980s, failure rates in transition training for the new *glass cockpit aircraft* were at an all-time high [27]. Early concerns with cockpit automation focused around question such as how to reduce the amount of pilot workload, and how much information does not the pilot needs.

Regarding the training, recent intensive research on training, and a better understanding of the problems produced by *glass cockpit pilots*, revealed that it is the nature of training that has to be changed rather than its duration. The glass cockpit was introduced in the 1970s (Boeing's 757, and 767 commercial aircrafts), and consists of several CRT or LED displays integrated by computers which replaced the multiple independent mechanical instrument used by that time. Cockpit technology has changed in several ways which require new practice and training methods.

A very powerful addition to the flight deck is the so-called FMS which extended the manual control (originally via a 2-axis-stick, then through a yoke, and finally through a fly-by-wire) to automatic handling a variety of tasks including navigation, flight path control, and aircraft systems monitoring [28]. The most advanced FMSs currently in operation are highly autonomous and reliable, and can perform long sequences of actions without pilot input (authority). The FMSs allow the pilot to select among several automation levels, provides advice on navigation, and detects and diagnoses abnormalities or faults. According to Wiener, who studied extensively the results of cockpit automation, the effect of automation on pilot workload is not an overall decrease or increase, but rather a redistribution of the workload [29]. For this reason he introduced the term *clumsy automation.* Wiener has distributed two set of questionnaires to Boeing 727 pilots 1 year apart to collect information on pilot workload, pilot errors, crew coordination, and pilot training for automation. Most of the pilots (more than 55%) replied that they were still being surprised by the automation after more than 1 year of line experience on the aircraft. Similar conclusions were also drawn by Sarter and Woods [30] who sampled a group of pilots of B737-300/400 which are equipped with a different glass cockpit. These pilots have also indicated the nature of and the reasons for the automation surprises (which can be regarded as symptoms of a loss of awareness of the status and behavior of automation, i.e., of a loss of mode awareness). Mode errors appear to occur due to a combination of gaps and misconceptions in pilot model of the automated systems. The above discussion shows that, even with the sophisticated cockpit automation such as the *FMS*, the pilot has a high workload and can make erroneous actions. The sophistication and variety of new automation systems on flight deck have led to new control modes for the pilot to understand, which need careful consideration to minimize the pilot confusion over states and modes in automated cockpits [31].

Figure 8.26(a) shows a typical cockpit board with the principal instruments, and Figure 8.26(b) shows the view of a modern automated cockpit.

8.4.3 Air Traffic Control

The original method of controlling takeoffs and landings was the use of an ATC standing in a prominent place on the airfield and employing colored flags to communicate with the pilots. The waving of a green flag meant that the pilots were to proceed with their planned takeoff or landing. But if the controller waved a red flag, the pilots were to hold their position until the controller had determined that it was safe to continue. This early type of air controller was difficult to control more than one aircraft simultaneously, and impossible to be used at night or during storming weather. The next attempt was the use of light guns. A light gun is a device that allows the controller to direct a narrow beam of high-intensity colored light to a specific aircraft. The gun was equipped with different-colored lenses to allow the controller to easily change the color of the light. The controller was operating the light gun from a *control tower* (a glassed in-room on the top of a hangar), or from a movable light gun station located near the arrival end of the runway. Light guns are still used today on most control towers as back-ups, either when the radios in the control tower or the aircraft are inoperative or when an aircraft is not radio equipped. The development of modern air traffic control systems started by equipping a control tower with radio transmitting and receiving equipment (at Cleveland, 1936).

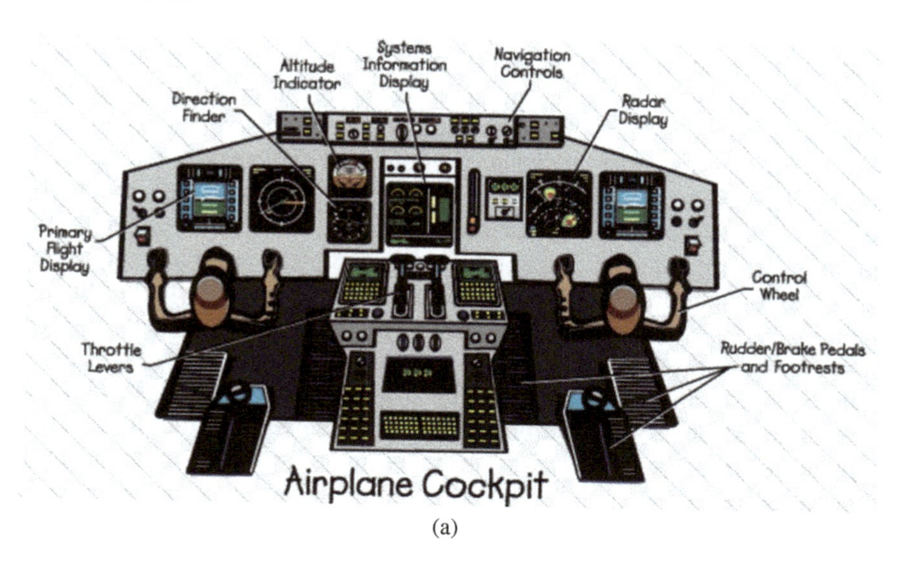

(a)

(b)

Figure 8.26 (a) Principal airplane cockpit instruments and (b) A modern cockpit example.

Source: (a) https://gr.pinterest.com/hadarmenashe/car-dashboard-cockpit and (b) www.wallpapersafari.com/cockpit-wallpaper

Today, radio has become the primary means of pilot-controller communication in the air traffic control system. The radio equipment has considerably changed since 1936, but the basic principles of radio communication remain unchanged. The earliest type of radio communication was one-way, i.e., traffic controllers could communicate with pilots but not vice versa. An interim solution for two-way communication was the use of receiving equipment in the control towers and transmitting equipment in the aircraft. To eliminate the so-called *navaid (navigation aid) interference* the aircraft transmitters used a different frequency than the ground-based navaids. This two-frequency system is called a *duplex communication system* (Figure 8.27(a)). The radio frequency bands allocated to aeronautical communications are determined by international agreements. These frequency bands exist mainly in high frequency (HF), very high frequency (VHF) and ultra-high frequency (UHF) spectrums.

The duplex system has some drawbacks which led to the development of a radio system that would allow pilots to communicate with controllers using one discrete frequency. This is known as *simplex communications* (Figure 8.27(b)). Simplex communications are today used in every ATC worldwide.

The **ICAO** developed standardized world's aviation systems and suggested procedures for aviation regulatory agencies [32]. These standards are known as 'International Standards and Recommended Practices' (and classified individually as ICAO Annexes). Every country member of ICAO

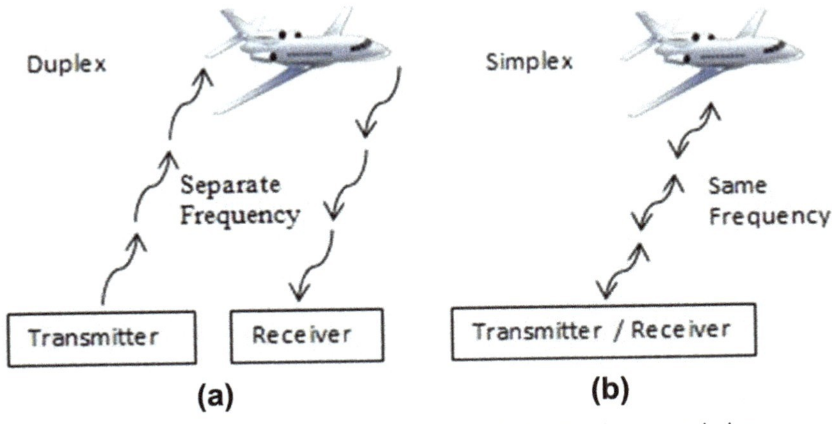

Figure 8.27 (a) Duplex transmission and (b) Simplex transmission.

has agreed to generally abide by these ICAO Annexes, unless they must be modified to meet national requirements. The adoption of these procedures has allowed pilots to fly over the world using a unique language (English), unique navigation aids (VOR: *VHF omnidirectional range*, ILS: *Instrument landing system*, NDB: *Non-directional radio beacon*, and MLS: *Microwave landing system)*, and the same procedures. ICAO requires that every country publish manuals describing its ATC system and any deviations from ICAO standards. ICAO recommends three types of aircraft operations, namely VFR: *Visual Flight Rules*, IFR: *Instrument Flight Rules, and* CVFR: *Controlled VFR).* Controlled VFR flights are separated by controllers as if they are IFR, but the pilots are not IFR rated and must remain in VFR conditions. The ICAO agreements specify that each nation will control its own sovereign space but will permit ICAO to determine who shall provide air traffic control service within international space. ICAO is only a voluntary regulatory agent, and so international ATC has been delegated to those member nations willing to accept this responsibility. ICAO has divided the total world airspace into *flight information regions* (FIRs), which identify which country controls the airspace and determines which procedures should be followed. For the purpose of '*en route*' ATC, each FIR identifies, normally, one major air traffic control facility. Typically, the boundaries of each FIR follow the geopolitical boundary of the concerned country. ICAO uses unique four-letter commercial air-ports' identifiers for ATC, whereas IATA *(International Air Transport Association)* uses three-letter codes primarily for travel agents and airline personnel.

Other additional facilities newly developed for cooperation with ATCs are the following:

- *URET: User Request Evaluation Tool* (a computer program which probes for potential conflicts of selected flight paths, processing real-time flight plan and track data from the host computer via a one-way interface).
- *PRAT: Prediction/Resolution Advisory Tool* (a decision support system that performs conflict prediction and resolution assistance for the ATC).
- *WARP: Weather and Radar Processor* (for collection, processing and dissemination of next-generation (NEXRA) and other weather information to controllers, traffic management specialists, area supervisors, pilots, and meteorologists).
- *ITWS: Integrated Terminal Weather* System(a fully automated weather prediction system providing enhanced information on weather hazards in the airspace within 60 nautical miles of an airport).

- *PRM: Precision Runway Monitor* (designed to face the problem that during instrument meteorological conditions, airports with parallel runways spaced less than 4,300 feet apart cannot conduct independent simultaneous instrument approaches due to equipment limitations).

The free Flight Operational Concept: Under the free-flight mode, pilots operating under **IFR** will be able to select their aircraft's path, speed, and altitude in real time. The ATC system will only be involved only when it is necessary to provide positive aircraft separation. A flight plan established by the pilot, is essentially a contract with the air traffic controller, and so any modification (required for example in case of thunderstorm) should be renegotiated with an air traffic controller. Under the free-flight concept, the pilot will be free to select the aircraft's rout, speed and altitude, and to make alterations with no ATC's preapproval (provided of course that the traffic density does not preclude a free flight).

In military aviation there is an additional factor (in comparison with commercial aviation), namely the need to deal with an enemy. Therefore, additional automation is required (including advanced radar and optical systems) for safe fast maneuvering and weapon firing. Still more advanced automation systems and facilities are required for UAVs (unmanned air vehicles) which are remotely controlled by an operator (pilot) on the ground. Figure 8.28 shows a typical air traffic room in an airport with two landing/take off lanes.

Figure 8.28 A typical air traffic control room.

Source: www.dxschool.org/air-traffic-control-usa

Other very important applications of automation are:

- *Automation in ship transportation* (performed by ships that require very sophisticated techniques, e.g., automatic roll stabilization, tracking of depth and obstacles using sonars and localization in latitude and longitude for navigation).
- *Automation in automobiles* (for enquiry control, transmissions, instrumentation, in-vehicle comfort, and in-vehicle entertainment).
- *Automation in railway systems* (these systems require high quality level of serviceability, reliability, and safety which are accomplished using automation though microcomputers).

Particular subsystems of railway automation are [6]:

- Train traffic control subsystem.
- Automatic train operation subsystem.
- Electric power supply control subsystem.
- Information transmission subsystem.
- Automatic car inspection subsystem.
- Supporting business management subsystem.

8.4.4 Automated Driving Vehicles

The dream of humans to achieve automated driving vehicles (i.e., intelligent vehicles autonomously driven or self-driven) goes back to the 1920s. Today, after 100 years, research has led to important results that are expected to help in fulfilling this dream [41–45]. An intelligent vehicle system is able to sense the environment and provide information or assist the driver in inputs to the vehicle system to maintain stability, traction, avoid collisions, or simply reroute the vehicle around a traffic jam. Driving systems are distinguished in:

- *Active systems* (Here the term active refers to the vehicle or system itself. The car may make decisions and execute tasks on its own without driver input).
- *Passive systems* (Here, the system keeps 'watch' and/or intervene to help the driver in reacting to a situation).

Active response and stability control can: (i) handle over-steering situations without driver involvement, (ii) provide input to the wheel velocity and lateral acceleration sensors, and (iii) evaluate and correct conditions by applying brakes to one or more wheels regardless of the driver pedal position.

Passive response merely hinders the effect a driver can have on the handling characteristics of the car and recognize a situation without acting on behalf of the driver by applying brakes or closing the throttle.

SAE (*Society of Automotive Engineers*) International (www.sae.org/autodrive) has developed and released a new standard (J3016) for the "Taxonomy and definitions of terms related to on-road motor vehicle automated driving systems". This standard provides a harmonized classification system and supporting definitions which:

- Identify six levels of driving automation from 'no automation' to full 'automation'.
- Base definitions and automation levels on functional aspects of technology.
- Describe taxonomical distinction for step-wise progression through the levels.
- Match uptoday's industry practice.
- Eliminate fuzziness and are useful for a wide variety of disciplines (engineering, legal, media, and public discourse).
- Educate people by clarifying for each automation level what role (if any) drivers have in performing the dynamic driving task while they are in an automated driving system.

The fundamental definitions included in the J3016 document are [45]:

- *Dynamic driving task* (i.e., operational aspects of automatic driving, such as steering, braking, accelerating, monitoring the vehicle and the road, and tactical aspects such as responding to events, determining when to change lanes, turn, etc.).
- *Driving mode* (i.e., a type of driving scenario with appropriate dynamic driving task requirements, such as expressway merging, high speed cruising, low speed traffic jam, closed-campus operations, etc.).
- *Request to intervene* (i.e., notification by the automatic driving system to a human driver that he/she should promptly begin or resume performance of the dynamic driving task).

Figure 8.29 shows the evolution of vehicle automation from no-automation to full automation starting from the level-1 function-specific automation in 2014–2017 and going to the full self-driving automation anticipated for 2025–2030 [41–47]. On the left of the figure the increase of automated vehicle control is shown by the increasing-size red '+' symbols upwards, while on the right of the figure the decrease of operator control is shown by the decreasing-size symbols '+' upwards.

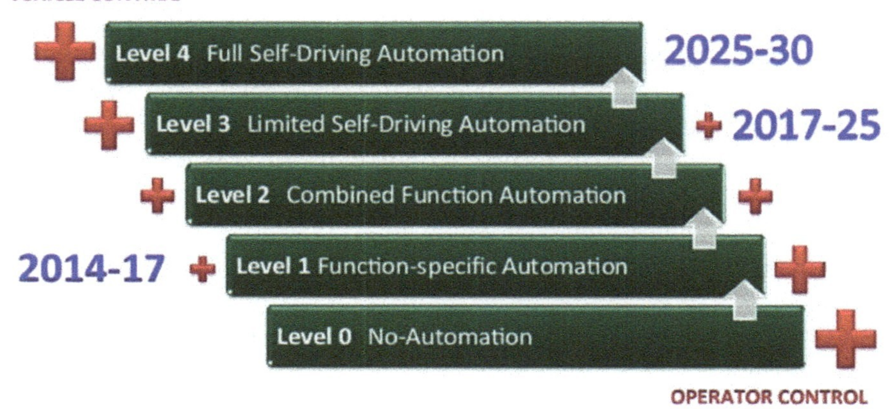

Figure 8.29 Evolution of vehicle automation 2014–2030.

Source: www.slideshare.net [47].

Figure 8.30 shows the milestones needed to be passed on the way the final goal of fully automated vehicles is met according to **SAE**, **NHTSA** [44] and **FHRI**.

Vehicle safety starts with the capability to avoid crashes. Any other capability that follows a crash is secondary. Vehicle safety is distinguished in:

- **Active safety** that provides features helping the driver to avoid a crash. Active safety systems include ABS (anti-lock braking system), BA (braking assistance system), EBD (electronic braking distribution system), ESC (electronic stability control), and AEB (autonomous braking system).
- **Passive safety** with features that minimize injuries when crash cannot be avoided. Passive safety includes airbags, seatbelt pretensions with force limiter and emergency locking retractors, and strong reinforced body structure.

Some higher-end cars may also include pre-crash and post-crash features. Autonomous braking is a priority feature of self-driving vehicles. An automated vehicle needs to precisely position itself along the road and to know its lateral position too. It must also be aware of lane boundaries, lane center, and where the car is supposed to drive. Self-driving cars must be capable to see around corners and know the environment before they reach an intersection. Traditional maps are not sufficient for these features.

SAE level	NHTSA level	BAST level	Steering, braking & acceleration	Monitoring of driving environment	Fallback performance	System capability
No Automation	0	Driver only	Human	Human	Human	none
Driver Assistance	1	Assisted	Human and system	Human	Human	
Partial Automation	2	Partially automated	System	Human	Human	
Conditional Automation	3	Highly automated	System	System	Human	
High Automation	3/4	Fully automated	System	System	System	
Full Automation		–	System	System	System	All driving modes

Figure 8.30 Vehicle driving automation milestones adopted by ASE, NHTSA, and FHRI.

Source: https://www.schlegelundpartner.com/cn/news/man-and-machine-automated-driving

A map with sufficient granularity is needed to enable the automated car to know which lane it is in on a multi-lane highway along with the lane location. To this end, online sensor and localization technology able to identify a vehicle and its surroundings on a map with very high resolution (say 10–20 cm) must be used. Such surroundings include lane marking types, intersection, signs, and other advanced safety features.

These scenarios and stages of development are subject to several legal and ethical problems which are currently under investigation at a regional and global level. Most advanced country is USA, while European countries are somewhat behind USA. The general legislation in USA (primarily determined by NHTSA and the Geneva Convention on road traffic of 1949) requires the active presence of a driver inside the vehicle who is capable of taking the control whenever it is necessary. Within USA each state enacts its own laws about automated driving cars. So far only four states (Michigan, California, Nevada, and Florida) have accepted automated driving software to be legal. In Germany the Federal Ministry of Transport has already allowed the use of driving assistance governed by corresponding legislation. Most major car manufacturers are planning to produce autonomous driving technologies of various degrees. For example, Google is testing a fully autonomous prototype that replaces the driver completely, and expects to commercialize its technology by 2020. Actually automakers are proceeding towards full autonomy in stages, currently most of them are at level 1 and only a few introduced level 2 capabilities. In the following we give the overall picture of today's vehicle autonomy (2016–2025).

- **By 2016**: Mercedes plans to introduce 'Autobahn Pilot'.
- **By 2017**: Mobileye expects to release, in co-operation with Delphi, a turnkey autonomous driving system suitable for rapid adoption by a variety of automakers.
- **By early 2017**: US DOT aims to publish a rule governing vehicle-to-vehicle communication, and GM expects Cadilac to have first adoption of the technology.
- **By 2018**: Tesla Motors hopes to have developed mature serial production of full self-driven cars (Level 4).
- **By 2018**: NISSAN expects to have a feature that can allow the vehicle maneuver its way on multi-lane highways.
- **By 2020**: Volvo envisages their vehicles will effectively be 'crash-free'.
- **By 2020**: GM, Mercedes Benz, Audi, Nissan, BMW, Renault, Tesla and Google all expect to sell vehicles that can drive themselves at least part of the time (Level 3).
- **By 2021**: Mobileye expects to bring, with the partnership of BMW and Intel, a fully autonomous vehicle in serial production.
- **By 2024**: Jaguar envisages to release the autonomous car.
- **By 2025**: Daimler and Ford expect autonomous vehicles on the market. Ford predicts it will have the first mass-market autonomous vehicle, but without specifying a target date.
- **By 2025**: Most GM vehicles will have automated driving functions as well as vehicle-to-vehicle communication technology.

Figures 8.31(a,b) show two typical examples of automated driving vehicles, and Figure 8.31(c) shows a driver riding in a moving self-driving car.

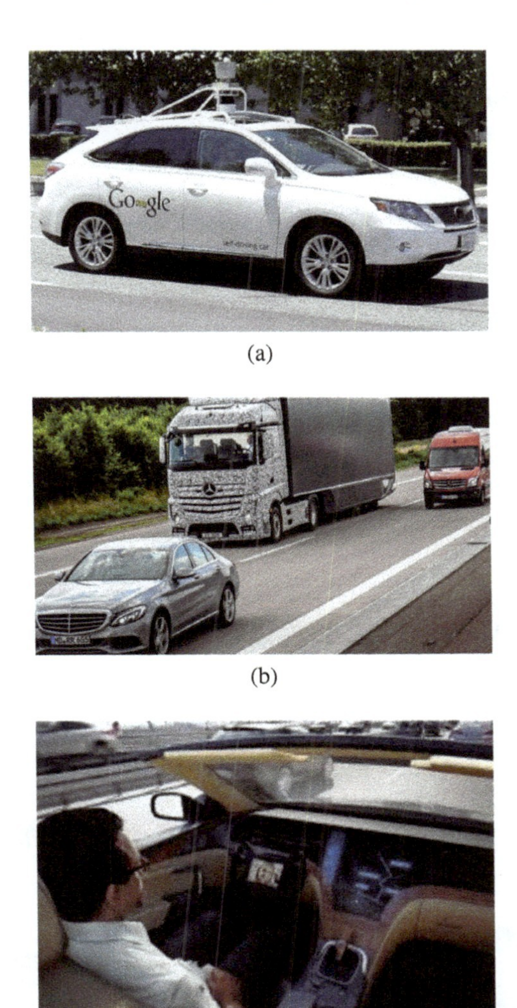

(a)

(b)

(c)

Figure 8.31 Automated vehicles: (a) Google self-driving car prototype, (b) Mercedes-Benz self-driving truck (which is planned to be on the market by 2025) in a high-way test cruise, and (c) A driver snapshot in a moving automated driving car.

Source: (a) www.coolestech.com (how-google-self-driving-car-works), (b) www.inhabitat.com (Mercedes-Benz elf-driving future truck completes its first autonomous test on the autobahn) and (c) http://360.here.com (2014/07/28/3-steps-automation-vp-connected-driving-ogi-redzic).

Figure 8.32(a) shows the basic sensors used in automated cars, and Figure 8.32(b) shows the types of environmental data gathered by self-driving car multiple sensors, namely:

- **LIDAR**: For edge detection.
- **RADAR**: For point detection.
- **Camera/Vision sensor**: For vision target.
- **Ultrasound**: For vehicle maneuvering assistance.
- **GPS**: For vehicle localization.

Specific tasks that a fully automated vehicle has to perform with the aid of the above sensors (without human driver involvement) include (Figure 8.32(b)):

- Traffic sign recognition.
- Lane departure warning.
- Cross traffic alert.
- Emergency braking.
- Pedestrian detection.
- Collision avoidance.
- Adaptive cruise control.
- Rear collision warning.
- Park assistance.

Figure 8.32(c) shows the types and functions of the vision systems and range sensors mounted on self-driving cars.

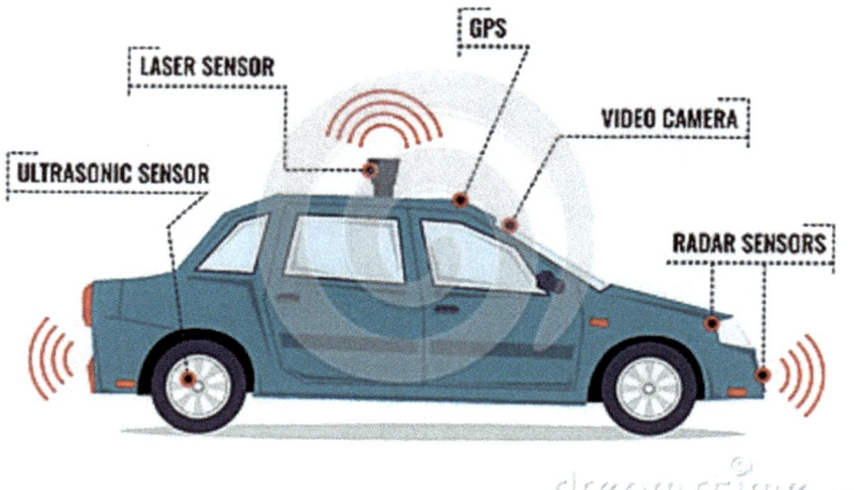

(a)

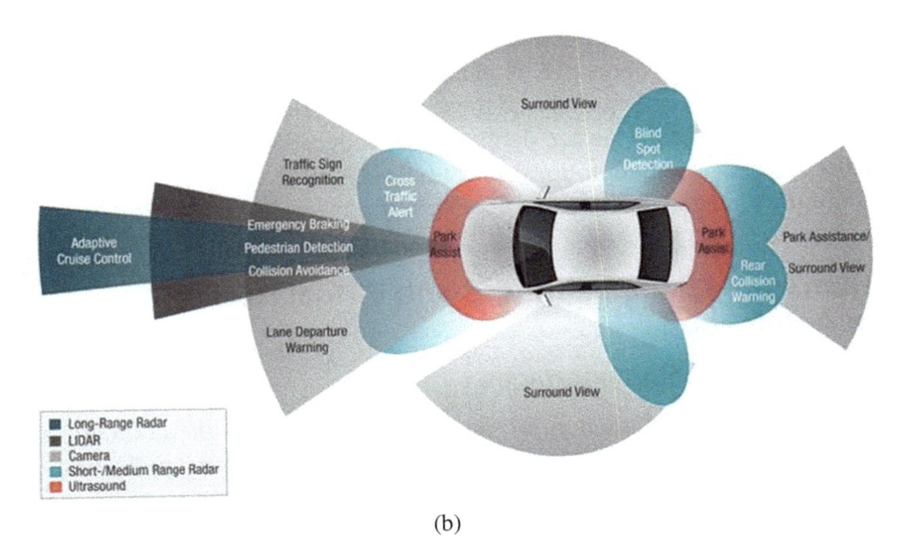

(b)

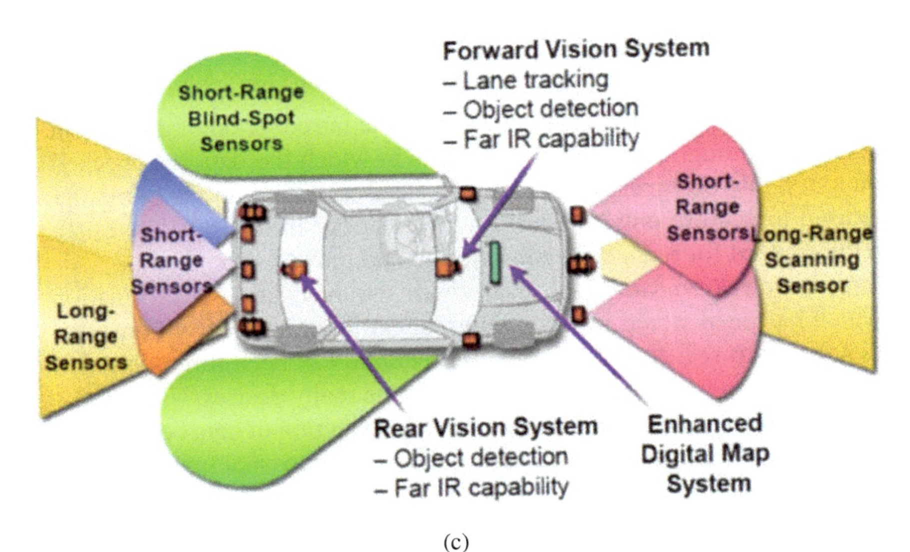

(c)

Figure 8.32 (a) Basic self-driving car sensors, (b,c) Multiple sensors gathering data for automated cars.

Source: (a) https://www.dream.com; (b) www.slideshare.net (Vivek Mara, Autonomous Driving, MST Seminar Automotive Sensors), and (c) www.inquisitr.com (Article: Autonomous Vehicles are the Wave of the Future, says Stephen Diago).

Statistical studies have revealed that about 90% of road accidents are due to human error. Proponents of automated vehicles claim that human errors will disappear with the use of these vehicles. But actually this is not expected to be exactly so (the accidents may be reduced considerably but not eliminated), since automation does not remove the human from the driving task. It simply changes the human role within it.

8.5 Human–Machine Interfaces and Virtual Reality in Automation

8.5.1 Human–Machine Interfaces

Communication and interaction of humans (operators) with automation systems is performed through proper interfaces. Automation applications (robotic systems, medical applications, process and power systems, etc.) can be divided into two main categories:

1. Applications concerned with pure information exchange.
2. Advanced applications that make use of the information exchange.

In the first category the following four basic applications are included:

- *Information Acquisition:* These deal with the introduction of data into a computer through an interactive process in which the data are checked on entry, any occurring errors are corrected at once, and the computer is used to acquire only the desired information. In this way there is a reduced need for redundant transcription and transmission of data.
- *Information Retrieval:* This deals with the recovery of information which has been stored in a computer, again in an interactive conversational way to secure that information required has been fully identified and that the user is helped to find what she/he really desires. Usually, the retrieval process is a prerequisite for the introduction of data because the originators of information may ask questions before making a decision about what is to be entered into files.
- *Editing:* The computer provides useful assistance in the preparation of stored text material for publication. The main part of this process consists of accepting raw text input and allowing the author to make alterations-adding or deleting material and moving portions of it around. Also, non-textual or highly structured material can be edited such as computer programs or market catalog records, etc.

- *Instruction:* Here, the instructional process might be considered that of asking the user (e.g., a student) questions, the answers of which are already known, and comparing her/his responses with the known information. Then, the computer acts differently according to whether she/he can or cannot answer correctly the questions. In information acquisition the computer asks for data it does not already have, and in information retrieval it works with the user to ascertain what is wanted.

In the more advanced applications the human and computer are jointly acting to produce a useful output. The outputs may be a product directly usable (e.g., an electronic circuit diagram, or a decision about an enterprise, or a computer program, etc.). Here, the human machine interaction is much closer to human–computer symbiosis, than in the pure information exchange applications. Three such applications are the following:

- *Interactive Programming:* The production of computer programs is one of the most effective applications of human-machine communication. Interactive programming is very frequently used interchangeably with the time-sharing term, but two are different: The former is usually an application of the later, but they can also exist without the other and in many other situations it is very difficult to keep these concepts separate.
- *Computer-Aided Design:* The process of design is, in an abstract viewpoint, the creation of an information structure or model which usually is highly complex, and expresses relationships which cannot be easily verbalized (e.g., color, shape, etc.). The *design* is different from the *product*, since it is information descriptive of the product. The computer is used to provide valuable help in evolving the design (i.e., the information set), and then use is made of the design separately to develop the product. In many cases the computer is also used to develop the product. For example, a computer can assist in the design of a textile fabric and then can control the process machine directly to produce the fabric previously designed. Today's design departments are usually equipped with advanced CAD systems which increase the design efficiency to rationalize the product-development process.
- *Computer-Aided Decision Making:* Real decisions in government, business, or industry are not an abstract selection between a *yes* or *no* – they are as much as the design of a plan or strategy as it is the selection from among a set of well-defined and discriminated alternatives. Thus computer-aided decision making should largely restricted to situations where the system and decision variables are relatively few and are

well defined. Unfortunately, real management or engineering has the opposite situation, i.e., many ill-defined variables to work with. Thus, management and automation systems of today usually employ combinations of information retrieval and simulation. The machines recover information on demand and make requested tests, but do not decide for the human; rather, they assist her/him to make a decision by supplying needed information and sometimes reminding her/him of what is to be dome or to be decided. The design of *advanced human-machine interfaces* (HMIs is actually a *multidisplinary* problem needing the cooperation of experts on human cognition, display technologies, graphics, software design, natural language processing, artificial intelligence, etc. One can say that design of an HMI is more an art than a science. As we already know, the goal of designing efficient HMI components in automated systems is to improve operational efficiency and overall productivity while providing a safe, comfortable and satisfying front-end for the operator or user. To this end, the *capabilities, the limitations* and the '*idioms*' of the human operator should be analyzed and taken into account for the HMI design.

Figure 8.33(a) illustrates the HMI/Interaction concept, and Figure 8.33(b) shows the general requirements of designing and building human-machine interfaces in practice. The human side involves the human cognition, sensors, and actuators. Similarly, the machine side includes processing units, sensors, actuators, and displays.

Human–machine interfaces are distinguished in:

- *Basic HMIs* (keys and keyboards, pointing devices, touch screens, light pens, graphic tablets, track balls, mousses, and joysticks).
- *Advanced HMIs* (graphical user interfaces/GUIs, static and dynamical visual displays, intelligent HMIs, natural language HMIs, multi-model HMIs, force sensing/tactile based HMI; HMI for knowledge-based systems).

Figure 8.34 shows the three principal HMI modes (auditory, visual, and tactile) that are typically used in intelligent/autonomous robots and other applications.

Auditory interfaces deal with information acquired by different audio signals. The functions involved are: *Speech recognition, Speaker recognition, Natural language dialog, Auditory emotion analysis*, and *Human-made noise/sign detection*. Historically, speech and speaker recognition have been the main focus of work in audio interfaces. Natural language dialog involves

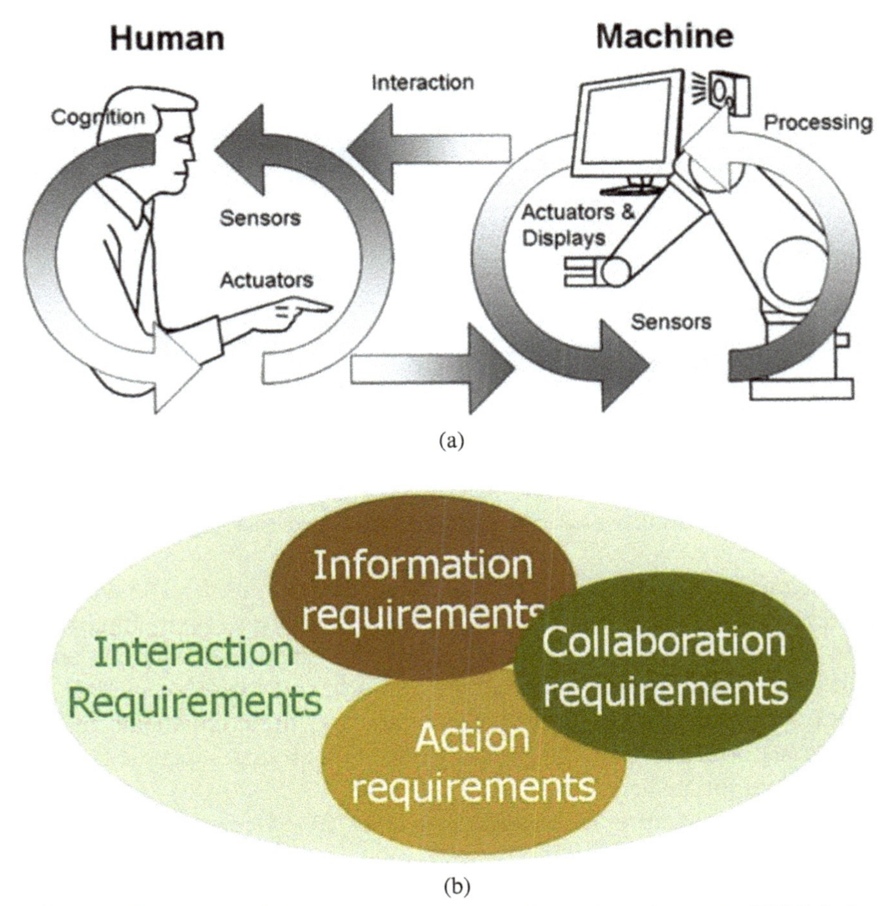

Figure 8.33 (a) The HMI/interaction concept, (b) General requirements of HMI design.
Source: www.doyouknow.in (/Articles/Technology/Introduction-To-Human-Machine-Interface).

communication with the machine through a kind of verbal language (e.g., a small subset of English). Basic issues that must be considered in natural language interfaces are:

- Ease of learning.
- Conciseness.
- Precision.
- Semantic complexity.
- Words and lexicons.
- Grammar and sentence structure.

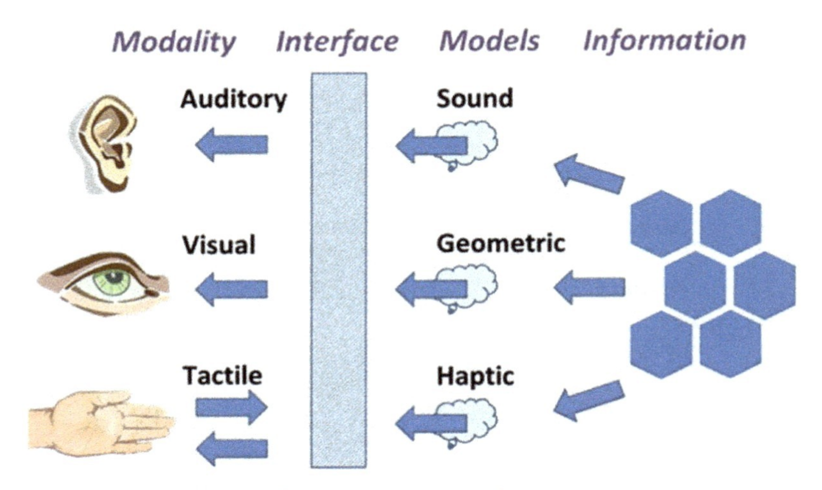

Figure 8.34 Human–machine interface modes.
Source: www.globalspec.com (Article: Human-Machine Interface Software Information).

The more natural is the language used the easier the learning is. The desire for conciseness is usually in conflict with the user friendliness. Many English sentences are ambiguous and should be used with great care. This is so because English does not use parenthesis as do artificial logical languages. Words, lexicons and grammar/sentence structure are the components that enable the transformation of language statements in a program-specific form that imitates appropriate actions.

A possible architecture of *natural language interfaces* is shown in Figure 8.35 [51]:

This architecture involves the following subsystems:

- ***Analysis* subsystem**: The linguistic analysis of the inputs (commands, queries) translates the natural language expressions into logic propositions using a suitable syntactic-semantic parser.
- ***Evaluation* subsystem**: This subsystem is responsible for further interpretation of the propositions and for the reference semantic interpretation. To this end, the natural language interface must have access to all processed information and the world representation inside the robot.
- ***Generation* subsystem**: This subsystem translates selected propositions into natural language descriptions, explanations, and queries. To this end, an incremental generator may generate the surface structures, using a referential semantics which connects verbal descriptions to visual and geometric information (e.g., localization expressions).

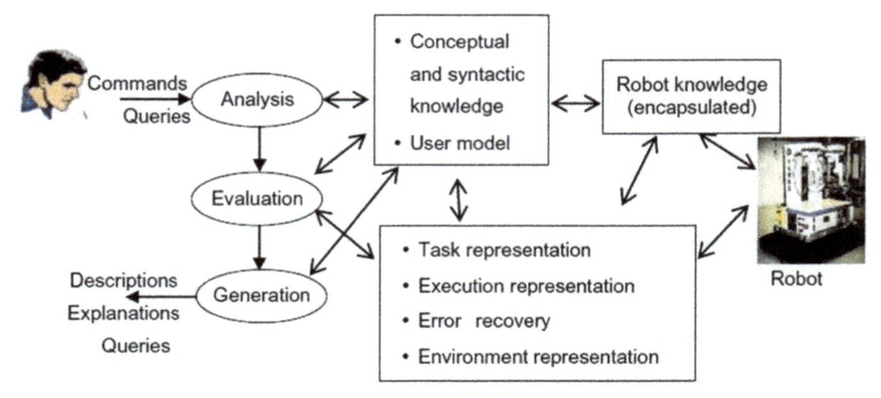

Figure 8.35 Architecture of natural language HMI (robot).

Visual Interfaces: *Visual perception* is fundamental for human-robot interaction, and includes the following aspects [53]:

- Facial expression recognition.
- Body movement tracking.
- Gesture recognition.
- Gaze detection (eyes movement tracking).

Facial expression recognition deals with the visual recognition of emotions. Body movement tracking and gesture recognition have different purpose but they are mainly used for direct interaction of human and robot on a command and action basis Gaze detection is usually an indirect form of interaction which is mostly used for better understanding of the human's attention, intent or focus in context sensitive situations. The vision-based interface uses one or more video cameras for data acquisition and starts with image-preprocessing operations. Figure 8.36 shows a possible architectural design of vision-based human-robot interaction [52].

The system involves two stages. The first stage performs the *detection of face* in image sequences using skin color segmentation. The second stage performs *eye tracking* in order to control the robot in accordance with the human's intention. *Image pre-processing* is used in order to take into account different lighting conditions and contrast, and involves contrast and illumination equalization, and filtering. *Illumination equalization* means the achievement of invariant illumination irrespectively of whether the environment is sunny or cloudy. To compensate for the lighting conditions and to improve the image contrast we use '*histogram equalization*'.

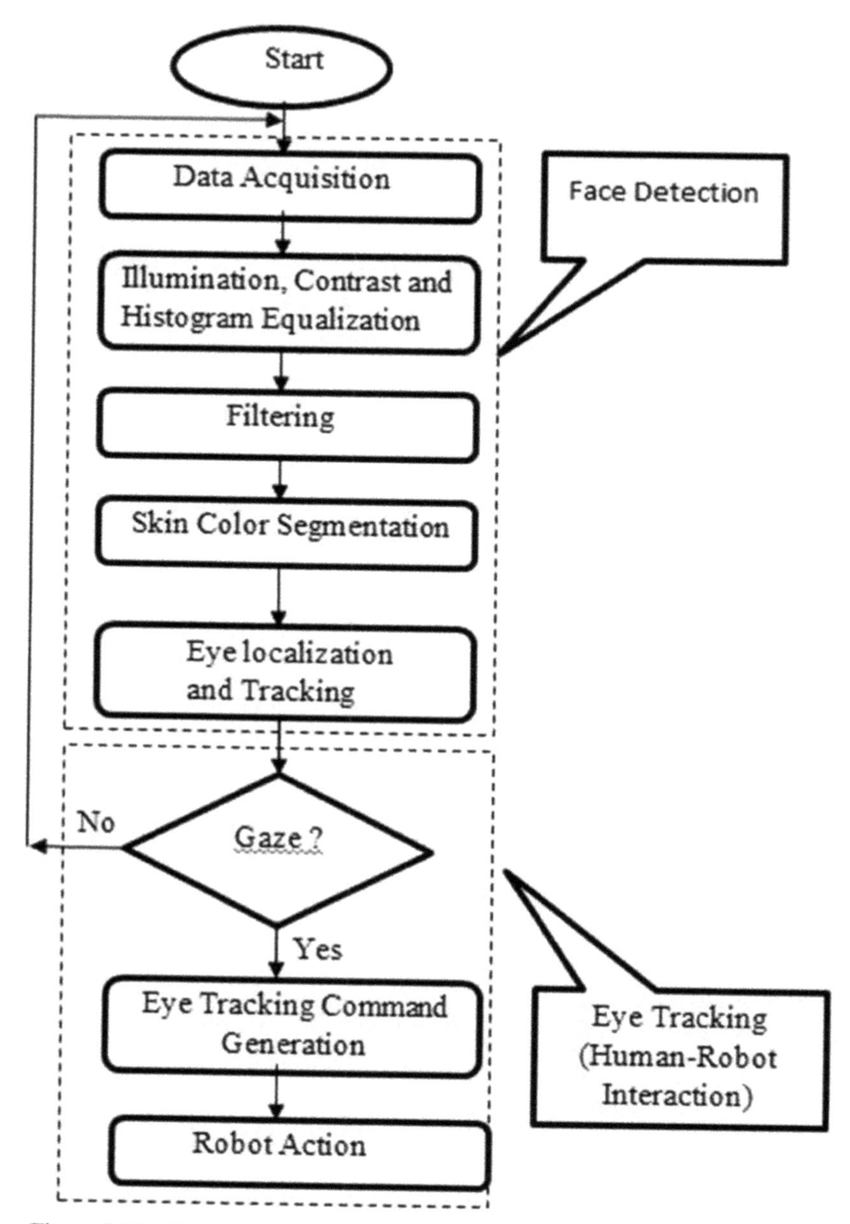

Figure 8.36 Representative architecture of vision-based human–robot interfaces.

The *filtering process* is used to clean-up the input images from existing noise. The detection of face is achieved using skin color segmentation. Once the face is found in an image, the eyes are then searched around the restricted areas inside the face. (e.g., using the so-called '*gravity center*' of the eye) [53].

Based on power requirements sensors used in advanced HMIs are distinguished in:

- *Active sensors* that use external power (excitation signal) for operation.
- *Passive sensors* that directly generate electrical signal in response to external stimuli.

Based on sensor placement, sensors are distinguished in:

- *Contact or touch sensors* (force sensors. tactile/haptic sensors).
- *Non-contact sensors* (vision sensors, ultrasound sensors, radar, etc.).

Tactile interfaces are realized using tactile/haptic sensors. A tactile sensor measures the parameters of a contact between the sensor and an object. The contact forces measured by a sensor can provide detailed information about the state of a grip. Texture, slip, impact, etc. generate force and position signatures that are important for identifying the conditions of a human or robotic manipulation.

In general, haptic/tactile interfaces are distinguished in:

- *Ground-based interfaces* (joysticks and hand controllers).
- *Body-based controllers* (exoskeleton flexible gloves, rigid links, and jointed linkages).
- *Tactile displays* (shape changers, vibrotactile, electrotactile, etc.).

Some typical examples of haptic sensors are depicted in Figure 8.37.

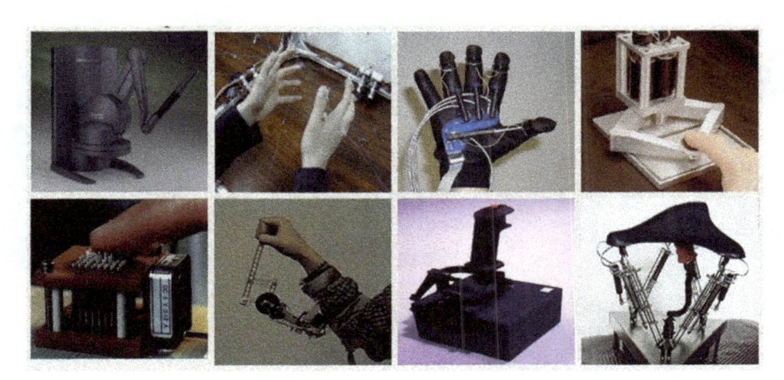

Figure 8.37 Eight examples of haptic sensors.

Source: www.slideplayer.com (/slide/6669939).

Figure 8.38 shows how HMIs have evolved over time since the1970s. Methodologies for the design of HMIs are presented in Helander et al. [33].

Figure 8.39(a) shows the structure of a HMI system through human haptics and machine/computer haptics (this is a common structure of telerobotic/teleoperation systems), and Figure 8.39(b) illustrates the general teleoperation/telemanipulation principle. The human operator sends his/her commands to the slave device (robotic manipulator, etc.) through the master device which is identical to the slave device, and the slave device applies the received commands to the environment (the object which is manipulated). The environment returns sensor feedback to the master device/operator in order to perform the required correction action, which is then sent to the slave device so as to close the remote feedback loop. As discussed earlier, one of the major problems in networked control systems, such as the telerobotic ones, is the existence of time delays (that normally are varying and unknown) in both directions of the information flow (from the master robot to the slave robot, and vice versa). Parameter and other modeling uncertainties also degrade the accuracy of the control and need special treatment. The state of art of the techniques developed over the years for facing both the communication delays and uncertainties (e.g., severe unknown parameter variations and other internal or external disturbances) is presented in [54]. These techniques

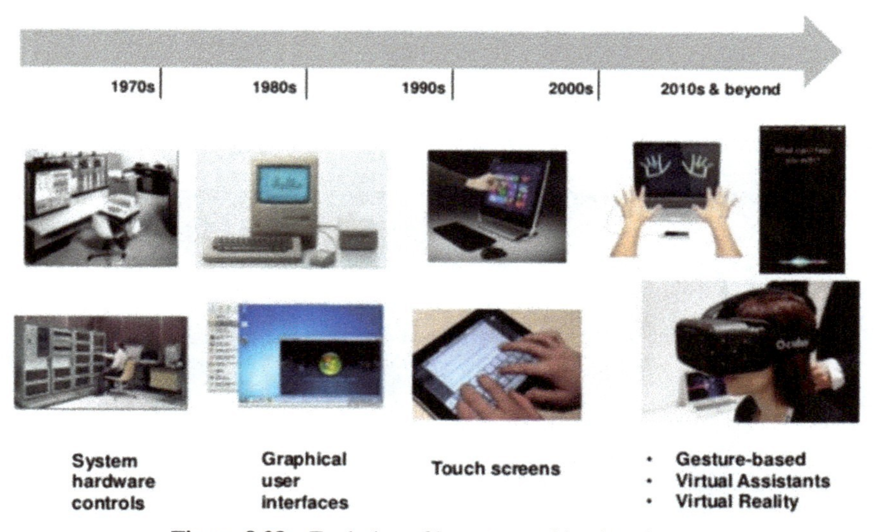

Figure 8.38 Evolution of human–machine interfaces.

Source: www.slideshare.net (/Danielahler1/the-future-of-human-machine-interfaces.html).

include predictive control, adaptive control, sliding-mode stabilizing control, robust control, neural control, fuzzy control, and neuro-fuzzy control.

Basic ontological concepts of telerobotic/telemanipulation systems are:

- *Telerobot:* A robot working from a distance and performing live actions at a distant environment using sensors and actuators.
- *Teleoperator/telemanipulator:* A system (robot, etc.) that enables human operators to sense and manipulate objects at a distance (beyond their reach).

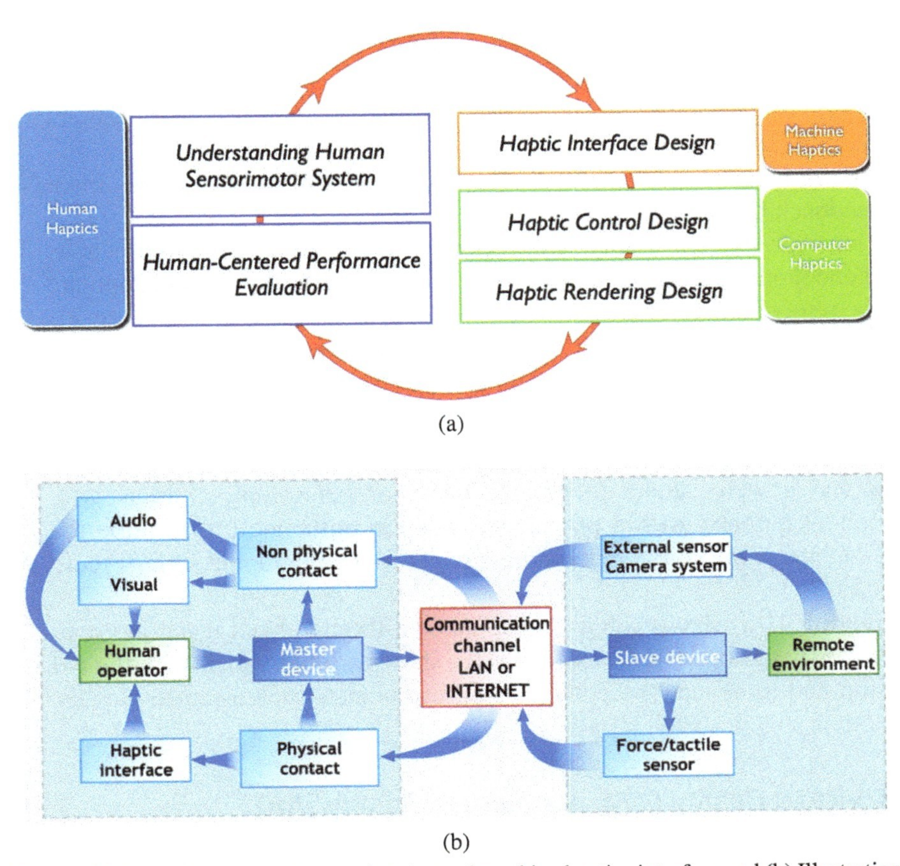

Figure 8.39 (a) Structure of human haptics and machine haptics interface and (b) Illustration of the general teleoperation principle (audio, visual, and haptic/tactile sensors and interfaces are in most cases employed).

Source: www.mogi.bme.hu (TAMOP/robot_applications/ch04.html).

- *Master or local site:* The site in which the human operator and the master robot are located.
- *Slave or remote site:* The site where the slave robot and the environment are located.
- *Master-slave system:* The complete system comprised by the master (local) and slave (remote) robots.
- *Manual or direct control:* Control of the robot motion by the human operator directly without the help of any device.
- *Supervisory control:* The operator controls the robot with high-level commands with the help of an intelligent and/or autonomous system.
- *Shared control:* Co-operated control which is performed partly by the human operator and partly by an intelligent/autonomous controller.
- *Unilateral control:* Control in which the operator cannot feel an accurate force from the remote site.
- *Bilateral control:* Control in which the operator gets accurate force feedback by the remote site.
- *Telepresence:* The ability of the operator to perceive the environment as if encountered directly (in addition to the operator's ability to manipulate the remote environment).
- *Transparency:* The feature that specifies how close the perceived mechanical impedance (force–torque velocity ratio) comes to recreating the true impedance of the environment.

Figure 8.40(a) shows a KUKA robotic writing system that uses a force/torque sensor and a web camera to get feedback for controlling a stylus tool. This robot equipped with a proper pen was reprogrammed to inscribe the entire *Martin Luther* bible onto endless paper in a calligraphic style (Figure 8.40(b,c)).

Figure 8.41(a) shows robotic grasping of a fragile object using a proper force sensitive gripper, and Figure 8.41(b) shows a robot which with the aid of vision and force/tactile feedback can grasp objects such as cups, glasses, balls, etc.).

8.5.2 Virtual Reality (VR), Augmented Reality (AR), and Mixed Reality (MR) Systems

Virtual Reality, **AR**, and **MR** are three technologies with a variety of applications in automation, robotics, engineering, medicine, military, art, movies, etc.

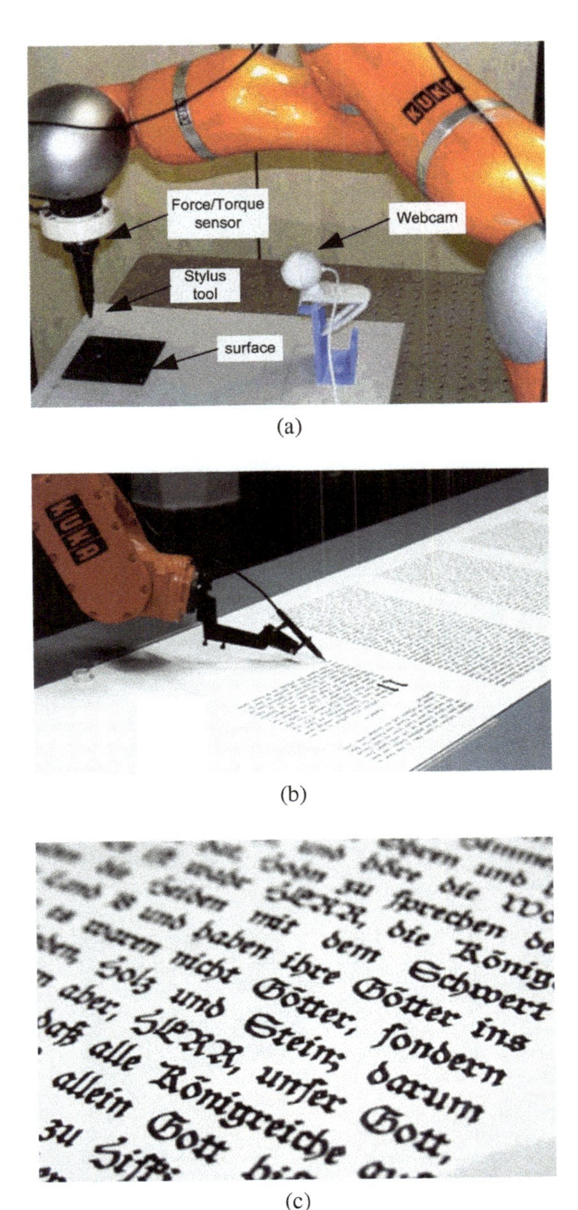

(a)

(b)

(c)

Figure 8.40 (a) A KUKA robotic system capable of writing on a surface using a stylus controlled via force/tactile and visual feedback, (b,c) The robot inscribing Martin Luther bible in a calligraphic style.

Source: www.nitin403.blogspot.gr (/2009/bible-writer-robot-house.html).

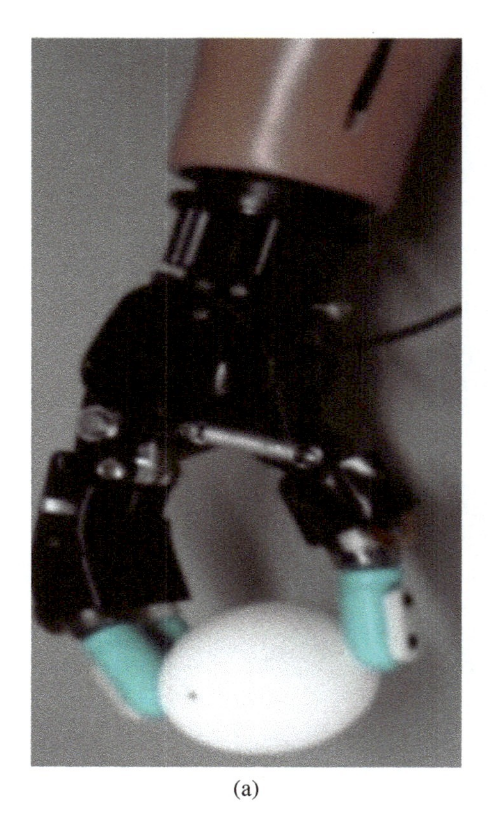

(a)

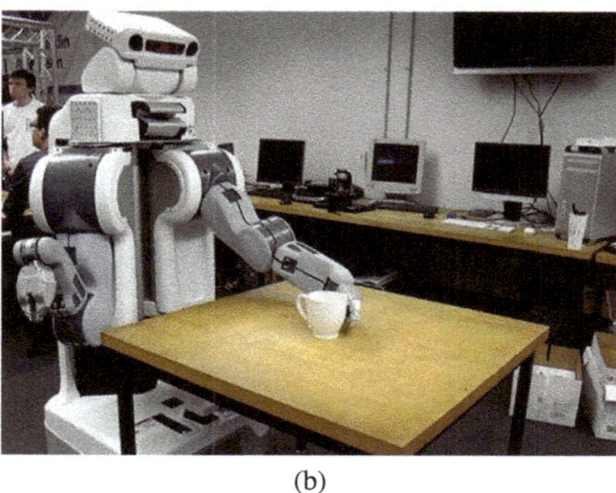

(b)

Figure 8.41 (a) Fragile object robotic grasping, (b) Robot seeing and grasping a cup with suitable cameras (eyes), fingers, and tactile/force feedback control.

Source: (a) [48], (b) www.slideshare.net (Article: G. Dedeoglu and G. Loeb, Robotic Design,: Frontiers in Visual and Tactile Sensing).

Virtual Reality is the creation of artificial/computer-based environments for users to inhabit placing them inside several experiences. The experiences in VR are totally immersive by means of visual, auditory, and haptic simulation such as the constructed reality is almost indistinguishable from the real deal. The user is completely inside it. VR typically requires a headed mount such as Oculus Rift Googles.

Augmented Reality is a VR with one foot in the real world blending the virtual and the real, and involves some type of artificial objects simulation in the real environment. Like VR, AR experiences typically involve some kind of devices through which one can view a physical reality whose elements are augmented (or supplemented) by computer generated sensory input such as sound, video, graphics or GPS data. In AR the real or visual can be easily told apart. AR provides a limited view, while VR is entirely immersive. AR applications are mostly dedicated on visual augmented reality and to some extend on tactile sensations in the form of haptic feedback.

Mixed Reality mixes the best of VR and AR to create a "hybrid reality". MR overlays synthetic content over the real world. MR is similar to VR. Their key difference is that in MR the virtual content and the real world content can interact to one another in real time. This interaction is facilitated by tools typically used in VR (e.g., Google's glass). Figure 8.42 shows typical devices used in virtual reality applications.

Figure 8.43(a) shows a number of typical mobile VR Headsets, Figure 8.43(b) shows the structure of a VR head-mounted augmented reality system, and Figure 8.43(c) shows the functions of a typical VR glasses headset (ProHT).

Figure 8.44 shows a snapshot of testing VR *Oculus* headset, and Figure 8.45(a,b) depicts two examples of VR gloves.

Figure 8.46(a) shows an example of virtual reality system that uses a headset combined with VR gloves, and Figure 8.46(b) shows a system where the Oculus Rift headsets are tested.

8.5.3 VR-based Medical Systems and Simulators

Virtual reality based systems also find important applications in medical education and surgery. They enable students and trained practitioners for better surgical preparation, and use of team training 3D systems. Figure 8.47(a–c) shows three examples of such VR medical systems. Surgical VR simulators have already facilitated training in basic laparoscopic surgery skills, using graphics that enable users to perform dexterous maneuvers as well as complete interventions.

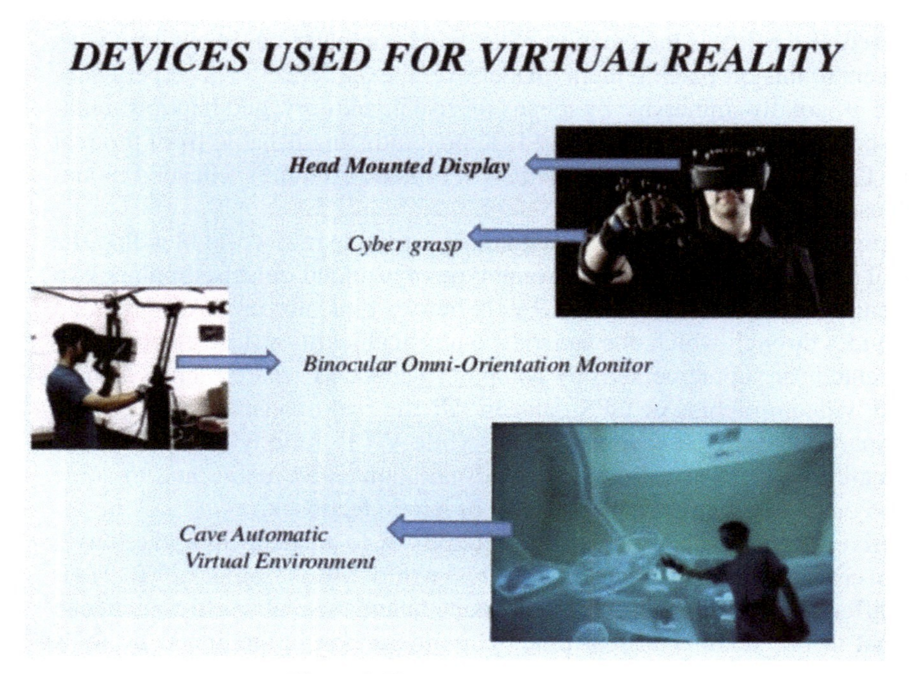

Figure 8.42 Typical VR devices.

Source: https://www.slideshare.net (PP Presentation: Anand Akshay, Virtual Reality).

8.6 Human Factors in Automation

Human factors **(HFs)** play a dominant role for the successful, safe and efficient operation of any modern technological system. In automation, '*human factors*' may be collectively defined as the study of relationships and interactions between equipment, processes and products, and the humans who make use of them. This interaction, on the part of the human involves issues of different nature: physical, psychological, cognitive, and thinking factors. As it is the case with many fields of science and technology, the field of human factors has begun under the demands of a war (here World War II), where it was required to match the human to the war equipment in terms of size, strength and sensing-cognitive capabilities. Taking into account the human factors, the automation designers try to improve continuously the machines and processes which the human is called to use such that to be more '*humane*' in many aspects.

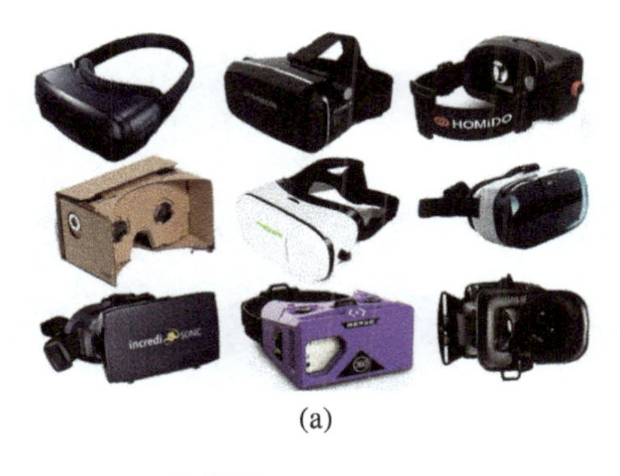

(a)

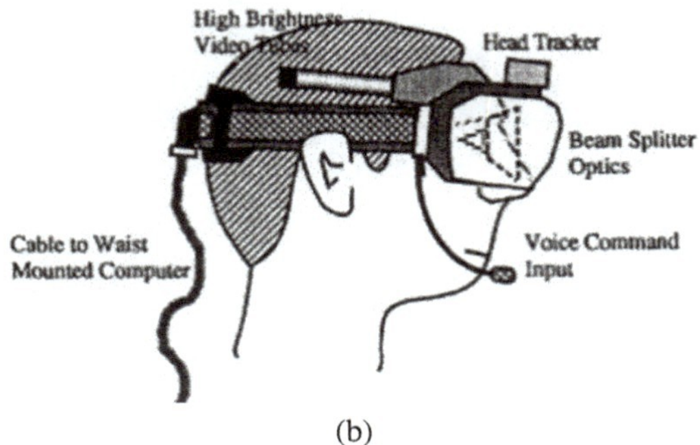

(b)

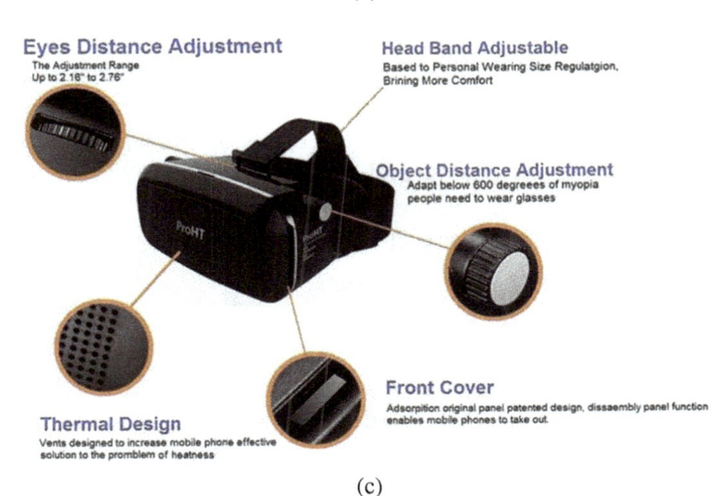

(c)

Figure 8.43 (a) Mobile VR headsets, (b) Components of an early head-mounted AR device, and (c) Typical VR glasses headset.

Source: www.moa.zcu.cz (Article: Virtual Reality Devices) and www.inlandproduct.com

Figure 8.44 Testing an Oculus headset.

Source: www.tested.com/tech/1215/why-virtual-reality-failed-and- why-there-still-hope/

The Human Features In the human factors field the human can be considered as a black-box, where her/his behavior is determined only by *stimuli* as inputs to the box and *response* as outputs. The dominant human features of concern are:

- *Physical* strength, sensory, and perceptual limits, etc.
- *Cognitive* e.g., a human cannot carry out in her/his mind large numbers of calculations without computation assistance.
- *Intellectual* e.g., does there exist a preferred way of information processing by the human which must be embedded in the design?
- *Motivational* e.g., can we comprehend human behavior without reference to her/his motivation, or is it necessary to take into account this motivation?

Human–Automation Relation In the past, the human's role in automation system was simply that of a pure executor (or controller). The operator was controlling a device, monitoring its performance, and modifying it via a specified sequence of actions. With the development of computerized systems the human's role has changed. The human is monitoring the system performance, but it is the computer software that controls individual equipment(s) to perform in desired ways. Today, the human is a partner of the machine, her/his role is to perceptually recognize the stimuli, and cognitively interpret the meaning of those stimuli. This means that the concept of error changes definition, which implies that the measurement process also changes.

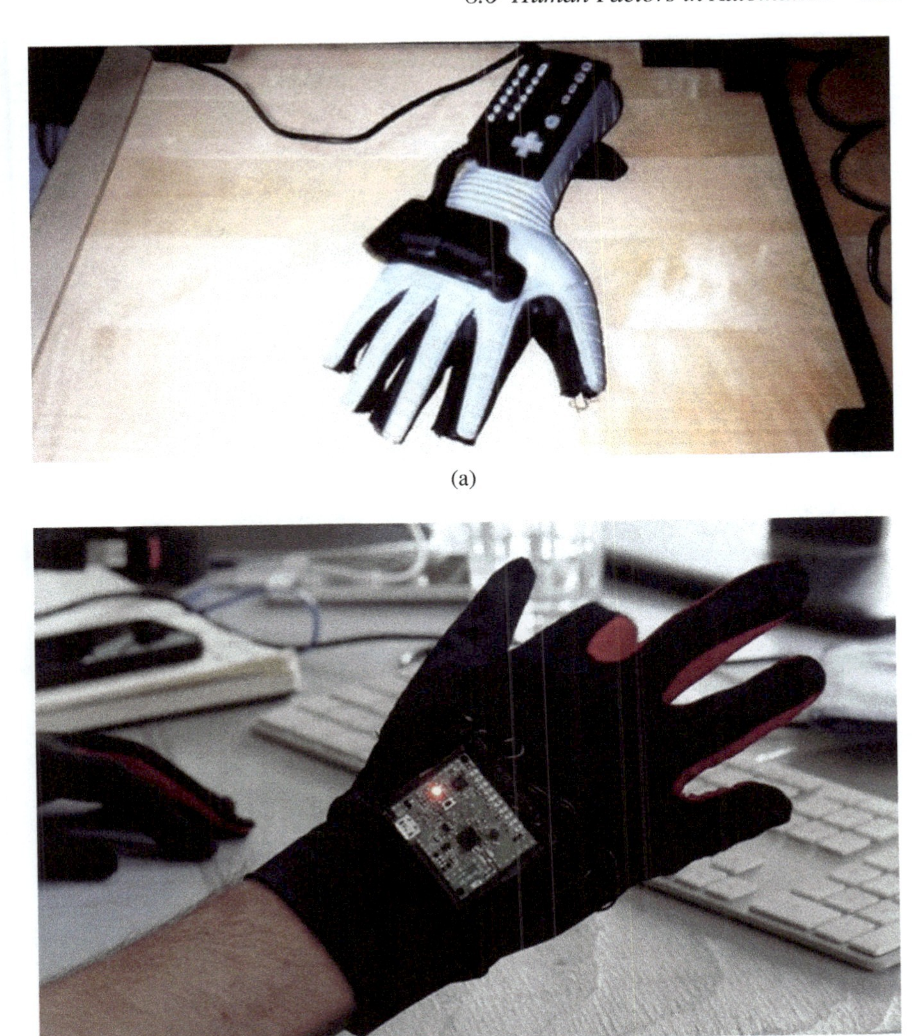

(a)

(b)

Figure 8.45 Two examples of VR gloves, (a) NES power glove, (b) The Manus glove.

Source: (a) mi.mu dev-blog: Data Gloves Overview, (b) www.tested.com/tech/ (Article: why-virtual-reality-failed-and-why-there-still-hope) and www.makery.info (Article: How to prototype in Virtual Reality, 23 June 2015).

It is clear, that automation affects people because it is a source of stimuli to which the human responds. This response is not shown as an overt behavior because the human is not a controller. The response can be a change in the

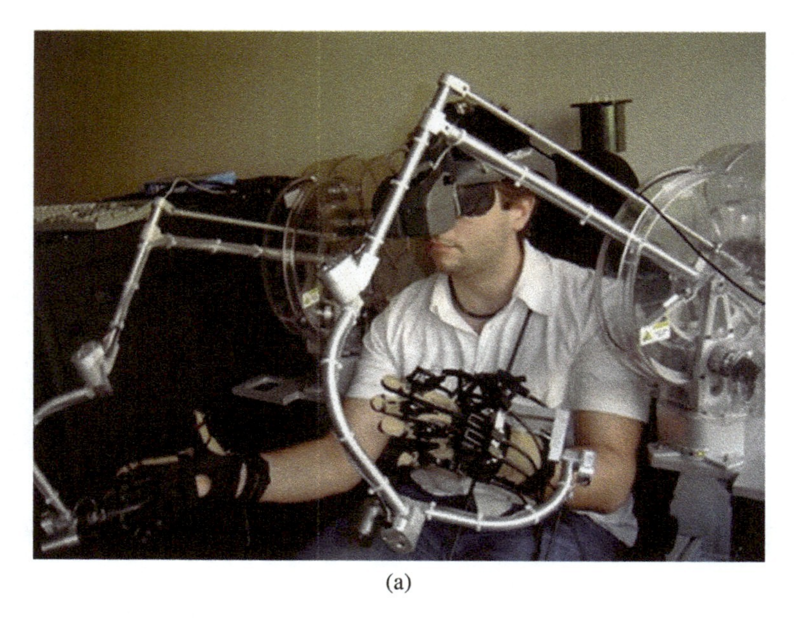

(a)

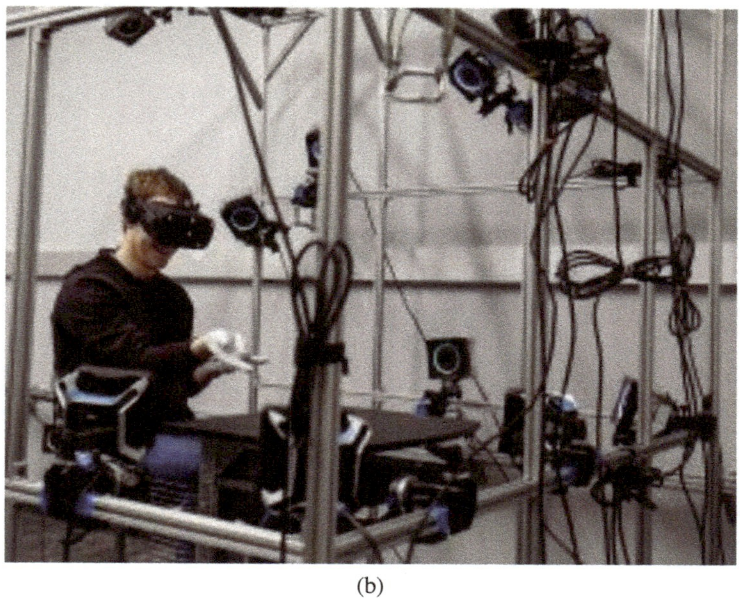

(b)

Figure 8.46 Virtual reality systems for robotic and other applications.

Source: (a) www.cyberglovesystems.com (haptic-workstation), (b) www.pcgames.com (Article: Mark Zuckerberg tests new Oculus Rift Gloves).

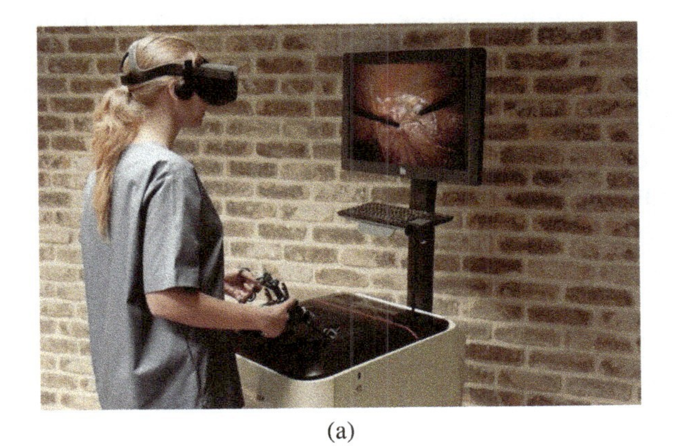

(a)

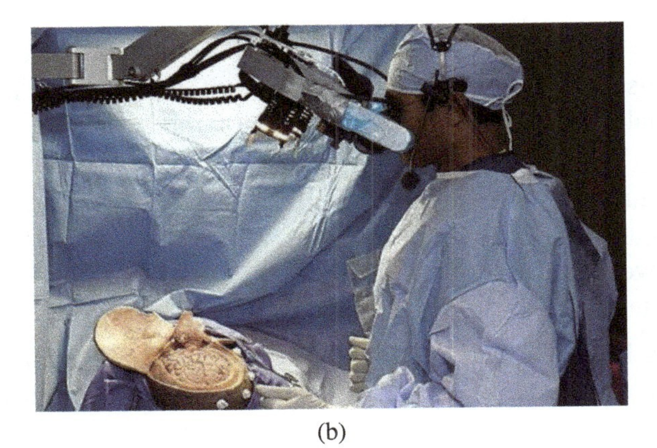

(b)

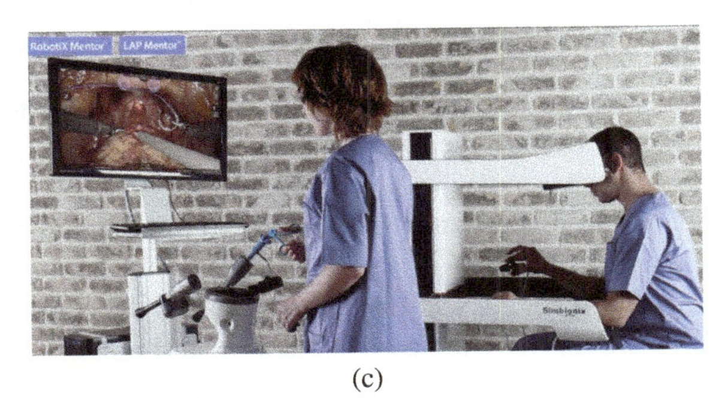

(c)

Figure 8.47 Three instances of VR-based medical training.

Source: (a) https://healthcaretraininganddeduction.com (Article: 3D Systems Combine VR Headsets, Simulation for Immersion in Surgical Training). (b) https://glozine.com (Article: Virtual Reality Finds its Way into Surgery). Also www.neurosurgerycns.files.wordpress.com and (c) www.hypergridbusiness.com

observer's attitude on her/his concept structure. Since one of the major goals of the human factors field is to make the humans (employees, etc.) happier with their technology, then the responses of them must be taken into account. Human factors consider human errors first in every piece of equipment and every system we design.

Examples of human factors are;

- Overload.
- Stress.
- Fatigue.
- Distraction.
- Situational awareness.
- Attention.
- Communication.
- Memory.

Figure 8.48 illustrates that human factors are affected by the following influences, actually lying at the cross-sections of the respective influence domains:

- Management systems.
- Equipment and facilities.
- Tasks and work processes.
- Environment.

Human factors in performance appraisal A performance appraisal is a process of evaluating an employee's performance on a job in terms of its requirements. It is the process by which the management finds out how effective it has been hiring and placing employees. Human biases/errors in appraisals include the following:

- *Halo effect*: Tendency to rate employee high or low on all factors.
- *Similarity*: Tendency to give special emphasis to qualities that appraisers perceive in themselves.
- *Recency*: Tendency to give greater importance to job behaviors near the end of the measurement period.
- *Central tendency*: (i) Tendency to rate all employees the same way, e.g., rating them all average, (ii) The unwillingness to assign extreme ratings.
- *Unclear standard*: An appraisal that is too open to interpretation.

Automation Today we look at the human (employee/worker) as more than a unit of production, since there is a converse relationship between automation and human. Of course, the human determines the type and extent of automation, but once developed the automation determines to a certain (sometimes

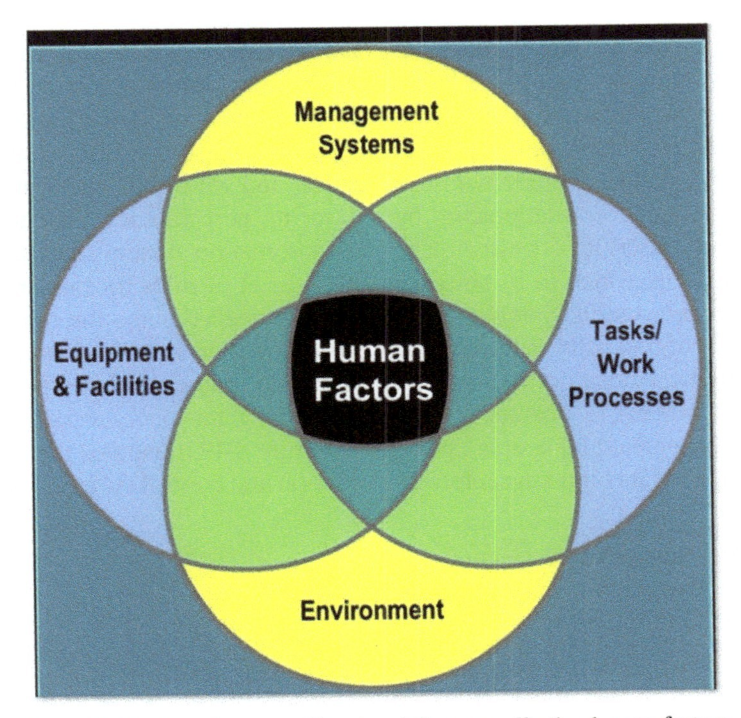

Figure 8.48 Ven diagram of four key influences affecting human factors.
Source: www.mideng.com (Article: Human Factors and Ergonomics).

high) degree how the humans behave. The elements of automation vary from the *molecular level* (resistors, transistors, microchips, etc.), where the human's involvement is just to repair or replace a component that fails, to the *simple tools* level where the human's attention is restricted to the tool at hand, and finally to *complete systems*, where the human's intervention becomes more sophisticated and is expressed by high level human functions. The three stages (degrees) of automation are the following:

- *Mechanization* (Mere replacement of a human performance with that of a machine).
- *Computerization* (Replacement of a human performance by a machine which is now a computer, hardware and software. The replacement is now more precise, more extensive, quicker, etc.)
- *Artificial Intelligence* (The degree of human replacement by artificial intelligence software is increasing but it is still very limited).

The automation (technology) is one of the primary stimuli of human behavior, since it does not only change the human role in immediate interaction with physical objects, but serves as a backcloth for almost all human actions. The human factors scientist is not a pure observer like sociologists, anthropologists, etc., but an *activist* whose education and role is to intervene in the human–automation relationship, by measuring performance, producing a stimulus and observing its effect. Complex automation systems inevitably need organization, which in its simplest form is a set of rules for humans to interact with and within systems. These rules are needed since the human-system interaction usually determines the efficiency of system performance. Simple equipment does not require an organization, but a multi-component system with humans to operate it does need some kind of organization to harmonize equipment and people operations. Figure 8.49 illustrates the three major functions of any human centered automation and control system. These functions are:

- **Perceive** (context assessment).
- **Select** (adaptation manager).
- **Act** (automation, HMI).

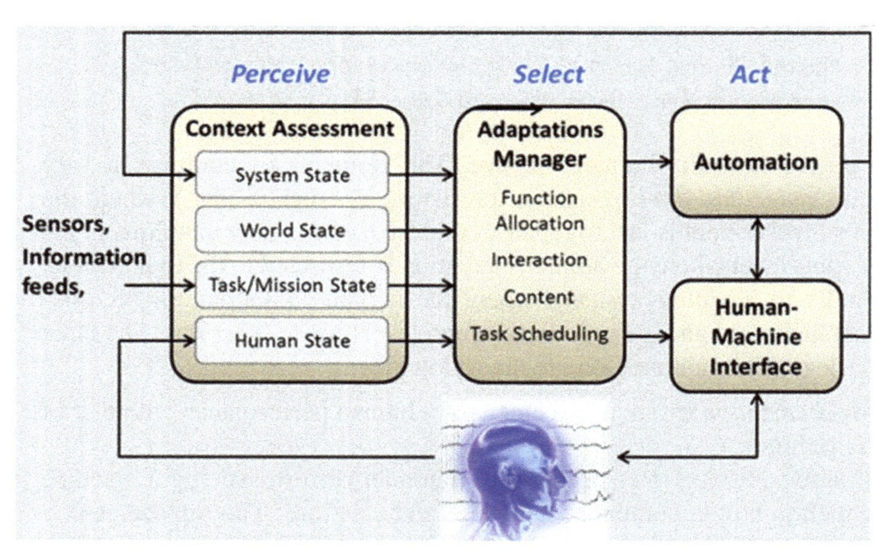

Figure 8.49 The three fundamental functions of automation and control systems, and their sub-functions.

Source: www.imse.iastate.edu/domeich/files/2012/12/image1.pngwww.s2is.org/Issues/v3/n3/papers/paper8.pdf

Figure 8.50 illustrates what is the beneficial role of **HCA** in factories and companies for the four principal players/actors involved, namely for:

- Customers, suppliers.
- Shareholders.
- Employees and partners.
- Society.

For activating and enhancing employees of an automated company, it is necessary and important to understand their day-to-day functioning and the overall contribution of the company to the society and world.

Three Key Human Factors in Automation The fact that human workload may be considerably increased in automated systems (contrary to the design goal) does not mean that automation is necessarily harmful, per se, but that a more careful implementation of automation is needed in modern complex large-scale systems. Three human factors that play a key role and

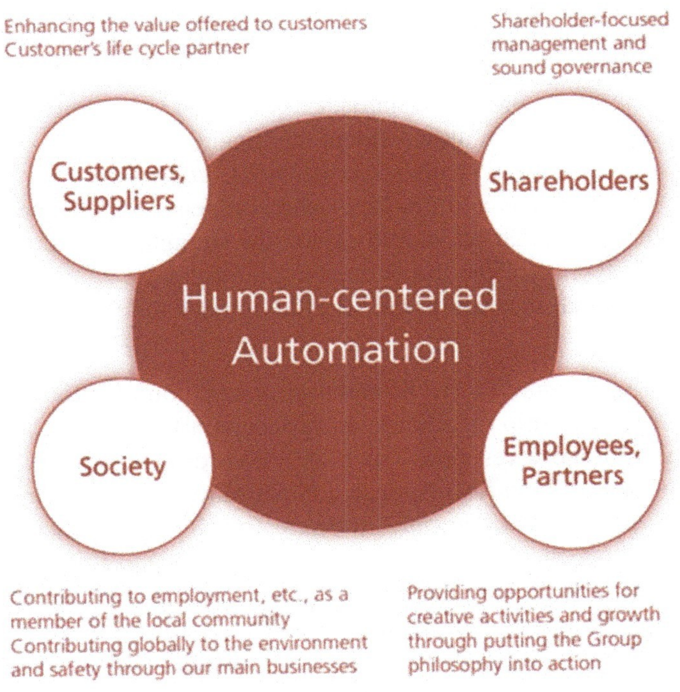

Figure 8.50 Benefits of human-centered automation for customers, suppliers, shareholders, employees, partners, and general society.

Source: www.azbil.com (/ir/management/interview).

should be considered in the design and implementation of flight decks (also applicable to other sophisticated automation systems) are the following:

- *Allocation of function.*
- *Stimulus response compatibility.*
- *Internal model of the operator.*

Allocation of Function: One of the basic questions in human factors engineering is whether a specific task is better to be assigned to a human or to automation. A first approach that was followed and applied to answer this question was based on lists of the functional merits and disadvantages of people versus machines. This approach, although logical, was not successful since machines and humans are not really comparable entities, even though the human factors professionals relate them in many details. The traditional *on–off* interface (automation is either *on or off*) does not lead always to less workload of the operator (pilot) as more automation is embedded. Thus, an intelligent interface is needed that would take into account both the workload of the operator (pilot) and the state of the system environment (weather, fuel load, and status of equipment), and recommend an optimal allocation of functions, where the pilot decides what flight automation should be used at any moment in the flight. Present flight deck automation is machine-centered (not human-centered) and has little resident intelligence.

Function/task allocation is also known as '*Compensatory Principle*' according to which humans and machines can compensate for each other's weaknesses. This means that we should exploit the strengths of both humans and machines differently. The fundamental premise here is to give the machines the tasks that they are good, and the humans the things that they are good at.

Stimulus-Response Compatibility: Stimulus-response compatibility is concerned with the relationship (geometric and conceptual) between a stimulus (such as a display) and a response (such as a control action) [34]. In aircrafts, an argument of stimulus response compatibility is the determination of the relative advantages of *moving airplane* (outside in) versus *moving horizon* (inside out) artificial horizon indicators. According to Kantowitz and Sorkin a better display has both moving. Two cases recorded in the literature with low stimulus–response compatibility are: The altitude deviation in the MD-80 under control of flight deck automation, and the vertical navigation (VNAV) functions of the FMS. In both cases the systems are difficult to be used correctly, but with improved training the frequency of error can be decreased. In general, automated systems which have low stimulus-response compatibility create extra work load to the pilot and lower trust in flight deck automation.

Internal Model of the Operator: This model describes the operator's internal representation of the controlled system (i.e., her/his conceptual understanding about the system components, processes, and input/output quantities). Unavoidably, the internal model varies considerably as a function of individual operators, tasks and environments. The internal model is used by the operator as a basis for planning future activities, deviating hypotheses about. It must be remarked here that the humans are poor monitors of automation. The human's physiological capabilities are not suitable for continuous monitoring and vigilance. The bandwidth of the human nervous system is not sufficiently wide to face sudden information load which may arise in critical situations of automated systems. We have many examples of this, such as nuclear reactors, chemical process plants, commercial and military aircrafts, etc. The human fails statistically at the low end of the bandwidth (less than 1 Hz), and fails almost surely at frequencies higher than 1 Hz. Older studies of human vigilance produced the arousal theory of vigilance, which postulated that the level of physiological arousal falls during a vigilance task, leading to the traditional vigilance reduction over time. Newer studies on arousal have linked arousal to the deployment of attention resources. The results of these studies show that vigilance tasks (even very simple) can impose considerable mental workload in the form of decision making and problem solving. *Thus the claim that automation always reduces workload is not correct.* As we already mentioned, automation changes the pattern of workload across work segments, and may increase the human monitoring work load because of the demand to monitor automation. In general, monitoring and vigilance are considered by humans as '*unstimulating*' and '*undesired*' tasks. Nevertheless, in all cases the human operators are assigned the task of monitoring and vigilance, and system developers are trying to achieve systems compatible with the human capabilities. Full automation of the monitoring process, is not a general solution, since automated monitors will increase the number of alarms which are already high in many cases, and human response to multiple alarms raises many human factors concerns. On the other hand, to protect against failure of automated monitors, a higher-level system that monitors the automated monitor is required, a process that could lead to infinite regress.

8.7 Conclusions

Automation is one of scientific/technical fields that play a dominant role in the development of modern society and the improvement of human quality of life. Three issues that need to be carefully examined for achieving a high degree of success and effectiveness when the concept of automation is applied in large-scale and complex modern systems, and guarantee safe operation in high risk situations are the following:

- Feedback.
- Task allocation.
- Operator reliance.

Feedback

Proper feedback is needed for the appropriate monitoring of the human actions performed, and correction of errors, and keeps alert. It is noted that many of the current problems are a result of automation in the sense that the automation is inappropriately designed and applied [36]. As explained extensively through this book, good feedback for the appropriate operation (safe, reliable, efficient, etc.) of a system is always an essential aspect of all control systems. But, adequate feedback to the human operators is absent far more than it is present, whether the system be a computer, a process or power plant, an autopilot, can industrial robot, etc. Without appropriate feedback, people are indeed out of the loop; they may not know if their request has been received, if the actions are performed properly, or if problems occur.

Task Allocation

Task allocation influences the actual overall properties of the designed system both from the view point of human factors and the view point of productivity and quality of products. Prior to task allocation, the tasks must be carefully analyzed. Actually, there exist many popular techniques for performing task analysis [37]. Current task analysis methods start with listing the sequential steps, and proceed to the specification of the logical conditions and probabilities for the transition from one step to another, the specification of whether a human or machine performs each step and so on. According to the MABA–MABA list [38].

- *Men are better for:* (i), detecting small quantities of visual, auditory or chemical energy, (ii) perceiving patterns of light and sound, (iii) improving and using flexible procedures, (iv) Storing information for extended periods of time, and recalling appropriate parts, (v) carrying out inductive inference, and (vi) performing judgment.
- *Machines are better for:* (i) responding quickly to control signals, (ii) exercising large forces accurately and smoothly, (iii) stirring information briefly, and erasing it completely, and (iv) carrying out deductive reasoning.

However, in allocating tasks to humans and machines, the different nature of them has to be understood as far as possible.

Operator Reliance

The decision of an operator to rely or not rely on automation is one of the most critical decisions that must be made during the operation of any complex system. Victor Riley [39] mentions an aircraft crash during a test flight, killing all seven on board, due to only 4 s delay of the pilot to retake manual control. Actually, there are many cases in the history of automation which have shown that the decision to rely or not rely on automation has been a critical link in the chains of events that have led to an accident (aircraft, rail, ships, nuclear plants, electric power plants, medical systems, process control, etc.). The two extreme possibilities are: (i) the operator over-relies on automation and fails to monitor or examine its performance, or (ii) the operator may not rely at all in automation because she/he has high (possibly erroneously) confidence in her/his own abilities to carry out the jobs manually. Many aircraft accidents belong to the first possibility; the Chernobyl nuclear accident belongs to the second category. The issue here is to understand and study the factors and biases that affect the human reliance on automation, and incorporate them in designing the training program of the operators. With reference to this problem, Sheridan and Ferrel [40] have expressed strong concern about the changing roles of human operators and automation, and have incorporated the operator's trust in automation as one of the primary issues of *supervisory control functions.*

References

[1] Bagrit, L. (1965). *The Age of Automation, The New American Library of World Literature*. New York, NY: Mentor Books.

[2] Francoit, W. (1964). *Automation, Industrialization Comes of Age*. New York, NY: Collier Books.

[3] Dunlop, J. T. (1962). *Automation and Technological Change: The American Assembly*. Upper Saddle River, NJ: Prentice-Hall.

[4] Thomas, R. E., and Christner, A. A. (1964). *The Effect of Various Levels of Automation on Human Operator's Performance, in Man-Machine Systems, Wright Patterson Air Force Base*, Report AMRL TDR-63-76, Dayton, OH.

[5] Dieterly, D. L. (1979). Preliminary Program Plan: Automation's Impact on Human Resource/Management. Mountain View, CA: NASA-Ames Research Center.

[6] Tzafestas, S. G. (2010). *Human and Nature Minding Automation: An Overview of Concepts, Methods, Tools and Applications*. Berlin: Springer.

[7] Smith, D., and Dieterly, D. L. (1980). *Automation Literature: A Brief Review and Analysis*. NASA Technical Memorandum.

[8] Mueller, E. (1969). *Technological Advance in Expanding Economy: Its Impact on a Gross-Section of the Labor Force*. Ann Arbor, MI: University of Michigan.

[9] Ratner, R. S., and Williams, J. G. (1973). *The Air Traffic Controller's Contribution to ATC System Capacity in Manual and Automated Environment's*. Terminal Operations, vol. 3, FAA RD-72-63. Washington, DC: Systems and Research Development Service.

[10] Noll, R. B., and Simpson, R. W. (1973). *Analysis of Terminal ATC Systems Operations*. Technical Report TR-73-74, dot tsc-103-73-4. Burlington, MA: Aerospace Systems, Inc.

[11] Sheridan, T. B. (2002). *Humans and Automation: Design and Research Issues*. New York, NY: Wiley.

[12] Tzafestas, S. G., and Pal, J. K. (eds). (1990). *Real-Time Microcomputer Control of Industrial Processes*. Boston, FL: Kluwer.

[13] Mukherjee, A. (2013). Process automation versus power system automation. *Int. J. Sci. Eng. Res.* 4, 88–96.

[14] Skrokov, M. R. (ed.). (1990). *Mini and Microcomputer Control in Industrial Processes*. New York, NY: Wiley.

[15] Driscoll, J. W., Sirbu, M., Alloway, R., Hammer, M., Harper, W., and Khalil, M. (1980). *Office Automation: A Comparison of in House Studies*. Cambridge, MA: MIT, Center for Policy Alternatives.
[16] Zisman, M. D. (1978). Office automation: evolution and revolution. *Sloan Manag. Rev*. 19, 1–16.
[17] Allison, Q. (1971). *Essence of Decisions: Explaining the Cuban Missile Crisis*. Boston, MA: Little Brown.
[18] Driscoll, J. W. (1979). People and the automated office. *Datamation* 25, 106–116.
[19] Hunt, V. D. (1983). *Industrial Robotics Handbook*. New York, NY: Industrial Press, Inc.
[20] Nof, S. Y. (ed.). (1999). *Handbook of Industrial Robotics*. New York, NY: Wiley.
[21] Taylor, B. H., et. al. (1996). *Computer Integrated Surgery*. Cambridge, MA: MIT Press.
[22] Katevas, N. (ed.). (2001). *Mobile Robotics in Health Care*. Amsterdam: IOS Press.
[23] Helaf, S., Mokhtari, M., and Abdulrazak, B. (eds). (2008). *The Engineering Handbook of Smart Technology for Aging, Disability, and Independence*. Hoboken, NJ: Wiley.
[24] Tzafestas, S. G. (2016). *Sociorobot World: A Guided Tour for All*. Berlin: Springer.
[25] Sarter, N. B. (1996). "Cockpit automation, from quantity to quality, from individual pilot to multiple agents," in *Automation and Human Performance; Theory and Applications*, eds R. Parasuraman and M. Mouloua (Mahwah, NJ: Erdbaum), 267–280.
[26] Edwards, E. (1977). Automation in civil transport aircraft. *Appl. Ergon*. 8, 194–198.
[27] Wiener, E. L. (1993). "Crew coordination and training in the advanced technology cockpit," in eds E. L. Wiener, B. G. Kanki and R. L. Helmreich, San Diego, CA: Academic.
[28] Billings, C. E. (1997). *Aviation Automation: The Search for a Human-Centered Approach*. Mahwak, NJ: Erlbaum.
[29] Wiener, E. L. (1993). "Crew coordination and training in the advanced-technology cockpit," in *Cockpit Resource*, eds E. L. Wiener, B. G. Kanki and R. L. Helmreich (San Diego, CA: Academic).
[30] Sarter, N. B., and Woods, D. D. (1992). Pilot industrial with cockpit automation: operational experiences with the flight management system. *Int. J. Aviat. Psychol*. 2, 303–321.

[31] Sarter, N. B., Woods, D. D., and Billings, C. E. (1997). "Automation surprises," in *Handbook of Human Factors and Ergonomics* ed. G. Salvendy (New York, NY: Wiley), 1926–1943.

[32] ICAO *Annexes to the Convention on International Civil Aviation*. Montreal, QC: ICAO.

[33] Helander, M. G., Landauer, T. K., and Prablu, P. (eds). (1997). Amsterdam: North-Holland.

[34] Kantowitz, B. H., Triggs, T. J., and Barnes, V. (1990). "Stimulus-response compatibility and human factors," in *Stimulus-Response Compatibility*, eds R. W. Proctor and T. Reeves (Amsterdam: North-Holland), 365–388.

[35] Kantowitz, B. H., and Bitner, A. C. Jr. (1992). "Using the aviation safety reporting system as a human factors research field," in *Proceedings of the 15th IEEE Annual Aerospace and Defense Conference*, Piscataway, NJ, 31–39.

[36] Norman, D. A. The problem of automation: Inappropriate feedback and interaction, not over-automation.

[37] Sheridan, T. B. (1997). "Task analysis, task allocation, and supervisory control," in eds M. Helander, T. K. Landauer and P. Priblu (Amsterdam: North-Holland), 1295–1327.

[38] Fitts, P. M. *Human Engineering for an Effective Air Navigation and Traffic Control Systems*. Columbus, OH: Ohio State University Research Foundation Report.

[39] Riley, V. (1996). "Operator reliance on automation," in eds R. Parasuraman and M. Mouloua (Mahwah, NJ: Erlbaum), 19–35.

[40] Sheridan, T. B., and Ferrel, T. B. (1974). *Man-Machine Systems: Information, Control and Decision Models of Human Performance*. Cambridge, MA: MIT Press.

[41] Lipson, H., and Kurman, M. (2016). *Driverless Intelligent Cars and the Road Ahead*. Cambridge, MA: MIT Press.

[42] Jurgen, R. K. (2013). *Automation Vehicles for Safer Driving*. Warrendale, PA: SAE International.

[43] Meyer, G., and Beiker, S. (eds). *Road Vehicle Automation*. Berlin: Springer.

[44] National Highway Traffic Control and Safety Administration: Automated Vehicles Policy (2013). Available at: http://nhtsa.gov/staticfiles/rulemaking/pdf/Automated_Vehicles_Policy.pdf

[45] https://www.sae.org/misc/pdfs/automated_driving.pdf

[46] https:www.schlegelundpartner.com/news/man-and-machine-automated-driving

[47] Low, T. (2014). "Human factors and vehicle automation: the good, the bad, and the ugly," in *Proceedings of the National Road Safety Conference*, Brighton.

[48] Cadwell, D. G., et al. (2012). "A high performance tactile feedback design and its integration in teleoperation," in *IEEE Transactions on Haptics*, Vol. 5.

[49] Carsten, O., et al. (2012). Control task substitution in semi-automated driving: does it matter what aspects are automated? *Human Factors* 54, 747–761.

[50] Brown, B., and Laurier, E. (2012). "The normal natural troubles of driving with GPS," in *Proceedings ACM Conference on Human Factors in Computing Systems* (Austin, TX: ACM Press), 1621–1630.

[51] Laengle, T., Lueth, T., Stopp, G., and Kamstrup, G. (1995). *KANTRA: A Natural Language Interface for Intelligent Robots*, Intelligent and Autonomous Systems (IAS-4) (Amsterdam: IOS Press), 365–372.

[52] Chen, J. Y.C., Haas, E. C., and Barnes, M. J. (2007). "Human performance issues and user interface design for teleoperated robots," in *Proceedings of the IEEE Transactions Systems, Man, and Cybernetics-Part C: Applications and Reviews*, 37, 1231–1245.

[53] Gonzalez, R. C., and Woods, R. E. (2002). *Digital Image Processing*. London: Prentice Hall.

[54] Tzafestas, S. G., and Mantelos, A.-I. (2013). "Time delay and uncertainty compensation in bilateral telerobotic systems: state of art with case studies," in Engineering Creative Design in Robotics and Mechatronics, eds M. K. Habib and P. Davim (IGI Global), 208–238.

9

Societal Issues

A collection of human beings... becomes a society only... when one individual has an effect, immediate or mediate, upon another... If there is to be a science whose subject matter is society and nothing else, it must exclusively investigate these interactions.
Georg Simmel

All human beings are born free and equal in dignity and rights.
Universal Declaration of Human Rights

All that is valuable in human society depends upon the opportunity for development accorded to the individuals.
Albert Einstein

9.1 Introduction

All fields considered in this book (systems theory, cybernetics, control, and automation) have societal implications which are overlapping and/or complementary. Systems theory offers the tools for systematically studying society and aiding its development. Cybernetics studies living beings, technological systems, and societal systems focusing on their communication/information and feedback control processes. In particular, cybernetics drives human evolution in its social forms through a very important process, namely conversation. Cybernetics is a theory of responsive (or adaptive) systems, in which reciprocal action-response (feed forward-feedback) cycles form the basic activity of communication. Feedback control is inherent in all living systems, and is embedded in all man-made systems and social systems. Actually, automation systems perform a plethora of operations and activities that can be monitored and controlled at several structural and operational levels. They, involve a variety of successful models, and control, supervision and communication techniques. All these systems in one or the other way influence the operation and growth of social systems.

The purpose of this chapter is to study the influence on society of systems, cybernetics, control, and automation at a detail deemed sufficient for the reader to understand and appreciate it. Specifically, after a quick look at the question of 'what is society', the chapter does the following:

- Discusses the role of systems theory in the society development and shows how its environments (internal, external) are involved in this development.
- Investigates the relation of cybernetics and society (sociocybernetics, behavioral cybernetics, economic cybernetics).
- Discusses the general impact of control on industrial systems, and the impact on two principal societal systems, namely management systems and economic systems.
- Presents key issues of the impact of automation on society (namely, productivity and capital formation, quality of working conditions, and unemployment), including the impact of robotics, and the impact of office automation on organizations.

9.2 What is Society?

The term *society* has its origin in the Latin *societas* from *socius,* and the French *societe* which means companion, chum, comrade, associate or fellow partner. At Dictionary.com one can find several alternative meanings of the word society. Some of them are the following:

- A group of humans broadly distinguished from other groups by mutual interests, participation in characteristic relationships, shared institutions, and a common culture.
- An organized group of persons, associated together for religious, benevolent, cultural, scientific, political, patriotic, or other purposes.
- A body of individuals living as members of a community.
- A highly structured system of human organization for large-scale community living that normally furnishes protection, continuity, security, and a national identity to its members.
- An organization or association of persons engaged in a common profession, activity or interest.
- The totality of social relationships among humans.

- In biological terms: society is a closely integrated group of social organisms of the same species exhibiting division of labor.

According to sociologist *Jenkins* [1] the term society refers to critical existential issues of humans, namely:

- The way humans exchange information, including both the sensory abilities and the behavioral interaction.
- Often community-based performance and phenomena cannot be reduced to individual behavior, i.e., the society's action is 'larger than the sum of its parts'.
- Collectives have usually life spans, going beyond the lifespan of individual members.
- All aspects of human life are tied in a collective sense.

According to *Herbert Spencer* "Society is an organism of functionally interdependent parts evolving through structural differentiation" (*The Study of Sociology*).

According to *Alston* [2] *'Society must be understood as a combination of functional, cognitive, and cultural systems'*. This view is illustrated by his 'triangle heuristic model of society' (Figure 9.1).

- **Cognition** is interpreted to be the issues that help people understanding the difference between *'what is'* versus *'what ought to be'*.
- **Culture** includes the social imperatives such as group, values, status, role, authority, ideology, etc.

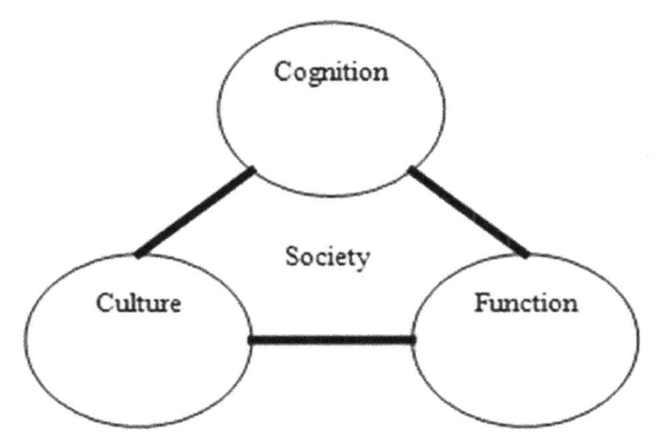

Figure 9.1 Alston's heuristic model of society [2].

- **Function** includes the institutional aspects of society, namely: norms (i.e., rules that govern activity and behavior), moral statements, and set of obligations and expectations. *Norms* are combined to produce *'roles'*. A role set specifies the individual. *Groups* match similar persons and interests, and interact by means of institutions and institutional complexes. *Institutions* are formed at the proper levels, namely community, local state, nation, international, as it may be required.

Society is sometimes separated from *culture.* According to *Clifford Geertz 'society is the organization of social relations, while culture emerges from symbolic forms (beliefs, ideologies, etc.)'*. In political science, the term *society* usually includes all the human relations, typically in contrast to the rulers (government, etc.) within a territory (State). An *ideology* links individual perceptions to those dominating in the overall society, and provides a base for consensus at the group level. In other words ideology is a way of looking and perceiving the world, shared by the members of a community. A society changes as any one of its elements changes persistently. Change results in stress, but it is needed for adaptation and adjustment to new internal and external conditions. When the cognitive, cultural and functional changes, accumulated over time, are no longer compatible with the ideology that served to interpret them, and solutions to social problems are no more possible within the existing systems of information and social organization, then *revolutionary change* takes place (as, e.g., the revolutionary change occurred in Western societies between the medieval period and the modern era). The study of *human society* is still continuing. Sociologists investigate the human behavior from different viewpoints such as cultural, economic, political and psychological perspectives, both qualitatively and quantitatively.

Basic issues/factors included in these studies are the following:

- General individual and group behavior (Genetic inheritance and social experience/organization factors and issues).
- Cultural factors and norms (tradition, beliefs, etc.).
- Social conflict factors (internal and external).
- Social trade-off factors (material or economic).
- Economic factors.
- Political factors.
- Professional and scientific society factors.
- Interaction of societies (cultural, civil, technological, etc.).
- Immigration factors.
- Social change/evolution factors.

Some free online information sources on these issues of human society diachronically are provided in [3–5]. Figure 9.2 shows members of a society attending an outdoor public speech, and Figure 9.3 shows a community protest.

For a social system to be sustainable there are several factors and requirements that must be satisfied. If some of them are not satisfied, the social

Figure 9.2 Society members attending a public speech.

Source: www.metabolomics2012.org

Figure 9.3 Members of a community protesting against tuition hikes.

Source: www.irregularities.com/2012/03/06/68-people-arrested-while-protesting-tuition-hikes-in-california/

system may break down. Therefore, it is essential for our society to know all the causes of social system breakdown and take care to control them before they occur. Ten major causes are depicted in Figure 9.4. *Sustainable development* of a society is a concept first introduced in the 1972 Stockholm Conference on Human Environment and further advanced in several *United Nations* (**UN**) Earth Summits where most countries actively participated.

According to *Gro Harlem Bruntland*, 'sustainable development means meeting the needs of the present generation without compromising the ability of future generations to meet their own needs'.

According to *Muscoe Martin* (1995) 'sustainability means holding up or to support from below'. Thus a society must be supported from below by its members, present and future.

UN established a *Commission for Sustainable Development* to further develop the principles and practice of sustainable development and coordinate the implementation of them worldwide.

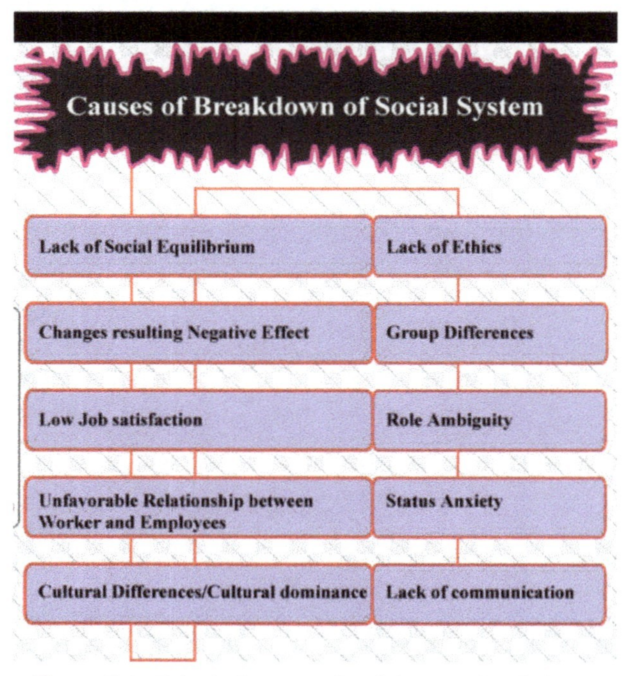

Figure 9.4 Principal causes of social system break down.

Source: www.pinterest.com

Figure 9.5(a) shows the principal players/actors of sustainable development (civil society, government, and business), and Figure 9.5(b) illustrates the essential dimensions of society (culture, polity, and economy) in relation to nature, the human being, and sustainable development.

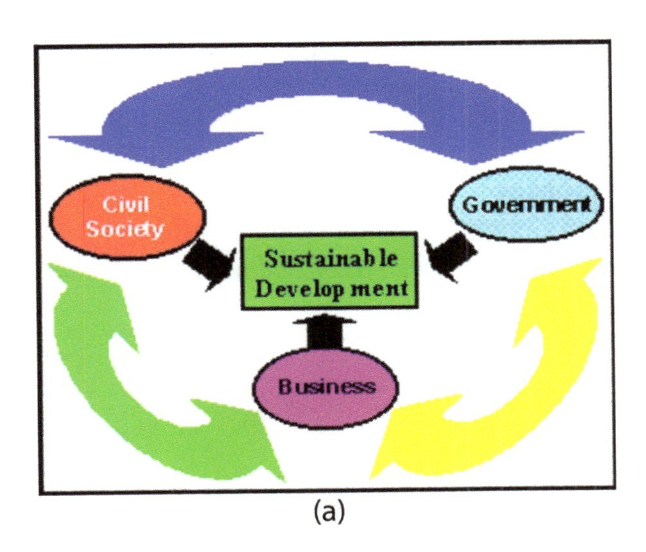

(a)

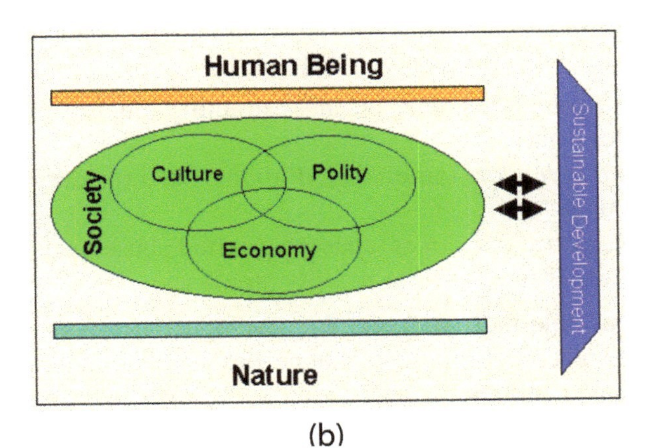

(b)

Figure 9.5 (a) Principal actors in sustainable development defining the domains where the key sectors of development are active. These domains provide the actors the substance for their dialog and interaction, (b) Primary dimensions of society with reference to human being, nature and sustainable development.

Source: www.psdn.org.ph (agenda21/unity.htm).

Figure 9.5(a) indicates that the substance of sustainable development lies in the harmonic integration of social cohesion equilibrium, strong and viable economy, and responsible and effective government to assure that development is a process of improvement.

General principles for sustainable development of a country, beyond sovereignty, include (Philippine Agenda 21):

- Viable and broad-based economy.
- Global cooperation.
- Gender sensitivity.
- Sustainable population.
- Community-based resource management.

Generally, sustainable development targets (not inclusively) to the following:

- Social equity.
- Job assurance.
- Poverty reduction.
- Good government.
- Peace and solidarity.
- Environmental protection and control.
- Ecological integrity (resource conservation, etc.).
- Overall 'quality of life' improvement.

Principal examples of indicators presently used, that tell us whether we are on the right path to achieve sustainability, are [35]:

- *Economic indicators* (income, business, training, quality of life, and human development).
- *Environmental indicators* (air quality, drinking water quantity and quality, and land use).
- *Societal and cultural indicators* (abuse, e.g., child abuse, racism perception, and volunteer rate for sustainability processes).

The selection of suitable indicators and the development of a sustainability program is a large and complex process that needs the collaboration of public and private bodies and entities.

9.3 Systems Theory in Society Development

In this section we will outline some issues concerning the influence of systems theory to *community/society* development [7]. Social development is a very complex process and cannot be described in a clear, unique, and accurate organized way. Social development involves a large number of elements and sub-process. Therefore, systems theory is a good approach to the study of society development, and the organization of the information that refers to it can be effectively done through the use of systems theory. All systems theory concepts (open and closed systems, feedback, system boundary, transfer of energy and information, and environment apply to the society concept). Many of the central concerns in social development, e.g., assessing power and influence, understanding the dynamics of intergroup relationships, and considering the changes involved in planning development activities, can be well understood and described by systems theory. In particular, the description of various environments related to a system, and the concepts of entropy and energy can be profitably used in explaining social development, as summarized below.

Energy: *Energy* can be thought as representing various things and influences that pass across the boundaries of systems. Such influences on the society systems are: food in order to survive, psychological energy, social power, etc. This *'energy'* which is often in the form of *'information'*, is usually the main product of human relationships, and is a basic necessary element in the functioning of societal systems. Social energy can be of several forms, and different people in a society hold various amounts of these kinds of power. Some types of influences are able to help societies progress, while other types can be unhelpful. Specifying the beneficial types of energy (influences) and understanding how to help communities gain access to and control them is one of the principal aims of society work.

Entropy: *Entropy* S is a fundamental concept of thermodynamics, and has several different interpretations [6]. According to the second Law of Thermodynamics entropy in nature is increasing ($\mathrm{d}S \geq 0$). Taking the view that entropy represents a measure of randomness, increasing entropy means increasing randomness (reducing order). As a result, systems tend to 'run down', and to progress to a stage of reduced coherence and eventually completely random order (disorder).

Families can be regarded as systems. Husband/wife bond is its central unifying force. If there is not adequate amount of social energy of the proper kind (in families this is the so-called 'love') to exchange between a husband and wife, there is likely to be some form of breakdown in their ordered relationship, and the bond between them could be weakened. In this case a unified family system could become less harmonious (husband and/or wife may carry on their separate lives), and if things go this way further, the family may be broken down. This can be generalized to all societies (communities).

Examples of energy or influence that can act as negative entropy are food, affection, education, medicine, friendship, or anything else which helps in sustaining or improving the conditions and unity of the members of the society.

Environments: There are four kinds of environments related to a system. These are the following (Figure 9.6) [7]:

- The system under study, and the internal environment of which it is aware (*Environment* **1**).
- The deeper internal environment of which the system is not aware (*Environment* **2**).
- The system's external environment of which it is aware (*Environment* **3**).
- The system's distant external environment of which it is not aware (*Environment* **4**).

The environments are the context in which any system exists. Energy or influences may be able to flow across the boundary from any environment to any other to change the balance of any part of the system. A short discussion of the above environments for a one industry town follows, where the system state (#1) is the total population and geographical area of the town [7]:

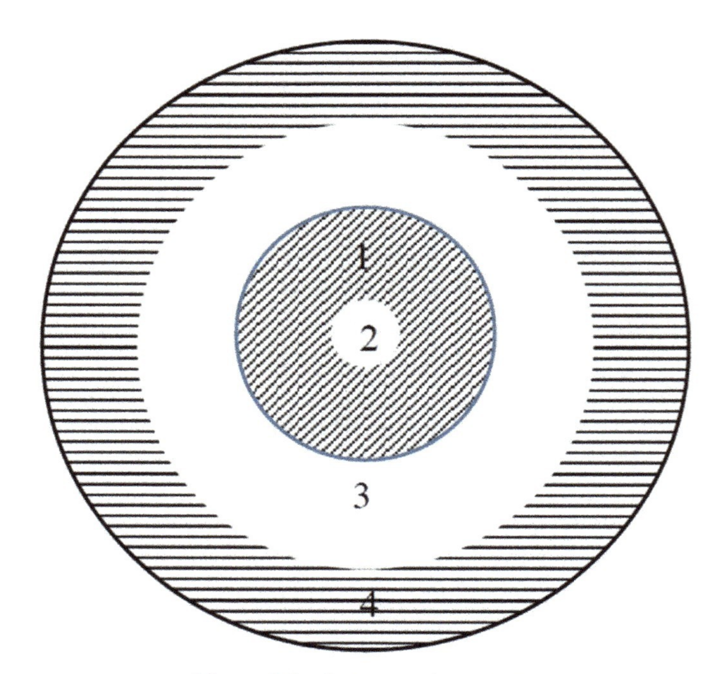

Figure 9.6 System environments.

Environment 2 The *deeper internal environment of which the system is not aware:* This environment could, for example, be an internal struggle among senior industry's managers for control of the operation. Even if the population of the town may not know about the struggle, the result of such a controversy will likely explode into the system and produce changes in the way the society operates.

Environment 3 *The external environment:* Changes in the external environment, e.g., a forest fire on the edge of town, could cause very large changes in the internal balance of the system.

Environment 4 *The distant external environment:* Changes in environments that the people of the town cannot see, e.g., a technological breakthrough in a remote laboratory which might greatly increase demand for mining product of the town, might eventually produce changes in the system. Decisions of political or economic nature that are made at a large distance from a society, are typically affecting the operation of the society. A fundamental goal of the society development is to educate people of the society how to best control the many influences from distant society's environments.

In summary, system theory can assist to solve many problems of society development, such as:

- Assessing the society.
- Selecting development goals.
- Planning a strategy to reach these goals.
- Carrying out activities to achieve goals.
- Evaluating progress and use the results in subsequent evaluations.

Clearly, all of the human activities/processes that define a society arise from communication, namely:

- The relationships among society's members are defined through communication.
- Linkages/interactions among subsystems depend on communication and information flow.
- All feedback processes involve communication.

9.4 Cybernetics and Society

As we have seen in Chapter 5, cybernetics concepts can be used to study social systems (macrosystems, mesosystems, and microsystems) and the activities taking place therein. This interconnection of cybernetics and society has led to the unified field called *'sociocybernetics'* which is based on the fact that both social systems and cybernetics are amenable to the application of *general systems theory* concepts, particularly second-order cybernetics concepts.

Three quite different avenues for applying cybernetics to the study of society, and its parts (social systems and processes) are the following [10]:

- The addition of social or sociological variables and components to control and information systems' models that are used to analyze systems from a purely technological viewpoint.
- The use of concepts and theories from cybernetics/GST to identify, describe, and explain the structure and behavior of social systems.
- The application of systems and system dynamics methodology (particularly in the areas of model construction and computer simulation) to the study of social structures and processes [11].

Important questions that provide a fruitful area of research include:

- Is planning of large-scale social systems realistically possible?
- Is equality among the composing units of social systems possible?
- Is hierarchical control a prerequisite for governing large-scale systems?

Partial answers to these questions can be found in Ashby, Braten and Buckley [12–14]. Now, let us have a look at *behavioral cybernetics* and its role in economic systems [15,16]. *Voevodsky* [15] hypothesizes that there are major cybernetic limits beyond which an economic system cannot continue to operate. These are limits where the losses will exceed the gains. If these limits are examined with respect to the communication processing that takes place within the dynamic systems, additional cybernetic factors emerge to explain why economic systems behave the way they do [16]. Behavioral cybernetics maintains that the response of all organism systems is the sum of two actions, namely:

- Maintenance of internal balance.
- Maintenance of external balance.

Internal balance is achieved through the following three-step communication process:

- Input processing.
- Response programming.
- Transformation processing.

The balance achieved and maintained at each step, influences the capability of the system to exist dynamically. The internal balance may also be influenced by the way in which external balance is achieved.

External balance establishment involves the way in which the inputs are obtained and the outputs are transmitted. In economic systems the actions of internal balance and external balance maintenance are called respectively *acquiring and marketing*. The establishment of internal and external balance in economic systems can be aggregated into a three step operation, namely:

- Acquiring.
- Producing.
- Marketing.

The overall economic balance is determined by the capability of the system to properly perform each of the above three processes. Two other major external factors that are important in behavioral cybernetics and useful in explaining system behavior are:

- External source of inputs.
- External consumer of outputs.

Economic cybernetics argues that the acquirability and marketability of an economic system, coupled with the input and market availability are cybernetic boundaries for producibility. This is shown in Figure 9.7 [16], where the area enclosed by *acquirability and marketability* constitutes the only zone of *economic order and production*.

To maintain economic stability, producibility must be adjusted to this zone. Economic stability can be maintained by increasing or reducing producibility. If the specific capability is exceeded it must be altered to compensate for the degree of producibility that exists. The overall conclusion

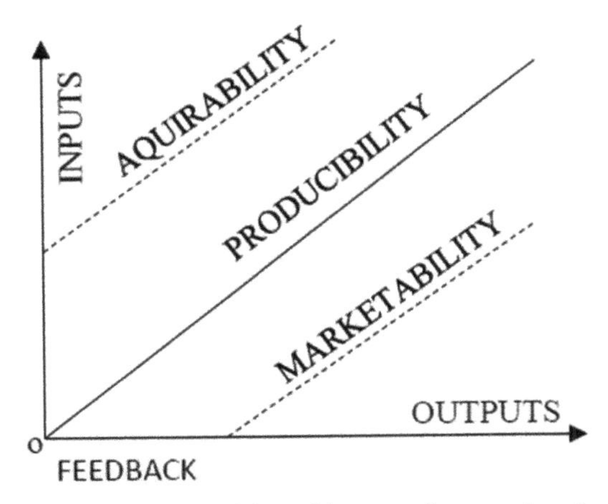

Figure 9.7 Pictorial representation of the stable zone of economic order and production.

is that improved efficiency, and acquirability, producibility and marketability are the only real corrective actions possible for economic systems in a non-monopolistic environment. Input-output analysis can also be employed for the prediction of the behavior of the system and for the prevention of failures [17].

In Jones [18] it is argued that cybernetics is a description of *systems in conversation,* i.e., it is about systems *'talking'* to each other. In other words, cybernetics is engaged in processes through which information is exchanged or communicated between each system or each element in a particular system (say a body or society). A crucial point is that this information has to be recognizable such that to change something within the *'receiver'*, while the response or the feedback the receiver provides also has to be recognized by the sender such that to change something in the original *'sender'*. A general sense conversation takes place between all types of systems (particularly *'living systems'* or things that show signs of life). This concept of 'living things' broadens the spectrum of things that behave in a cybernetic way (e.g., human, or animals that are subject to metabolic processes, or ecological systems where predator prey processes take place). Social systems are also cybernetic systems, because they involve feedback structures that allow exchanges of materials and information or adaptation processes to face changes in the proximal environment. In Jones [18], a number of societal and other examples that are treated using the principles of thermodynamics are provided.

9.5 The Impact of Control

Control is everywhere. Nuclear reactor power plants, aircraft and spacecraft, automobiles, trains, and ships, homes and buildings, enterprises, chemical plants, etc. are complex large scale systems that show the ubiquity of control technology. Many man-made systems of modern society would simply not be possible without control. The advancement of control technology was promoted by new related technologies, e.g., novel sensors, new computer hardware and software, novel computational, simulation, and optimization algorithms, fast and precise vision systems, sophisticated interfaces, communication networks, etc. Some decades ago, discussion on the impact of control technology on society has been limited to a few industrial systems (process plants, electric power plants, etc.) but today these traditional domains of control have been supplemented by many others. Successful applications of control are not the result solely of control expertise. Deep domain knowledge has always been necessary. The connection of classical and novel domains has been established and strengthened as a result of the broadening of the control applications. A comprehensive study of the impact of control technology on modern life was presented in Samad and Annaswammy [8]. The first part of this study presents detailed discussions on the role of control in both traditional and emerging application domains. These include the following:

- Aerospace.
- Process industries.
- Power plants.
- Automotive.
- Robotics.
- Biological systems.
- Renewable energy and smart grids.
- Ecological systems.

The second part of the study highlights significant particular advancements in the field in a way which can be profitable both for the control professionals and stakeholders. A few examples of these successful control stories are:

- Mobile telephones in the tuning of feedback loops across the globe.
- Collision avoidance systems in air traffic management that rely on estimation and control algorithms.
- Advanced control is now widely implemented in printers, copiers, and other devices.
- Optimization and control technology implemented on railroads has reduced enormously fuel consumption.
- Control technology has revolutionized stability and traction control, and automotive safety.

Control technology has widely influenced our lives. Some typical examples of this influence are listed in Table 9.1.

In the following we will briefly examine the control issues of two typical societal systems namely:

- Management control systems.
- Economic systems.

Management Control Systems

The term **MCS** was coined by *Anthony* [99] as a system that collects and employs information to evaluate the performance of various resources involved in organizations. These resources include:

- Physical/technological resources.
- Economic resources.
- Human resources.

Table 9.1 Control-aided way vs. old way of handling societal tasks and applications

Old Way	Control-Aided Way
Accidents and time loss in working in dangerous areas (e.g. mines, volcanoes, etc.)	Controlled robots with on-board cameras now perform these dangerous tasks (including data collection for space travel and below the earth's surface)
Climbing stair in skyscrapers	Elevators and escalators increase accessibility
Several surgical operations were fatal and risky because of human error	Robotic and programmed surgical operations achieve precision and accuracy
Manual labor to assemble automobiles	Machine assembly using robots (arms and cranes)
Washing dirty dishes in sink by soaking them in soap water	Dishwasher with no human intervention except loading and unloading influence control and jobs services

A **MCS** considers the overall organization as an integrated system, and includes the study of the influence of the resources in implementing the organization's strategies towards the achievement of its goals. **MCS**s involve always decision making methods and procedures, and optimization algorithms. The concepts, sub-problems, and techniques of the general decision making and control design of management systems are shown in block diagram form in Figure 9.8 [7]. Decision making is usually a selection process among available alternative solutions or procedures. Control is the process of using the decision making results for the initiation and implementation of the corrective actions. Management control is the control that refers to organizations and business enterprises. The feedback control is the process by which the desired goals are achieved.

Actually, the structure of Figure 9.6 embraces both technological and managerial systems. The technological and managerial (behavioral) operations are interacting and must be considered as a totality. This structure involves three general stages, namely: (i) *Problem formulation,* (ii) *Selection process* (optimal design/plan), and (iii) *implementation of the design,* which are divided in several steps as follows:

Problem formulation

- System view.
- Modeling.
- Requirements/design goals.

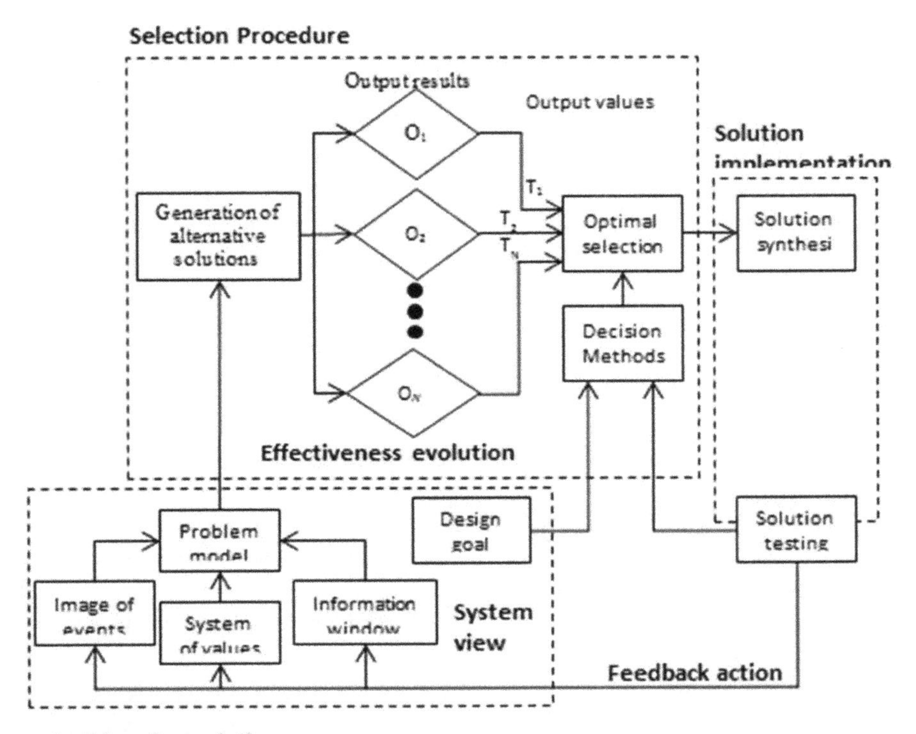

Figure 9.8 Structure of the general feedback decision and management control model.

Selection procedure

- Generation of alternative solutions.
- Evaluation of the effectiveness of solutions.
- Optimal solution selection.

Solution implementation

- Synthesis/implementation of the selected solution.
- Testing of the system.
- Control.

The ***system view*** reflects all our experience and information from the past, and involves the *view of facts* (from the past), the *information window* (that keeps only the data which are relevant to the problem/system at hand), and the *system of values* upon which the system design will be based.

The ***modeling step***, which constitutes the core of the optimal decision making, is the process of constructing or developing a model of the system which will be based on the *system view.* Here, system is any union of cooperative and interacting components. The mathematical models of the systems contain differential equations, algebraic equations, and the constraints of the variable/quantities involved. As we have seen in Chapter 6, system dynamic models can be described either as transfer functions in the complex frequency domain (if the system is time-invariant) or as state space equations in the time domain (for both time-invariant and time-varying systems). Thus the *control design* can be made either by classical control techniques or by modern control techniques.

The ***design goals*** vary from case to case and may involve minimum energy consumption, maximum economic (or other) profit, minimum time control, desired final state achievement, decoupling/no interacting control, etc.

The ***generation of alternative solutions*** involves alternative procedures, alternative plants/programs/schedules/price policies, alternative design/ selection processes, etc.

The ***evaluation stage*** requires the availability of proper decision models and quantification of the inputs and outputs.

The ***optimal solution selection*** can be performed by two-valued and multi-valued decision theory and by dynamic/static optimization theory (see Chapter 6: Modern Control).

The ***synthesis and implementation*** of the selected solution may reveal practical problems, not anticipated at the beginning. Therefore, very often we correct the solution using *feedback procedures and controllers* based on the input/output measurements. In all cases a proper detection/measurement mechanism is needed.

Feedback control is applied at several levels (enterprise level, national level, international level) and for several processes of the society. Some examples of such processes are:

- Resource allocation.
- Inventory of products and raw materials.
- Economic transactions.
- Human factors and psychology.
- Legal and ethical frameworks.

By 1998, *Anthony's* original definition of management control as 'the process by which managers influence other members of the organization to implement the organization strategies', included *human behavioral issues*. Management control systems involve both formal and informal information subsystems and processes. Therefore management control depends on issues like *'cost accounting'*, *'financial accounting'*, *'regulatory compliance'*, etc. According to *Simons* [19] 'taking an isolated view of features such as accounting systems or managerial behavior will leave the theory weak in both explanatory and predictive power'. Based on empirical evidence, Simons looks at management control systems from the viewpoint of how managers use control systems under different conditions taking normative design rationality as given. He also explicitly links, in his framework, controls with both strategy formulation and implementation. In quantitative procedures, management employs the models and techniques of *operation research* **(OR)** [20], and also economic theory techniques. Some techniques used in MCS are:

- Program management techniques.
- Capital budgeting.
- Just-in-Time (JIT) techniques.
- Target costing.
- Activity-based costing.
- Total quality management (TQM)
- Benchmarking.

Anthony in 1995 has actually defined management control as separated from strategic control and operational control. Langfield-Smith argues that this is an artificial separation that may no longer be useful in an environment where employee empowerment has become popular. Actually, management control involves formal and informal controls, output and behavior controls, market, bureaucracy and clan controls, administrative and social controls and results, and action and personnel controls.

Stafford Beer is recognized the first to apply cybernetics to management defining cybernetics as '*the science of effective organization*'. He coined the term **'postwid'** which is an abbreviation of the definition of a system's *de facto* purpose as 'the **p**urpose **o**f a **s**ys**t**em is **w**hat **i**t **d**oes'. He has studied what is called '*cybernetic factory*' as an example of scientific thinking model in an organization, and provided the pictorial representation of the cybernetic factory shown in Figure 9.9. An organization/company is considered to be a system in an environment, and also a control system [33]. It includes a homeostatic ultra-stability loop that involves the company as a black box which is transformed to a homomorphic model, and the environment as another black box also transformed to an associated homomorphic model.

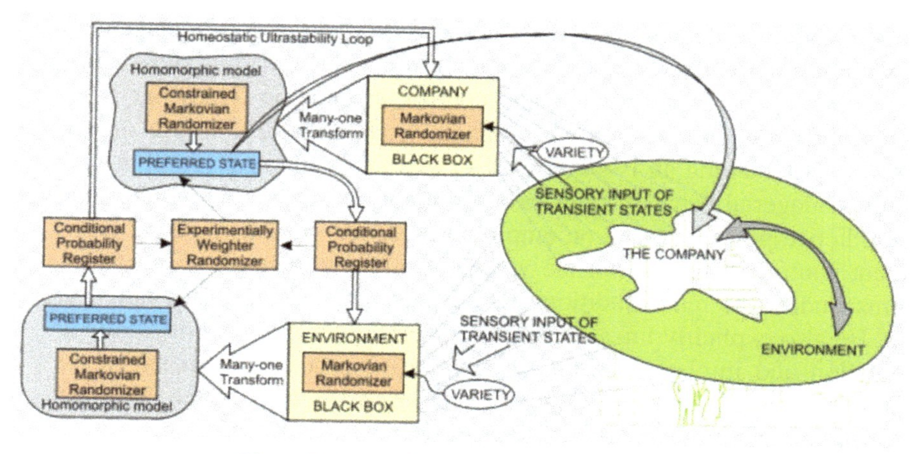

Figure 9.9 Beer's cybernetic factory model.

Source: www.newworldencyclopedia.org (Entry: cybernetics).

There is also a loop starting from the preferred state as input, and including sensory feedback from the transient states. The variety involved in the system is modeled by Markovian randomizers.

The cybernetic factory model has been applied by Beer and other cyberneticists to steel industry, operational research problems, entrepreneurial systems, organizational cybernetics, viable system theory, etc.

Viable system model The **VSM** was developed by Beer in order to enable organizations and companies achieve *agility* and thus be able to face unexpected real-time market and other events. The VSM looks at a company as if it is literally a living being and describes how it should be structured to operate most effectively in its environment. According to VSM theory, a system is a viable system if it is able to create, implement, and regulate its own operating policies such as to sustain itself over time. To do so it needs to be free to design and execute its own operations within pre-specified performance ranges and areas of responsibility. VSM explains in great detail how several systems connect to one another to create a greater one, a whole which emerges from simpler parts and can do things that the individual parts cannot do themselves. Emergence is a feature that enables a whole system to behave consistently. Business theory uses the holistic synergy concept $(2 + 2 = 5)$ to signify that the whole is greater than the sum of its parts [34].

Figure 9.10 gives an abstract pictorial representation of the VSM structure, where systems 3. 4 and 5 constitute a meta system that performs

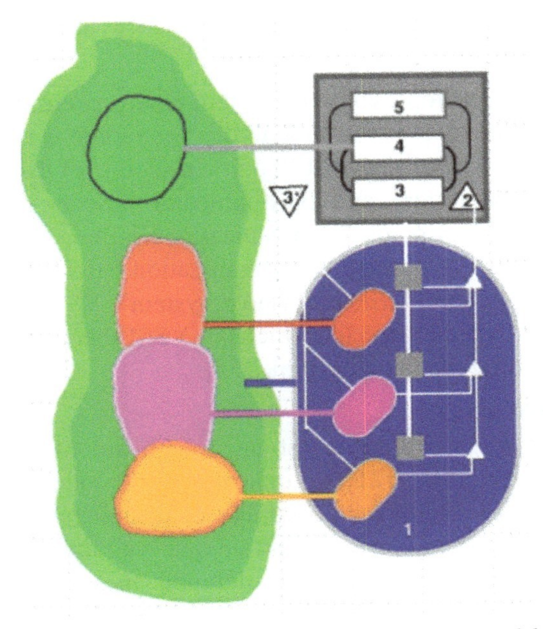

Figure 9.10 Abstract structure of viable system model.

Source: [34] (viable system model).

planning and control processes. Very broadly, these systems can be described as follows [www.synchro.com viable system model]:

- *System 3*: **Control** (control is achieved through bilateral communication between subunit and meta system).
- *System 4*: **Intelligence** (this is a two-way link between primary activity and external environment, and is fundamental to adaptability).
- *System 5*: **Policy** (provides clarity about the overall direction, purpose, and values of the organization, and needs to be highly selective in the information it receives).

Systems 1 and 2 perform implementation and coordination functions, and system 3* carries out the auditing function. Specifically:

- *System 1* (**Implementation**) involves the operating units for getting things (products, services, etc.) done. System 1 is actually a microcosm of the entire system and needs to be as autonomous as possible. Each operating unit involves its own systems 1–5 in order to be a viable system itself.

- *System 2* (**Coordination**) coordinates the interfaces and the value-added operations of the organization.

The green area on the left represents the system environment. Stafford Beer points out that 'all systems are vital to viability, then there is no meaning to be *more important*', and advocates that 'the first principle of the organization is that managerial, operational, and environmental varieties diffusion tend to equate'. Another key principle of organizations is the *recursive system principle* which says "In a recursive organization structure any viable system contains and is contained in a viable system'. VSM is a recursive system at each level of recursion (like fractals). VSM theory finds a variety of applications including the law discipline [34].

Economic Control Systems
Over the years, **ECS** have been investigated with aid of *'theories'* and *'models'*. *Kevin Hoover* (1995) [21] has indicated that "model is a ubiquitous term in economics, and a term with a variety of meanings". Similarly, the term 'theory' is not interpreted in a broad unique way among economists. According to *Klein and Romero* [22] after defining *'model', "theory'* has a higher normative status than *'model",* and also "a theory does not require a 'model', and a 'model' is not sufficient for a 'theory". But these authors leave the term 'theory' undefined. According to *Goldfarb* and *Ratner* [23]: "A widespread use of 'theory and model is that '*theory*' is a broad conceptual approach (as in '*price theory')* while *'models',* typically in mathematical (including graphical) form, are applications of a theory to particular settings, and/or represent explorations of different sets of assumptions conditionally allowable by the theory approach". In this way, for example, *'price theory models'* have a fully understandable and standard meaning. This view of the relation of *'economic theory' and 'economic model'* is adopted in most classical textbooks on economics. Of course, this does not mean that the terms 'model' or 'theory' are able to carry all the weight of competing possible interpretations. According to *Kevin Hoover* the term 'model' has a variety of meanings, and so we have many types of economic models, including the following opposing classes:

- Evaluative/interpretive models.
- Observational models.

Deming states: "Without theory, experience has no meaning. Without theory, one has no questions to ask. Hence without theory, there is no learning. Theory is a window into the world. Theory leads to prediction. Without prediction, experience and examples teach nothing. To copy an example of success, without understanding it with the aid of theory, may lead to disaster" (*Economics for Industry, Government, Education*, p.106, 1993). The application of *feedback control* is done using a 'model' which is formulated on the basis of a 'theory' and tested using real data. Currently, there are many textbooks and research books concerned with the application of formal control theories (linear, nonlinear, stochastic, adaptive, etc.) based on specific mathematical models. Economic models were formulated and applied at two levels:

- Microeconomic level.
- Macroeconomic level.

In all cases of feedback control, the main question is to find a (good) control goal. Usually this goal is related to an efficient (optimal or sub-optimal) use of relevant resources (time, capital, material, and human resources) in relation to the product of service provided. The above issues hold for both the microeconomic and macroeconomic levels. However, the tools that are appropriate on the two economic levels are not the same.

Economic models also include models that have been developed for the following aspects (Manuela Samek/slideplayer.com/slide/4785358):

- *Labor supply*: These models are built for explaining labor supply decisions, the influence of working hours' regulation, the taxation, and the welfare systems on labor supply decisions.
- *Labor demand*: These models are built for explaining how companies and industries decide if and how much labor to employ in the production process. These models may also incorporate adjusting costs to take into account the relation between employment protection legislation and labor market performance.
- *Wage determination*: These models are built for explaining why wages are not flexible, what variables affect wage bargaining and its effect on the wage dynamics and structure.
- *Human capital*: These models are built for explaining why individuals invest in education and training, and what the personal and social benefits and returns of this investment are.

An economic model can be expressed in the form of words, diagrams or mathematical equations depending on the user and the purpose of the model as shown in Figure 9.11.

Figure 9.12 shows qualitative curves of production and consumption quantities in the 'price-quantity' plane. Economic equilibrium occurs at the

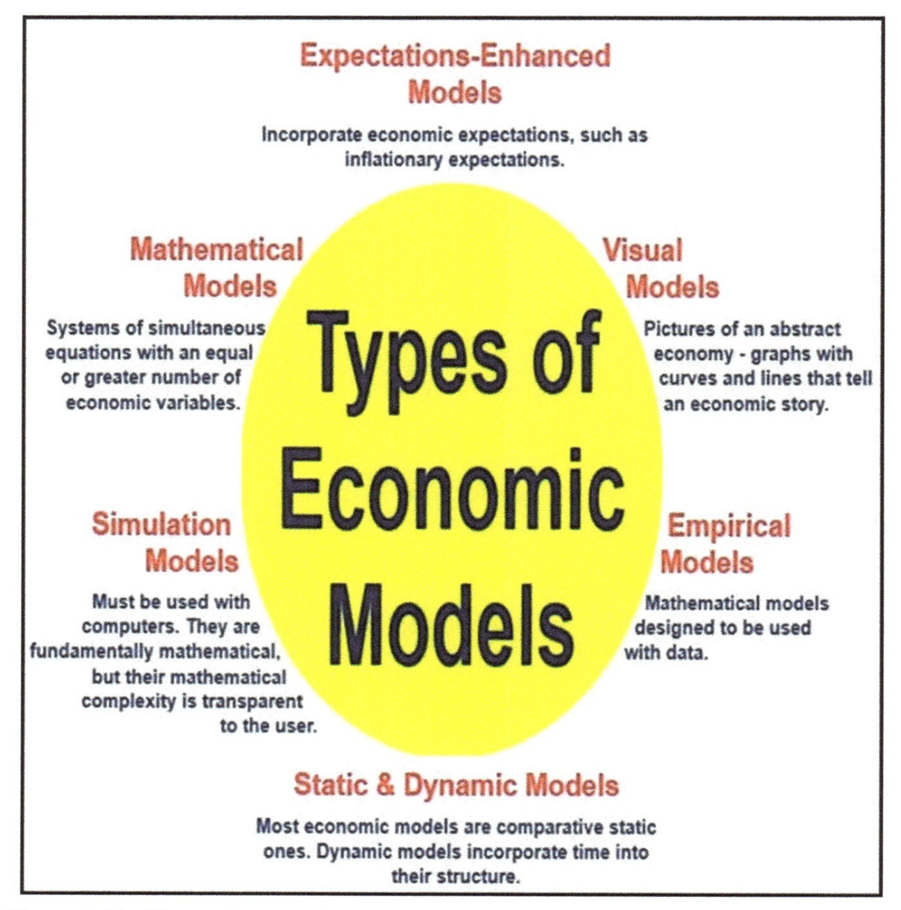

Figure 9.11 Types of economic models (mathematical, simulation, visual, and empirical models).

Source: www.marketbusinessnews.com/financial-gkossary/economic-model-definition-meaning

point where the supply curve crosses the demand curve. The corresponding quantity of goods is called '*equilibrium quantity*'.

The macroeconomic and microeconomic control tools are summarized as follows:

- *Classical macroeconomic control tools:* Legislations that guarantee certain minimal constraints (e.g., minimum wages), short-term economic policies that address current problems (e.g., price limiters), and monetary interventions of the central bank, etc.

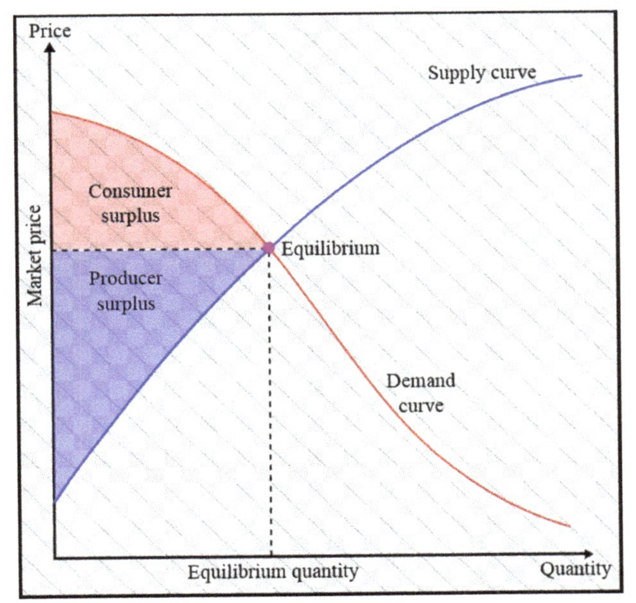

Figure 9.12 Production-consumption curves. Economic equilibrium occurs when production quantity is equal to the consumption quantity.

Source: www.lookfordiagnosis.com/mesh_info.php?term=Models%2C+Economic&lang=1

- *Classical microeconomic control tools:* These tools include managers that constraint the interactions of employees through a specified organization chart, the amount of allowed resources (capital, material, space, input), process control, i.e., real-time control procedures using rule sets (*Tayloristic rules),* supervision procedures, and output (product) measurements. On the microeconomic level, large companies perform their business processes within specified *organizational units* (business units) in order to create products or services that are supplied to the market. The steps followed in these business processes are determined by the *operational structure* of each company. Economic actions are based on negative feedback which assures the achievement of a predictable equilibrium for prices and market shares. Negative feedback tends to stabilize the economy because any major changes will be offset by the reactions they generate. According to conventional economic control theory the equilibrium marks the *'best'* outcome possible under the circumstances. Here, it is useful to note that economic systems also exhibit positive feedback effects that need to be observed and taken care.

These self-reinforcing processes may appear on all economic levels. Typically, increased production brings additional benefits. Producing more product units a company can obtain more experience in the manufacturing process and understand how to produce additional units even more cheaply. Thus, companies and countries that gain high volume and experience in a high-technology industry can exploit advantages of lower cost and higher quality that may make it possible for them to shut other companies or countries out.

The three major questions of the economic problem are:

- What to produce.
- How to produce.
- For whom to produce.

An economic system is characterized (non-exclusively) by the following:

- Government involvement.
- Financial institutions, firms, and households.
- The ownership of resources.
- Price control.
- Freedom of choice,
- Who gets the resulting profit?

A pictorial illustration of the economic cycle (*peak-recession-recovery-peak cycle*) is given in Figure 9.13(a), and an overall economic-operation diagram (*circular flow of income*) which refers to the five principal sectors of economy, namely:

- Households.
- Firms.
- Financial institutions.
- Government.
- Rest of the world

is shown in Figure 9.13(b). Economic processes included in Figure 9.11(b) are:

- Government transfers.
- Government purchases.
- Government borrowing.
- Taxes.
- Investment.
- Private savings.

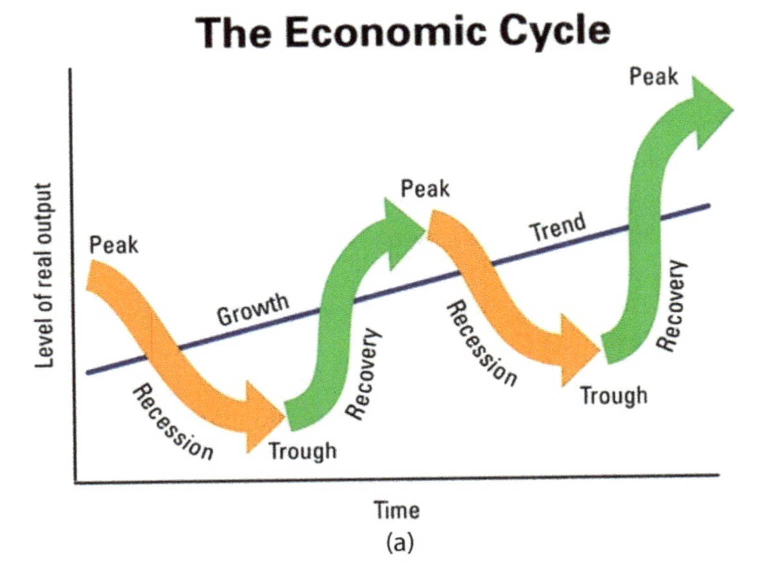

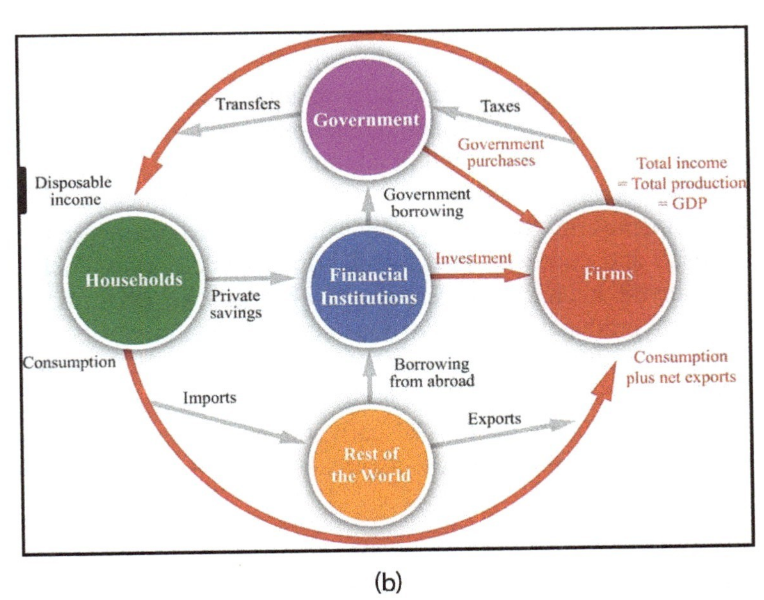

(b)

Figure 9.13 (a) A graphical representation of the economic cycle and (b) an illustration of the circular flow of income for the five principal economic sectors.

Source: https://saylordotorg.github.io (Macroeconomic Theory through Applications, and The Circular Flow of Income, Saylor Company, 2012).

- Imports.
- Exports.
- Consumption.
- Disposable income.

In general, economic systems are distinguished in four types:

- *Traditional economic systems* (family and community-based systems).
- *Command or planned economic systems* (all communist or dictatorship countries).
- *Market-driven (neo-liberal) economic systems* (USA and all capitalistic and free economies).
- *Mixed economic systems* (USA, Japan and most modern countries).

This distinction is based on the way these systems operate and are controlled.

Traditional economic systems tend to follow long-established patterns and so they are not very dynamic. The quality-of-life standards are static and the individual persons do not have much occupational or financial mobility. Typically, in many traditional economics, community interests have higher priority than private interests, although in some of them some kind of private property is respected under a strict set of conditions and obligations.

Command economic systems are fully controlled by the government, which decides how to distribute and use the available resources. Government regulates and control prices, wages, and, sometimes, what kind of work the citizens do. Socialism is the main kind of command economy.

P*ure-market economic systems* are controlled by the interaction of individuals and companies in the market environment that determines how resources are used and goods are distributed. Here it is the individual who selects how to invest his/her own resources, what jobs to perform, what services or goods to produce, and what to consume. In a pure market economy the government has not any involvement in the economic life. Today, in most developed countries, the economy is of *the mixed type.* Of course the kind of *mix* differs from state to state. Usually, the mix between public and private sectors is not static but changes adaptively according to the particular conditions each time.

9.6 The Impact of Automation on Society

Automation overlaps with control, computer technology, information systems, communication systems, and other technologies. Therefore automation and technology, and human life cannot be separated; modern society has a cyclic co-dependence on automation and technology. Humans use automation and technology to travel, to communicate, to do business, and to live in comfort. Automation must integrate, in an appropriate and successful way, technical, organizational, economic, cultural, social, and human attitudes with a view to enhance **QoL**.

The typical approach for operating industrial automation systems is the so-called **HITL**. This is an old concept that allows operation to be performed at several automation levels between fully human controlled and fully automated control. The operator of the system plays a crucial role having the task to carry-out complex operations of supervision, critical control, repair, maintenance, and optimization. Operators have to optimize performance, making the most of human judgment while facing human restrictions, such as sensitivity to information overload or systematic cognitive bias. The overall goal is to get safe and efficient systems with products matching the human needs and preferences. HITL is the typical practice in process control, power systems control, manufacturing systems, aircraft control, etc.

Society-in-the-loop (**SITL**) is a scaled up version of **HITL**. Although HITL automation is concerned with embedding the judgment of individual humans or groups in the optimization of conventional automation systems, SITL is about incorporating the judgment of society in the governance of societal outcomes, e.g., an intelligent algorithm that controls huge numbers of self-driving cars, or a set of new filtering/brainwashing algorithms that influence the political beliefs and preferences of millions of citizens, or algorithms that suggest how the allocation of resources and labor in the total economy should be made. The question is 'what is the HITL equivalent of these governance algorithms?' Here is where a shift is made from HITL to 'society in the loop'. SITL suits more to functions of large communities and particularly to government and governed people. Modern government is the outcome of an implicit agreement (or social contract) between the ruled and the rulers, aimed at fulfilling the general will of citizens. Similarly, SITL can be regarded as a way to introduce the general will into an '*algorithmic social contract*'.

Figure 9.14 shows the process of transferring *humanities/human values* into *artificial intelligence/intelligent automation* systems through the *society-in-the-loop* involvement. The driving force is represented by the *societal expectations*, and the system corrective feedback is implemented by the *evaluation function*. SITL constitutes one of the ways of co-evolution of society and automation/technology.

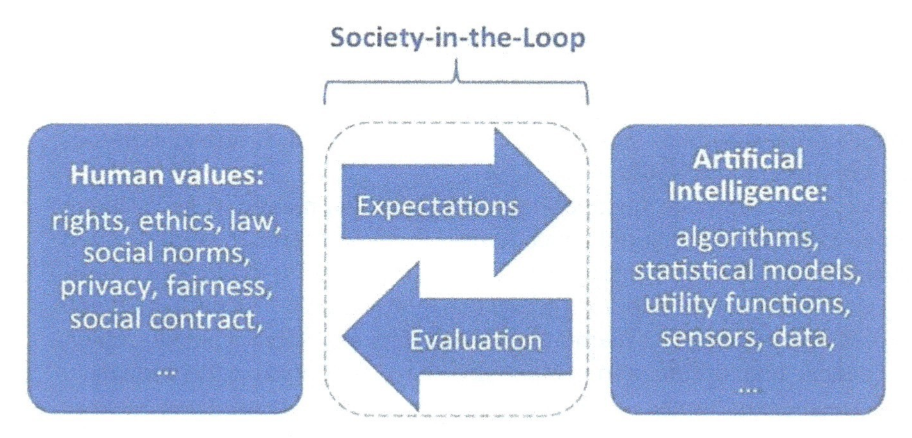

Figure 9.14 Illustration of the 'society-in-the-loop' concept which refers to the integration of human values in artificial intelligence and automation.

Source: MIT Media Lab. (Iyad Rahway, Society in the Loop), https://medium.com

Automation is an ever-growing technology with an enormous gamma of subjects and applications. However, automation and technology may have both positive and negative impact on society, and has caused many concerns. A study on these issues is presented in Martin et al. [25] where the term *'appropriate automation'* is used to cover technical, organizational, economic and social/cultural issues.

Overall considerations addressed in Martin et al. [25] are:

- Human–machine incomparability.
- Responsibility of engineers.
- The social dimension of automation technology.
- Humanization of technology versus human engineering.
- Cultural aspects of automation.
- Cultural impact on automation.

In Kutay [26], the economic impact of automation technology is assessed. It is argued that a new conceptual framework is needed to analyze the economic impact of automation technology, and the benefits and costs associated with automation technology (strategic benefits, reduction in labor costs, etc.). The paper first demonstrates the inadequacy of current economic analysis techniques to assess the benefits of automation technology, and then presents a new methodology that can be integrated to an expert system for assessing the economic impact of various types of automation technology.

In the following we will briefly discuss a number of crucial issues of the automation that are of long time concern by the human and society. Environmental issues of automation are discussed in Tzafestas [27]. The issues considered here are the following:

Productivity and Capital Formation
Productivity is a complex concept not uniquely defined and measured. Furthermore, even after some specific definition is chosen, industrial (and office) productivity depends on many interacting factors. Therefore, productivity improvements cannot be attributed to any single methodology or technology. Robotics, for example, is an input to productivity ratio P which is defined as:

$$P = \text{(Units of output)/(Units of input)}$$

and represents both capital and technical knowledge. Human labor is another input to P. 'Human–computer' as a united entity is a third input to P. What combinations of inputs to productivity ratio should be adopted is a social issue of great importance.

Capital formation is another issue of automation related to productivity. Economics often attribute the capability to create new investment capital to the growth of productivity. Two social questions about capital formation are the following:

- Is the capital available to fund the construction of new systems and the modernization of existing ones that will use automation technologies, sufficient?
- Is there sufficient capital to fund research and development by business men who wish to develop new types of automation equipment?

The answers to these questions depend on the legal and economic status of each state and on how automation is perceived by investors and managers to be a promising technology in which to invest.

Quality of Working Conditions

- Working conditions are improved if automation is used for jobs that are dangerous, boring or unpleasant, and if the new jobs created by automation are better.
- Productivity increases may also, in the larger term, result in a shorter, and more flexible scheduled work week.
- Equipping an employee with a job helper (e.g., a robot extender) not only ceases job stress but also opens job opportunities to people with handicaps or other limitations.

Of course, whether the above benefits are realized depends, in part, on the specific ways in which industry and administration uses automation. Many people have expressed concern that automation increases the possibilities for employer surveillance of employees, and that automation could be used by employers to 'downgrade' jobs that require working with automated systems.

Unemployment

Unemployment was the most important issue in discussion about the social effect of automation four or five decades ago, but now is at an acceptable equilibrium level due to increasing generation of new jobs. In any case, automation and related technologies can affect labor in several ways such as:

- The effects of automation on the relative proportion of machines to humans (the 'capital-labor ratio') in a given industry.
- The need of expert workers with particular job skills and abilities in a certain industry.
- The extent of change in production numbers and prices in the countries in which automation and new technology are introduced.

To assess the effects of automation to future labor levels, a reference line is needed against which job loss or gain can be measured. This reference line might be a projection of current trends, but must also take into account the *virtual unemployment and virtual employment issues.* Virtual unemployment represents the jobs which have been lost if a given plant or organization has not responded to market demands by automating. Virtual employment represents the jobs which were not explicitly eliminated, but that would have existed with automation not adopted.

Today, industrial automation and robotics exhibit more opportunities than disadvantages within manufacturing community, since as a whole are expanding among more industries due to the integration of robots. A recent global study has anticipated that robot driven job creation in North America could reach 1.5 million by the end of 2016 (Figure 9.15).

Automation and new technologies are changing the skills required to adapt to the new workplace. Jobs are affected. Some jobs will disappear and others will be in high demand. Already jobs exist that had never heard some years ago, e.g. the data scientist specialization which is in enormous demand. To keep-up with this disruption, business companies have to encourage their staff learn new skills. According to Economic World Forum, this change from 2015 to 2020 for the ten top skills is as shown in Table 9.2.

Looking particularly at *robotics*, the following positive/negative technological and societal impacts have been documented and recorded in the literature.

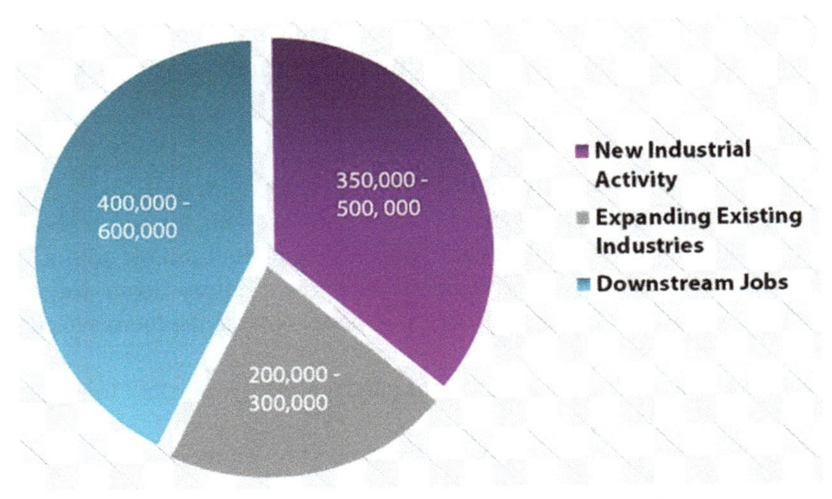

Figure 9.15 Global robot-driven job creation 2012–2016.

Source: www.octopus.com (Industrial Robots and Manufacturing: A Catalyst for Loss or Growth?).

Table 9.2 Change from 2015 to 2020 for the ten top skills

2015	2020
1. Complex problem solving	Complex problem solving
2. Critical thinking	Coordinating with others
3. Creativity	People management
4. People management	Critical thinking
5. Coordinating with others	Negotiation
6. Emotional intelligence	Quality control
7. Judgment and decision making	Service orientation
8. Service orientation	Judgment and decision making
9. Cognitive flexibility	Creativity

Advantages

- *Mechanical Power:* Humans can lift about 45–50kg, whereas robots can lift many tons.
- *Motion Velocity:* Humans can respond and act at 1 cps, whereas robots can at 1,000cps.
- *Reliability:* Robot work is much more reliable than human work.
- *Sensitivity: Robots* have much less sensitivity to environmental conditions (temperature, pressure, vibration) than humans.
- *Endurance:* Robots can work uniformly until breakdown from wear. Humans have much reduced endurance and monitoring capabilities (about 30min of monitoring).

- *Precision of Work:* Robots have definite precision of work, whereas human precision varies according to physical, psychological and training conditions.

Disadvantages

- Robots have incompatibility in terms of workspace with humans.
- Robots have incompatibility in terms of motion with human (robots move linearly or at acute angles and abruptly stop, but humans do not).
- Human safety: Some workers are at risk from injury from robots (maintenance workers, programmers, personnel outside the danger zone, etc.).
- Robots are difficult to operate (control panels differ from robot to robot: lack of standardization).
- Feeling of isolation among workers surrounded by many robots.
- Telepresence and virtual reality in telerobotic and other systems raise again old philosophical questions of being and existence in the field of ontology.

Some positive/negative impacts of general technology on society are the following [28]:

Positive impacts

- *Technology has improved communication:* Both society and organizations use communication to exchange information. People use technology to communicate with each other. Electronic facilities (radio, television, internet, social media have substantially contributed to the improvement of the manner we exchange ideas that can develop our societies). Small and large enterprises and organizations use Internet, computer-communication networks, and mobile communication technology to grow and improve their services to customers.
- *Technology has improved transportation:* Transportation plays a dominant role in our society. Both society and business have benefited from the new transportation methods. Transportation provides mobility for people and goods. Transportation uses automobiles, trains, airplanes, motorbikes, people, energy, materials, roads information, etc. As we have discussed in Chapter 8, automation has developed improved or novel transportation means for faster, safer, and more comfortable travelling/material transfer.
- *Technology has improved education and learning process:* Education is the backbone for the development of every society and society.

Most educational institutes and Universities started integrating educational technologies with the principal aim to improve the way students learn. Technologies such as smart white boards, computers, mobile phones, projectors and world-wide-web/internet are increasingly used in classrooms to boost student's motivation to learn. Visual education is becoming more popular because it has beneficial features for education in many subjects (mathematics, physics, biology, economics, and much more). The organizations and business have invested considerable amounts of money in various educational technologies that can be employed by both teachers and students.

Negative Impacts

- *Increased population:* Technology has improved the health and medical care facilities and is continuously contributing to the research carried out to solve most health problems affecting humans. For developed countries this is good, but it is bad for developing countries which have not the capacity to access these health care benefits brought by technology. Developed countries can control the growth of their population in order to match a planned population level. But this is not so in developing countries which have high population growth rates while health care is poor and food is scare.
- *Resource depletion:* Automation has the effect of increased natural resource consumption. Billions of customers/consumers of new technological products (computers, mobile phones, etc.) for their homes have increasingly high demands of natural resources (like aluminum) which will never return to the mother earth. This means that at one time there will not be available the natural resources required for the further development and sustainability of the coming human generations. Similarly, the intensive farming practices will deplete the soil, which calls for the development of commercial fertilizer to assure healthy harvest. But these artificial fertilizers have chemicals that are dangerous to the soil and human lives.
- *Increased pollution:* Air and water pollution is increased by technology which may be continuously worsening if no proper measures are taken place to produce *green* technological and automation equipment. Unfortunately, as engineers are working hard to create the best technologies for both society and business, they release harmful chemicals and gasses that are polluting our environment with the result of climatic change *(global warming)* **(GW)**. The increased pollution

caused by technology and automation, and the design methods of nature minding (green/pollution free) automation is thoroughly examined in Tzafestas [27].

In Cernetic et al. [29], several issues from research and development in the area of social impact of automation are presented. The paper starts by providing a summary from the extensive survey of literature partly in quantitative and partly in qualitative terms. Next the main findings from the literature survey are given. Finally, a view to the future is given by indicating three important avenues for further development. The following key social attributes are summarized:

- Employment.
- Work.
- Work organization.
- Culture.

Regarding the attribute of culture, the authors state that the definition and requirements of *'socially appropriate technology'* critically depend on cultural peculiarities of the country where this technology will be used [25]. An overall consideration of human-centered manufacturing from a broader social and political perspective is provided in Rodd [30], and an overview of the foundations of human-centered systems design is provided in Gill [31]. The suggestions for further research and development given in Cernetic et al. [29] include the following:

- Increase the intensity of working on various issues of 'social impact of automation'.
- Widen (extend) the areas of consideration in order to fully understand the underlying processes and to be able to derive appropriate solutions.
- Expand the consideration from mere automation towards information and communication technologies.
- Take a look at the future trends combining the possible future directions of development in **ACT** which also cover a part of **ICT**.

We close our discussion with a look at the impact of **OA** on the organizations [32]. The two major factors that motivate business organizations to consider automated office systems are:

- A crucial need to improve the productivity of both clerical and managerial office employers.
- The increasing complexity of organizational decision-making and information needs.

In Olson and Lucas [32] a narrow view of office automation is taken that concentrates on the administrative component of an organization's functioning. A crucial component of automated office systems under this view (focus) is their communication functions which are realized by communications technology. Automated office systems are based on interactive workstations connected to communications network. Each workstation involves some degree of functionality of three components, namely: communications, text processing, and personal applications. Personal applications (on-line calendar, scheduling programs, etc.) include the capability for streamlining individual administrative tasks and are used by individuals at their own discretion. In Olson and Lucas [32] some potential long-term organizational implications of office automation are discussed in order to:

- Call attention to the need for research to increase our understanding of the potential effects of the technology.
- Alert practitioners to take a broad perspective when implementing these systems.

Office automation is expected to increase organizational productivity through redefinition of office work rather than increased efficiency of current office functions. The potential changes in the nature of work discussed in Olson and Lucas [32] are expressed by the following propositions.

Effects of Office Automation on the Nature of Work

- "Automated office systems, especially text processing functions, can improve the quality of written documents produced (e.g., reports)".
- "Automated office systems, especially text processing functions, can permit increased specialization of skills to support administrative and clerical tasks".
- "Automated office systems, especially communication functions, can alter the physical and temporal boundaries of work".

Effects of Office Automation on Individuals

- "Automated office systems can affect the role identification and stress of office workers, especially secretarial and clerical workers".
- "Automated office systems can affect the perceived status and job satisfaction of office workers, especially secretarial and clerical workers".
- "Changes in the physical and temporal nature of work supported by automated office systems can affect the worker's feelings of identity with organizational goals and criteria for promotability, especially for professional and managerial workers".

Effects of Office Automation on Organizational Communications

- "Automated office systems, especially communications functions, can lead to improved efficiency of communication 'for all office employers'.
- "Automated office systems, especially communications functions, can lead to an increase in the total volume of communications by organization members".
- "Automated office systems, especially communications functions, can affect the total volume of communications between departments".
- "Automated office systems, especially communications functions, can lead to a decrease in the amount of face-to-face contact between a manager and secretary, between colleagues, and between superiors and subordinates".

Effects of Office Automation on Management Process

- "Automated office systems, especially communications functions and personal applications, can affect manager's perceptions of the degree of rationality, flexibility, and free space of their work."
- "Automated office systems, through their effect on the physical and temporal nature of work, can affect methods for monitoring and controlling work".
- "Automated office systems can be utilized to help increase the span of control managers".

Effects of Office Automation on Interpersonal Relations

- "Automated office systems, especially communications functions and personal applications, can reduce the quantity and quality of social interaction and social reinforcement in the office".
- "Automated office systems, especially communications functions, can affect the number of *sociometric* links within an organization, the volume of communications among existing links, and the volume of communications upward in the hierarchy".

Effects of Office Automation on Organizational Structure

- "Automated office systems can facilitate changes in the definition of physical organizational boundaries".
- "Automated office systems can help improve the ability of the organization to accommodate structural changes".

Detailed discussions on the above propositions on the effects of office automation on organizations, and their implications for the design and implementation of automated office systems are provided in Olson and Lucas [32].

The above propositions provide a good starting point for research on the impact of automated office systems.

9.7 Conclusions

This chapter has studied the societal aspects of systems theory, cybernetics, control, and automation in a unified way. Most fundamental key issues were considered, namely society development, sociocybernetics, behavioral/economic cybernetics, industrial control, and managerial/economic control, as well as the impact of automation on society (unemployment, working conditions, productivity, capital formation) and the impact of automation on organizations. The social consequences of technology are still continually studied with many psychological, behavioral, and efficiency of workers issues requiring further deep study and research. The field of social impact is actually multidisciplinary and needs the synergy of psychology, human factors, human–machine interaction/cooperation, sociology, technology, etc. Overall, the primary goal of the automation and technology social impact field is to design and implement-human-friendly, human-centered or human-and-nature-minding systems, which hopefully lead to higher standards and comforts for individual persons, families, and entire societies of any kind, as well as for the environmental pollution protection, and sustainable life, climate, and ecology.

References

[1] Jenkins, R. (2002). *Foundations of Sociology*. Basingstoke: Palgrave MacMillan.

[2] Alston, R. M. (1983). *The Individual vs. the Public Interest: Political Ideology and National Forest Policy*. (Boulder, CO: West View Press).

[3] Davis, K. (1949). Human Society. New York, NY: MacMillan Company.

[4] Blackman, F. W. (1926). *History of Human Society*. Availble at: http:/manybooks.net/titles/blackmanf3061030610-8.html

[5] Science for All Americans. (2013) *Online: Human Society*. Available at: www.project2061.org/publications/sfaa/online.chp7.htm

[6] Tzafestas, S. G. (2017). *Energy, Information, Feedback, Adaptation, and Self-organization: Fundamental Elements of Life and Society*. Berlin: Springer.

[7] Tamas, A. (1987). *System Theory in community development*. Available at: http://es.scribd.com/document/36961633/System-Theory-in-CD

[8] Samad, T., and Annaswammy, A. M. (2011), "Cognitive control," in *The Impact of Control Technology*, eds T. Samad, and A. M. Annaswammy (Newyork, NY: IEEE Control Systems Society)
[9] Rigdon, J. (1994). Technological gains are cutting costs, and jobs, in services. *Wall St.* 10, A1.
[10] Geyer, F., and van der Zouwen, J. (1991). Cybernetics and social science: theories and research in sociocybernetics. *Kybernetes* 20, 81–92.
[11] Cavallo, R. E. (1979). *The Role of Systems Methodology of Social Science Research.* Boston MA,: Nijhoff Publishing.
[12] Ashby, W. R. (1956). *An Introduction to Cybernetics*. London: Chapman and Hall.
[13] Braten, S. (1978). "Systems research and social sciences," in *Applied General Systems Research: Recent Developments and Trends,* ed. G. J. Klir (New York, NY: Plenum Press), 655–685.
[14] Buckley, W. (1967). *Sociology and Modern Systems Theory*. Englewood Cliffs, NJ: Prentice Hall.
[15] Voevodsky, J. (1970). *Behavioral Cybernetics: An Exploration*. Berkeley CA: San Jose State University.
[16] Duke, E. M. (1976). Cybernetic factors in economic systems. *Cybernet. Forum* 8, 21–24.
[17] Leontief, W. (1966). *Input-Output Economics*. New York, NY: Oxford University Press.
[18] Jones, S. (2013). "Cybernetics in society and art," in *Proceedings of the Nineteenth International Symposium of Electronic Art, (ISEA 2013)*, eds K. Cleland, L. Fisher and R. Harley (Technology and the University of Sydney: Sydney).
[19] Simons, R. L. (1995). *Levers of Control: How Managers Use Innovative Control Systems to Drive Strategic Renewal*. Boston, MA: Harvard Business School Press.
[20] Tzafestas, S. G. (1982). "Optimal modal control of production inventory system," in *Optimal Control of Dynamic Operational Research Models*. (Amsterdam: North Holland Company) 1–71.
[21] Hoover, K. (1995). Facts and artifacts: calibration of the empirical assessment of real business cycle models. *Oxford Econ. Pap. New Ser.* 47, 24–44.
[22] Klein, D., and Romero, P. (2007). Model Building versus theorizing: the pau of theory, journal of economic theory. *Econ. J. Watch* 4, 241–271.

[23] Goldfarb, R. S., and Ratner, J. (2008). Economics in Practice: follow-up theory and models terminology through the looking glass. *Econ. J. Watch* 5, 91–108.

[24] Taylor, F. W. (1906). On the art of cutting metals. *ASME J. Eng. Ind.* 28, 310–350.

[25] Martin, T., Kivinen, J., Rijnsdorp, J. E., Rodd, M. G., and Rouse, W. B. (1991). Appropriate automation: Integrating technical, human, organizational, economic and cultural factors. *Automatica*, 27, 901–917.

[26] Kutay, A. (1989). *The Economic Impact of Automation Technology*. Technical Report CMU-RI-TR-89-13A Robotics Institute, Carnegie Mellon University.

[27] Tzafestas, S. G. (2010). *Human and Nature Minding Automation: An Overview of Concepts, Methods, Tools, and Applications*. Berlin: Springer.

[28] Ramey, K. (2012). *Technology and Society: Impact of Technology on Society*. Available at: www.useoftechnology.com/technology-society-impact-technology-on-society [accessed November, 12, 2012].

[29] Cernetic, J., Strmcnik, S., and Brandt, D. (2002). "Revisiting the Social Impact of Automation", in *Proceedings of the Fifteenth Triennial IFAC World Congress*, Barcelona.

[30] Rodd, M. G. (1994) Human-centered manufacturing for the developing world. *IEEE Technol. Soc. Mag.* 13, 25–32.

[31] Gill, K. S. (1996). "Human machine symbiosis," in *Foundations of Human-Centered Systems Design*, ed. K. S. Gill (Berlin: Springer), 1–68.

[32] Olson, M. H., and Lucas, H. C. (1982). The Impact of office automation on the organization: Some implications for research and practice. *Commun. ACM* 25, 838–847.

[33] Beer, S. (1951). Sketch for a cybernetic factory, Cybernetics and Management (Chap. XVI), English University Press, 142–152.

[34] Freedom from Fear Magazine (2014). *Cybernetics and Law*. Available at: http://f3magazine.unicri.it

[35] Tzafestas, S. G. (2010). *Human and Nature Minding Automation: An Overview of Concepts, Methods, Tools and Applications*. Berlin: Springer.

10

Ethical and Philosophical Issues

Great people have great values and great ethics.
Jeffrey Gitomer

Ethics is knowing the difference between
what you have the right to do and what is right to do.
Potter Stewart

Action is indeed the sole medium of expression for Ethics.
Jeffrey Gitomer

10.1 Introduction

Ethics is concerned with the study and justification of moral beliefs. The term comes from the Greek word *ήθος* (ethos) which means *moral character*. Character is an inner-driven view of what constitutes morality, whereas *conduct* is an outer-driven view. Philosophers regard ethics as moral philosophy and morals as societal beliefs. Thus, some society's morals are not ethical, because they merely represent the belief of the majority. Other philosophers argue that the ethics has a relativistic nature, in the sense that what is right is determined by what the majority believe. This school of thought started in ancient Greece by Aristotle who argued that 'ethical rules should always be seen in the light of traditions and the accepted opinions of the community'.

Systems, cybernetics, control, and automation are overlapping fields of science and technology of interdisciplinary nature which have created extensive and crucial ethical concerns all over the years. They can be used for the good and the bad and so, besides legislation, they call for application of proper ethical theories and principles which complement the law and help to assure the growth, development, and survival/sustainability of modern societies.

The purpose of the present chapter is to investigate the ethical issues of **SCCA**. Specifically, the chapter:

- Provides a short discussion that addresses the general question '*what is ethics?*'
- Discusses the ethical issues and concerns of systems engineering starting with the question '*what is systems engineering?*'
- Discusses the ethics of systems thinking.
- Examines the ethics of cybernetics focusing on the ethics of 'cybernetic organisms' (cyborgs) and human 'implants'.
- Provides a general discussion of the ethics of control and automation, and outlines the basic elements of 'robot ethics' (roboethics) and 'management control ethics'.
- Presents an overview of systems philosophy and its evolution.
- Discusses the control and cybernetics philosophy.

10.2 What is Ethics?

As seen in Chapter 2, ethics is one of the principal branches of philosophy developed since the ancient times in parallel with the other branches (logic, metaphysics, epistemology, and aesthetics). Applied ethics is an area of ethics which is concerned with the study of the application of ethics in actual life. *Ethical systems (theories)* that can be used for ethical decision making include:

Kantian Deontological ethics system This is based on two forms of the categorical imperative, namely: (i) act only following moral rules that you can at the same time wish to be universal laws, (ii) act such that you always treat both yourself and others as ends in themselves, and never only as a means to an end. An act that you would not wish to be a universal moral law may be *wrong* or *not belonging in the moral domain.* The second imperative could be better formulated as: 'when enlisting others for your own ends, treat them with honesty and do not use coercion').

Utilitarian ethics system *(John Stuart Mill)* Utilitarianism is distinguished in *act utilitarianism* and *rule utilitarianism.* In the first, acts are to be judged according to their utility. In the second, moral rules are to be judged according to their net utility. Judging moral rules rather than actions is a better form of utilitarianism which can be a better foundation for ethics. In utilitarianism, weighing the consequences is a difficult issue. Mill defined the 'good' in terms of well-being (pleasure, or happiness), which is the Aristotelian '*ευδαιμονία*' (eudemonia=happiness).

Virtue or character-based ethics system This ethics system looks at the virtue or moral character of the individual who performs an action, rather than on ethical duties and rules or the consequences of an action, i.e., it is a person-based ethics. Virtue ethics gives guidance for the kind of features and behaviors a good person will try to achieve, and so it is concerned with the whole of a person's life rather than particular actions. According to virtue ethics (i) "A right act is the action a virtuous person would do under the same conditions", and (ii) "A good person is someone who lives virtuously; a person that possesses and lives the virtues".

An ideal ethical action can be made by a proper combination of the principles of *Utilitarianism* (beneficence). *Deontologicalism* (justice), and *Individualism* (virtue ethics).

Social contract system *(Thomas Hobbes, Jean-Jacques Rousseau, John Locke, John Rawls)* In social contract we accept rules that specify how we treat others, accepting the rules rationally for our mutual benefit. The rules are usually formulated in terms of *rights*, i.e., respects the rights of others. The concept of rights is very important and can be used either to specify what the moral domain is, or to delineate the limits of government. It is remarked that in real life what is right for one person may not be right for another person, i.e., it is all relative for the individual *(Subjective Relativism)*. Clearly, although it is recognized that individuals are unique, rational discussion of morality between individuals is impossible. Also, what is right in one culture or society may not be right for another culture, i.e., it is all relative to the culture or society *(Cultural Relativism:* proponent *William Graham Summer)*. Subjective and cultural relativism has been rejected by Quinn as a foundation for ethics.

Value-Based System *(J. Dewey)* This ethical system employs some *value system* which consists of the ordering and prioritization of ethical and ideological values that an individual or society holds [1]. Values are distinguished in (i) Ethical values which specify what is right or wrong, and moral or immoral. They define what is allowed or prohibited in the society that holds these values, (ii) ideological values which refer to more general or wider areas of religion, political, social, and economic morals. A value system must be consistent, but in real life this may not be so.

Case-Based System This is a modern ethics system that tries to overcome the apparently impossible divide between deontology and utilitarianism. It is also known as *casuistry* [2], and starts with immediate facts of a particular case. Casuists start with a particular case itself and then examine what are morally significant features (both theoretically and practically). Casuistry

Figure 10.1 The ingredients of ethics.

Source: www.goldfish-consulting.co.za/blog/ethics-0

finds extensive application in juridical and ethical considerations of law of ethics. For example, lying is always not permissible if we follow the deontological principle. However, in casuistry one might conclude that a person is wrong to lie in formal testimony under oath, but lying is the best action if the lie saves life.

Ethics in theory and actual everyday life involves all the above, namely values, moral principles, rules and regulations, rules of conduct, ethical practices, and research aiming at improving and enhancing ethics (Figure 10.1).

Figure 10.2 gives a practical 6-step scheme for resolving ethical problems and dilemmas.

The *evolution of ethics* is somehow a bridge between biology and human behavior. This bridge is realized by examining the cultural and biological feedback systems that inspire the evolution of social rules. When biology and culture are in conflict there result conflicting acts as a kind of informational feedback which tells people that there are crucial problems that need to be resolved. Thus, one can say that evolution of ethical systems is resulting from the struggle of human species (and societies) to survive. On the other hand, a whole web of related '*rule systems*' (e.g., state laws, professional codes, and customs) evolve to further human adaptation.

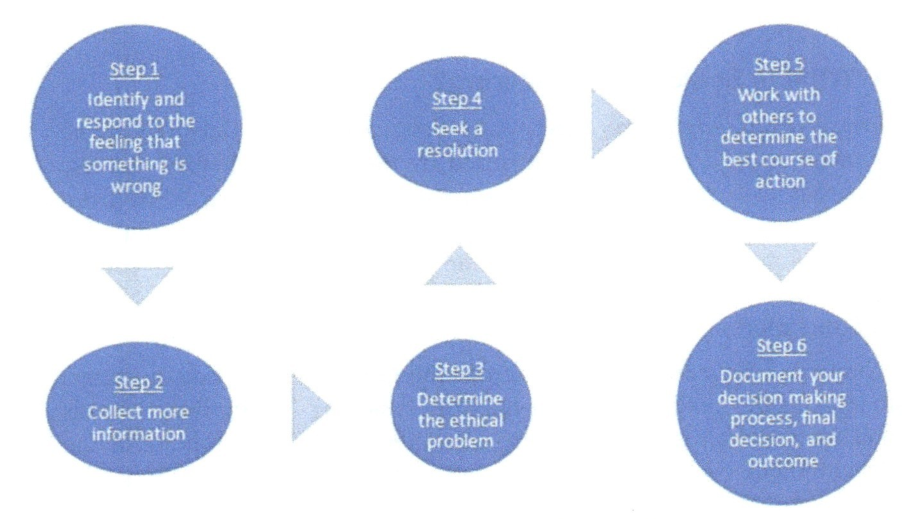

Figure 10.2 A 6-step scheme for ethical problem resolution in practice.

Source: www.keywordsuggest.org/gallery/425427.html

Ethical systems are agreed rules of conduct that come from past experience. *Moral laws* (informally known) evolve over centuries of time and are often influenced and expressed by human emotions. However, the terms '*moral*' *and* '*ethical*' are frequently used '*interchangeably*' [3]. As with philosophy, for anything we care to be interested in, we have an *ethics* that deals with ethical issues related to its design, technological implementation and use in the society. Therefore we have many particular ethics, e.g.:

- Ethics of computer science.
- Ethics of artificial intelligence.
- Ethics of robotics.
- Ethics of information.
- Ethics of cyberspace.
- Ethics of medicine, etc.

In the following sections we will briefly outline the following:

- Ethics of systems engineering.
- Ethics of systems thinking.
- Ethics of cybernetics.
- Ethics of control.
- Ethics of automation.

10.3 Ethics of Systems Engineering

10.3.1 What is Systems Engineering?

The term *systems engineering* **(SE)** was first used at the Bell Telephone Laboratories in the 1940s. It was inspired by the need to identify and use the properties of a system as a whole, and not as they appear to be from the particular properties of the parts. Systems engineering was initiated as only one approach, but today represents an entire discipline in engineering. Actually, there is no single definition of what the term 'systems engineering' means. Some representative definitions (meanings) are the following:

- "*Systems engineering* is an interdisciplinary approach and a means to enable the realization of successful systems. The customer needs and the required functionality are identified from the beginning of the development cycle, documenting requirements, then proceeding with the design synthesis and system validation while considering the complete problem" (*INCOSE*: International Council on Systems Engineering definition).
- "*Industrial and systems engineering* (often used interchangeably with '*industrial engineering*') is a term used to emphasize three key points namely: (i) components (including machines and people) that interact with each other to produce the overall behavior of the system, (ii) the system under study which is always a subsystem of a larger system and the interactions involved must also be considered, (iii) the systems include humans, and must be optimized as a whole".
- "*In technology management* (or engineering management) the systems engineering approach is set to account for manufacturability, installation, operations, maintenance, repair, coordination, and disposal of a system".

Figure 10.3 illustrates in a pictorial way the major engineering and technology management branches involved and contributing in the realization of systems engineering.

The philosophical roots of systems engineering are traced in Bertalanffy's 'General *Systems Theory*' and Forester's '*Systems Dynamics*' . The field of systems engineering has evolved over time to embrace a holistic concept of 'systems' and 'engineering processes'. Today, systems engineering is used in both the narrower and wider sense, and can bind scientists from different fields working in the system design and development into a unified group.

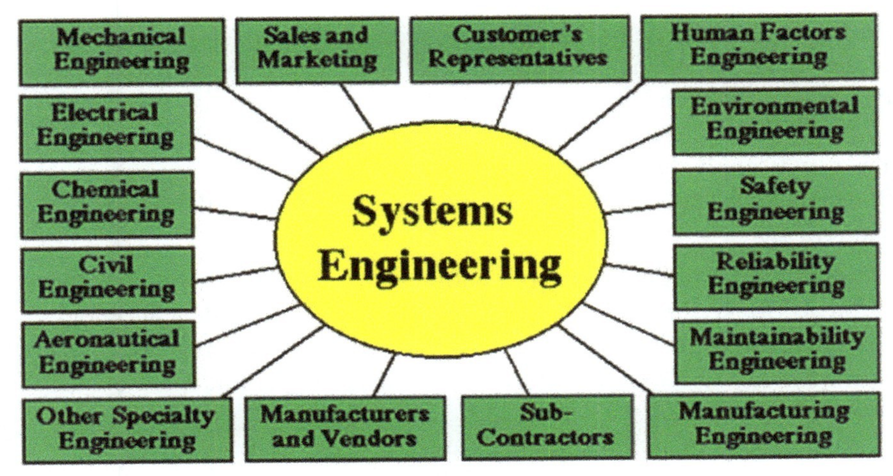

Figure 10.3 General pictorial representation of systems engineering.

Source: www.incose.org/AboutSE/Careers

10.3.2 Ethics of Systems Engineering

Ethics of systems engineering is distinguished in [4]:

- Personal ethics.
- Corporate ethics.
- Professional ethics.

Personal ethics refers to the way an individual relates to an organization and to the people this person works with. The two major ethical rules ('*golden rules*') that specify the way a person may perform are:

- *Treat the others in the way you would like to be treated:* This rule governs the interpersonal relationships. For example, effective systems engineering is based on team work. An engineer may be excellent, but if he can't work on a team the company will have little use for him.
- *Those who have the gold determine the rules:* This rule aims to satisfy the clients (external or internal). But in the process of assuring the cients' satisfaction the internal principles of the company must not be violated in any way.

Corporate ethics attempts to assure making a fair profit within the company's operational framework (suppliers, distributors, employees, and clients). The company tries to minimize wrong tendencies such as personal biases, and shortcuts to maximize profit.

Professional ethics helps the engineers to decide what to do in a situation where they think something is wrong. In situations that are not clear and pose difficult ethical problems this professional decision is not an easy task. In all cases where the engineering activity seems to be ethically wrong or illegal one must try to apply the golden rules.

Before taking any action an engineer must consider the following aspects [4]:

- *The law* (Refer to the Company's attorney).
- *Motives* (Make sure that you know, or refer to an independent attorney).
- *Actions* (Understand why you are doing what you are doing).
- *Ethics policy* (Follow the ethical code for employees adopted by the company).
- *Consequences of actions* (Make sure that you know well what the consequences of your acts are, and make your best to assure they are ethical).

An engineering or business company that considers seriously the ethical issues must do the following:

- Specify the actions that are not acceptable by the company.
- Develop a method for reporting violations.
- Adopt a consistent way of enforcement.
- Impose just punishment to violators.
- Ensure and show that justice has been carried out.
- Protect the reporters of violation.

Scientists and engineers are first of all obliged to obey the following universal deontological rules [5], both in their private life and their professional work in a company.

- Don't kill.
- Don't cause pain.
- Don't disable.
- Don't deprive of freedom.
- Don't deprive of pleasure.
- Don't deceive.
- Keep your promise.
- Don't cheat.
- Obey the law.
- Do your duty.

This is a multi-rule ethical system, and as in all multi-rule systems, it is possible to face a conflict between the rules. To address the rule conflict problem one may treat the ethical rules as dictating *prima facie duties* [6]. This, for example, means that if an agent gives a promise, it has the obligation to keep the promise. Rules may have exceptions, and moral considerations derived from other rules, may override the rule. As argued in [5] these rules are not absolute. A way for deciding when it is okay not to follow a rule is provided in [5]. This rule is the following:

'Everyone is always to obey the rule except when an impartial rational person can advocate that violating it be publicly allowed. Anyone who violates the rule, when an impartial rational person could not advocate that such a violation may be publicly allowed, may be punished'.

Modern technological developments that raise ethical issues in systems engineering include:

- *Data storage* (The costs of maintaining large databases with details about individuals are rapidly decreasing).
- *Data analysis* (Companies can analyze large quantities of data referring to people in the process of developing detailed profiles of individual performance).
- *Networking* (Data copying from one place to another and accessing personal data from remote locations is becoming much easier with the use of computer networks and the Internet).

The scope of engineering ethics should involve at minimum the following (Figure 10.4):

- Moral reasoning and ethical theories.
- Engineering as social experimentation.
- Responsibility for safety.
- Responsibility for employees.
- Engineers, managers, consultants, and headers.
- Rights for engineers.
- Global issues.

Professional societies have developed and released *ethics codes* that provide guidance for interaction between professionals such that they can serve both each other and the whole society in the best way, without the fear of other professionals undercutting them with less ethical actions [7].

The code of the *National Society of Professional Engineers* **(NSPE)** is stated as follows [8]:

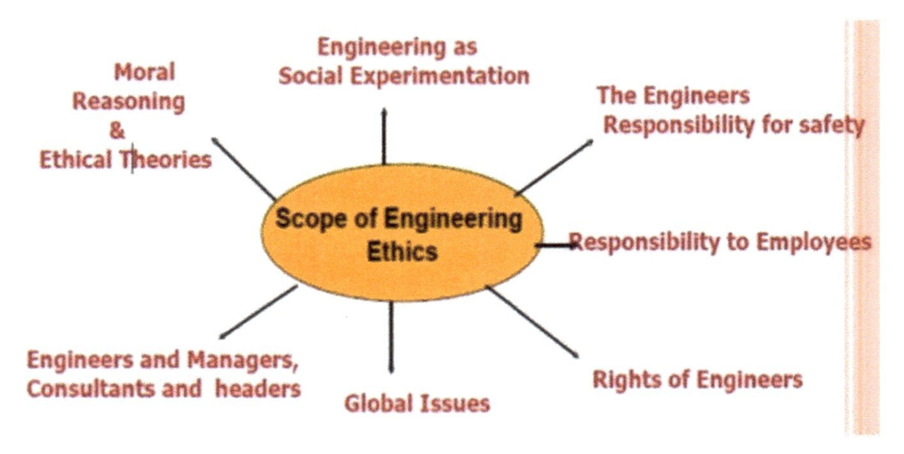

Figure 10.4 Scope of engineering ethics.

Source: https://www.slideshare.net/ (Mugiwaral/engineering-ethics-cases, Oct. 15, 2000).

'Engineers in the fulfillment of their professional duties shall:

- Hold paramount the *safety, health and welfare* of the public.
- Perform services only in areas of their competence.
- Issue public statements only in an objective and truthful manner.
- Act for each employer or client as faithful agents or trustees.
- Avoid deceptive acts.
- Conduct themselves honorably, responsibly, ethically, and lawfully so as to enhance the honor, the reputation, and usefulness of the profession'.

This code is addressed to the entire engineering profession with no reference to particular engineering specialties. The detailed code which includes: (i) Rules of Practice, (ii) Professional Obligations, and (iii) a Statement by the NSPE Executive Committee, can be found in [8].

The fundamental canons of the **ABET** (*Accreditation Board of Engineering and Technology*) code of ethics of engineers are [53]:

- "Engineers shall hold paramount the *safety, health and welfare* of the public in the performance of their professional duties.
- Engineers shall perform services only in the areas of their competence.
- Engineers shall issue public statements only in an objective and truthful manner.
- Engineers shall act in professional matters for each employer or client as faithful agents or trustees, and shall avoid conflicts of interest.

- Engineers shall build their professional reputation on the merit of their services and shall not compete unfairly with others.
- Engineers shall act in such a manner as to uphold and enhance the *honor, integrity and dignity of the profession.*
- Engineers shall *continue their professional development* throughout their careers and shall provide opportunities for the professional development of those engineers under their supervision."

Two other engineering codes are:

- The code of the Institute of Electrical and Electronic Engineers **(IEEE)** [9].
- The code of the American Society for Mechanical Engineers **(ASME)** [10].

The key ethical question in system design is the following:

- How to 'act ethically' in situations that involve new technologies, new objects of design, and new contexts?

Particular questions are:

- Do users have actually decision power? If yes what kind?
- Does the design method or process guide designers and researchers to develop their interests and attitude towards participants?
- Does a design method, tool or process recognize and encourage participants' abilities to learn?
- Does a design method, tool or process include participants' evaluations of what is being designed as well as the design process itself?

10.4 Ethics of Systems Thinking

A short discussion of systems thinking was given in Section 4.5. Here we will outline the ethical implications of systems thinking in organizational settings [11]. Philosophy may be regarded as a kind of '*conceptual engineering*'. 'Systems thinking' needs its own conceptual engineering in order to reflect on the structure of thoughts ideas contained in it. The relations of systems thinking to ethics are [12]:

- Thinking well is ethical.
- Ethics requires thinking well.

In any situation, as a general principle, it is better to think well than to think poorly [12]. People have a duty to seek the truth, and when it is matter of choice, one does not right when he/she does not think well. Because people have an ethical obligation to think well, the question is whether systems thinking helps people in their 'conceptual engineering', or in other words 'does systems thinking help them to think? The works of Talcott Parsons, Niklas Luhmann, Jay Forrester, and Russel Ackoff have provided important answers to these questions. Here, we will address the question: 'Does systems thinking helps us perform ethically?

In Blackburn [12] it is argued that, "since the claims of ethics are grounded in truth, i.e., 'moral judgments possess truth-value', to the extent that 'systems thinking' helps us to think about human organizations, it helps us to do ethics". 'Systems thinking' is a vehicle driving us to think better, but even so there will always be ethical concerns in organizations. Actually, no matter how well we think, we will always have to face the ordinary life controversies with other people. Conceptual engineering cannot resolve all of them, but even so it can assist us in our attempt to act ethically [13].

For both Kant's deontological ethics theory (*theory of duties*) and Mill's utilitarian (*consequential*) theory an important step is to identify the agents, i.e.:

- To whom does a person owe a duty?
- Whose happiness is likely to be affected?

'Systems thinking' enables us to perform this identification and increase the range of stakeholders in order to include everyone involved in the system, e.g., co-workers, managers, vendors, suppliers, distributors, customers, etc. When people perform tasks within a system, their behavior tends to be a response to the actual performance of the system. This is one of the insights ascribable to systems thinking.

In Hitt [13], 'Ethics and Leadership', *Williams Hitt* discusses the various tensions in the every-day operation of organizations, which put ethical dilemmas for the organizational leaders. These tensions are recognized by systems thinking to be part of the structural properties of organizations.

'Systems thinking' assists organizational leaders in revealing the closer correlation between what an employee does and what is responsible for. Therefore systems thinking dilutes responsibility only when it belongs to the system itself. According to *Wheatley* [14], systems theory encourages leaders to trust and involve their subordinates more, not less. The '*equifinality principle*' (a goal may be achieved in different ways) implies that once participants accept the goal and values of the system and hold themselves accountably to it, they should be free to find their own best way to perform. More, issues on the ethics of systems thinking are provided in Harter [11].

A partial list of required competencies for efficient and good thinking in applications of modern society is the following:

- *Creative thinking* (Use available supportive methods and tools, avoid obvious solutions with undesired side effects, surprise others and have fun).
- *Scientific thinking* (Quantify and measure, formulate and test hypotheses).
- *Nonlinear thinking* (Action and reaction need not to be closely related, try to find small changes with big effects, use policies as a leverage).
- *Operational thinking* (Think in the same way as events are already happening, be confident in units of measurement).
- *System as a cause* (To find a system cause the system's structure should be identified first, do not blame others).
- *10 miles view* (Look ahead in time, expand your perception, try to see both trees and forest).
- *Closed loop thinking* (Locate feedback loops of causal relationships, search for feedbacks in both policies and mental models).
- *Dynamic thinking* (Search for repeated behavior patterns over time, think in continual terms, try to perceive impacts of small changes).
- *Value-focused thinking* (Focus on the essential activities that must occur prior to the adoption of a decision policy).

'Thinking about values' possesses the following features [55] (Figure 10.5):

- Uncovers hidden objectives.
- Identifies decision opportunities.
- Creates and evaluates alternatives.
- Helps to get more productive information.
- Improves communication of the agents participating in a decision.
- Facilitates the involvement of multiple stakeholders.
- Enhances the coordination of interrelated decisions.
- Guides strategic thinking.

10.5 Ethics of Cybernetics

Ethics of cybernetics is a field in which both philosophers and cybernetics researchers have devoted considerable efforts that led to interesting results. Meta-theorists of cybernetics have made large claims about the impact of cybernetics, for example Turkle [15] and Introna [16]. Turkle's basic thesis is

Figure 10.5 Features of thinking about values [55].

that the virtual realities made possible and surely encouraged by technical advancements, manifest on the computer screen, concretely illustrate and instantiate a radical aesthetic relativism (freedom), concomitant de-centered and multiple selves (constructed selves) [17]. In other words, cybernetics has the ability to support a new form of selfhood, the *de-centered* and *multiple self* (selves). The multiple self is not accountable in the same way that the integral self of morality is held accountable. Therefore, according to Turkle, cybernetics liberates the traditional self for the freedom of multiple selves. Lucas Introna condemns information technology and supports the reverse point. He also credits cybernetics with a radical transformation, or the possibility of a radical transformation of morality. Essentially both Turkle and Introna argue that cybernetics cannot simply be viewed from the perspective of morality, but rather that morality must be viewed from the point of view of information technology. An extensive discussion of the above ethical views is provided in Cohen [17].

Here we will discuss the ethical issues which refer to some recent cybernetics advancements that directly influence human life, namely *cyborgs* (**cy**bernetic **or**ganisms) and *implants*. Cybernetics can be classified by:

- *Placement.*
- *Responsiveness.*
- *Function.*

Placement may be external modification (glasses), external replacement (glass eye), and implanted (fully cybernetic eye).

Responsiveness ranges from *inert* (peg leg), to *mechanical* and *interactive* responsiveness.

Function ranges from mechanical (running shoes), to physiological (insulin implant), and neurological (brain implant).

Not all cybernetic devices can be easily fit in these categories, while many will fit more than one of the categories particularly for function. All these device classes offer big challenges for the development of cybernetics, the biggest one being implanted cybernetics (especially interactive ones).

Cyborg and implant ethics *Cyborg and implant technology* has been made possible by the fact that the brain central nervous system bioelectrical signals can be connected directly to computers and robot parts that are either external to the body or implanted in the body. Clearly, making an artificial hand with the movement and sensors needed to create an entity that approaches the capability for movement and feeling of a normal functioning biological hand is an extremely advanced attainment. Of course, when one succeeds in decoding the brain's movement signals, then these signals can be directly connected to external electronic devices (mobile phones, TV sets, etc.) such that it may be possible to control electronic devices with the power of thought alone, i.e., without the use of articulated language or external motor device. This means that cyborg-type prostheses can also be virtual.

- *Cyborg and implant technology* is a two-sided concept; it can be used for the good or for the bad, and like other modern technologies may have negative and positive consequences.
- *Cyborg and implant ethics* includes the ethics of medicine, the ethics of medical robots, and the ethics of assistive robotics (especially *prosthetic ethics*).

An implant is anything implanted within the human body which includes devices beyond normal biological functioning. An advanced new class of implants is that in which interfaces are created between neural tissue and microprobes in order to achieve communication between a patient's nervous system and devices that replace or supplement a malfunctioning organ.

Examples of new neural interface implants and cyborgs which have particularly strong ethical implications are [20]:

- *Cochlear implants*: These are used in the everyday treatment of deafness. A cochlear implant is a device, which, activated by sound, directly stimulates the auditory nerve, therefore by-passing dysfunctional parts of the inner ear. Cochlear implants give rise to an 'artificial' sense of hearing which is inferior to natural hearing but yet sufficient for social functioning.
- *Prosthetic Vision for blind people*: This is based on essentially the same principles as cochlear implants, i.e., stimuli from technological sensors are relayed to nervous system via a nerve-implant interface. This interface can be placed as retina clips and chips implanted in the visual cortex of the brain.
- *Brain implants for bladder control.*
- *Brain implants that block tremor* in patients with tremor such in Parkinson's disease.

The primary ethical concerns surrounding the development of cyborgs and implants are focused on human dignity, human relations, protection of physical/bodily harm, and the management of the health and other personal data evaluation. In all cases the primary rules of medical and robotic ethics, namely autonomy, non-maleficence, beneficence, justice, truthfulness, dignity, and accurate evaluation, should be respected [21]. Implant ethics overlaps with several other sub-disciplines of bioethics, e.g. '*neuroethics*' that covers the ethical issues to which advances in neuroscience (including brain implants) may give rise [21]. A general discussion of ethical and moral concerns referring to advanced medical technologies is provided in Satava [22].

The drawbacks of cyborgs include the following:

- Cyborgs do not heal body damage normally, but, instead, body parts are repaired.
- Replacing broken limbs and damaged armor plating can be expensive and time consuming.
- Cyborgs can think the surrounding world in multiple dimensions, whereas human beings are more restricted in that sense.

A philosophical discussion about cyborgs and implants, and the relationship between body and machine is provided in [13]. A general scientific discussion about cyborgs and the future of mankind is given in [18, 19]. General ethical issues of organ or cell transplantation or implantation of various types of technological device include the following [20]:

- *Donation issues* (E.g., in organ transplantation surgery, issues like consent, compensation, and risk of exploration have been strongly and extensively discussed by ethicists).
- *End-of-life decisions concerning donation* (These are among the most pressing issues in transplantation ethics).
- *Distribution issues* (The severity of the problem is determined by the price of the interventions rather than whether or not it involves an implantation).
- *Improvement of function above normal levels issues* (Implant ethics has to deal with issues of organ normality and dysfunction, and with the admissibility of human enhancement).
- *Fear that non-voluntary interventions may be carried* (This can perhaps be done in the form of brain implants used to control other human beings).

A full discussion on the above ethical issues is provided in [20]. Four examples of human cyborgs are the following [23]:

Example 1 A person with extreme color blindness (achromatopsia), equipped with an 'eyeborg' (a special electronic eye), which renders perceived colors as sounds on the musical scale, becomes capable of experiencing colors beyond the scope of normal human perception. The name of each color is memorized and becomes a 'perception'. This way the device allows the user to 'hear' color (Figure 10.6).

Example 2 One of the first women that became cyborgs (Claudia Mitchell) was outfitted with a bionic limb. The limb is connected to her nervous system enabling her to control it with her mind (Figure 10.7).

Example 3 *Jens Naumann* has lost his sight (in both eyes) due to a couple of serious accidents. He became the first person in the world to receive an artificial vision system, equipped with an electronic eye connected directly to his visual cortex through brain implants (Figure 10.8).

Example 4 After losing part of his arm because of an accident at work, *Nigel Ackland* got an upgrade enabling him to control the arm through muscle movement in his remaining forearm (Figure 10.9). He can independently move each of his five fingers to grip delicate objects or pour a liquid into a glass. The range of movement achieved is really extraordinary.

Figure 10.6 Wearing an eyeborg the artist Neil Harbisson, born with achromatopsia, can see colors.

Source: http://www.mnn.com/leaderboard/stories/7-real-life-human-cyborgs

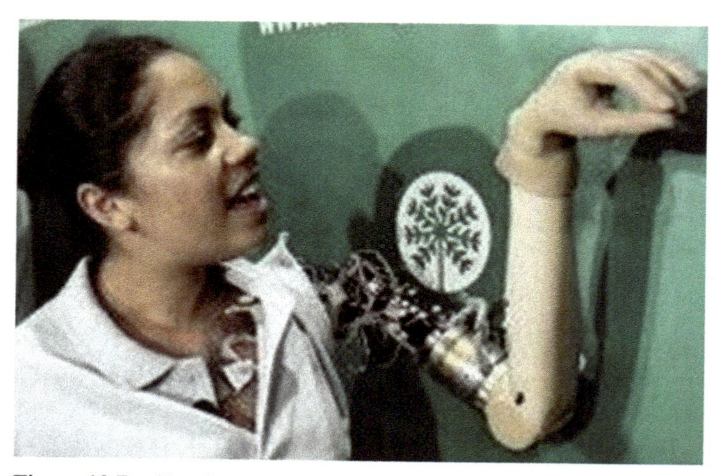

Figure 10.7 Claudia Mitchell can control a bionic limb with her mind.

Source: http://www.mnn.com/leaderboard/stories/7-real-life-human-cyborgs

Figure 10.8 Jens Naumann sees with a cyborg (electronic) eye connected directly to his visual cortex.

Source: http://www.mnn.com/leaderboard/stories/7-real-life-human-cyborgs

Figure 10.9 A cyborg controlling the arm and fingers though muscle move.

Source: http://www.mnn.com/leaderboard/stories/7-real-life-human-cyborgs

10.6 Ethics of Control and Automation

10.6.1 General Issues

Jointly, control and automation embrace all systems and applications that use feedback control and automation principles and technologies. Ethics of control and automation are overlapping considerably and belong to the ethics of technology (technoethics) [24].

As already seen, examples of control and automation systems include:

- Process control and automation.
- Power systems.
- Industrial systems.
- Robotic systems.
- Vehicle driving automation.
- Office automation systems.
- Land, sea, and air transportation systems.
- Organizational/economic systems.
- Environmental control systems, etc.

Therefore, ethics of control and automation involves the ethics of all the above system categories, including the ethics of modern technologies like information and communication technologies, etc. In practice, the ethics of control and automation is distinguished in the following subdivisions:

- The ethics of people involved in the design process.
- The ethics of the systems and technologies themselves.
- The ethics of people using the control and automation systems and products.

In general, *technology* is the branch of knowledge which is concerned with the development and employment of technical means and their interrelation with life, society, and the environment. It is the core to human development and a key focus for understanding human life, human society, and human consciousness. The term '*technoethics*' was coined (in 1977) by the philosopher Mario Bunge, in his effort to describe the responsibilities of scientists and technologists to develop ethics as a branch of technology. He argued that "the technologist must be held not only technically but also morally responsible for whatever he designs or executes: not only should his artifacts be optimally efficient, but, far from being harmful, they should be beneficial both in the short run and long term". He recognized a pressing need in society to create a new field called 'Technoethics' to discover rationally grounded rules for guiding science and technological progress [25].

Technoethics is a broad field and needs extensive descriptions to be fully presented. Here we will briefly discuss the ethics of two key branches of technoethics with reference to control and automation. Specifically, the following branches of technoethics will be considered; the first belonging to the 'hard' systems category and the second belonging to the 'soft' systems category.

- Roboethics (robot ethics).
- Management control ethics.

10.6.2 Roboethics

This branch involves the following sub-branches [26]:

- Medical roboethics.
- Assistive roboethics.
- Socialized roboethics.
- War roboethics.
- Automated car roboethics.

Medical roboethics involves both the medical ethics principles, and principles that are related to the use of medical robots (surgical robots, etc.). Medical ethics calls the doctors to uphold the following five fundamental principles (called ***ABCDE*** *principles*):

- ***Autonomy*** (A doctor serves to advise a patient only on health care decisions, and a patient's own choices are essential in determining personal health).
- ***Beneficence*** (A doctor is always acting in the best interest of the patient).
- ***Confidentiality*** (The conversations between doctors and patients are confidential. This is the so-called *patient-physician privilege).*
- ***Do no harm/Non-maleficence*** (This principle is based on the concept that it is more important not to harm than to do good).
- ***Equity and Justice*** (The medical community works to *eliminate disparities* in resources and health care across regions, cultures and communities as well as to abolish discrimination in health care).

In overall, medical ethics and roboethics appeal traditional practical *'more'* principles such as:

- Keep promises and tell the truth (except when telling the truth results in obvious harm).

- Do not interfere with the lives of other people unless they ask for this sort of help.
- Do not be so selfish that the good of others is never taken into account.

In robotic surgery three ethical issues that must be considered are:

- The duty of the manufacturer to ensure that operators of the robot are adequately trained.
- The duty of the hospital to allow properly credential surgeons to use the robots.
- The duty of the surgeons to act ethically following the codes of medical ethics.

Assistive roboethics is concerned with the ethics of robots designed for aiding '*People with Special Needs*' (**PwSN**) in order to assist them to improve their mobility and attain their best physical and/or social level. PwSN include persons with loss of upper limb control, persons with loss of lower limb control, and persons with loss of spatio-temporal orientation).

Assistive robotic devices include:

- Assistive robots for people with impaired upper limbs and hands (smart canes and walkers, etc.).
- Assistive robots for people with impaired lower limbs (robotic wheel chairs, etc.)
- Rehabilitation robots (upper limb and lower limb).
- Orthotic devices.
- Prosthetic devices.

The ethical principles of assistive robotics include the ABCDE medical ethics principles (autonomy, beneficence, confidentiality, do no harm/non-maleficence, equity and justice), and special aspects of doctor's/caregiver's responsibility (truthfulness, privacy, data integrity, clinical accuracy, quality, reliability). All assistive technology programs should incorporate ethics statements into administrative policies and comply with professional codes of assistive ethics (e.g., RESNA code [27]).

Socialized Roboethics is concerned with the ethics of socialized robots (robot social partners) which are designed to interact with humans socially and emotionally. They are used as companions, and adjust to therapy especially for vulnerable persons (e.g., children with autism, aged people with dementia, etc.).

Socialized robots raise a number of ethical concerns that belong to the psychological, social, and emotional sphere. The social and emotional (nonphysical) ethical issues that have to be addressed when using socialized robots include the following [28]:

- *Attachment* (This ethical issue arises when a user is emotionally attached to the robot. Attachment can appear in all kinds of user, children, adult and elderly, and can create problems, e.g., when the robot is removed due to failure or operational degradation).
- *Deception* (This risk can be created by the use of robots in assistive settings, especially in robot companions, teachers or coaches).
- *Awareness* (This issue concerns both users and care givers. They both need to be accurately informed on the risks and hazards associated with use of robots).
- *Robot authority* (A robot designed to play the role of a therapist is given some authority to exert influence on the patient. Therefore, the ethical question arises who actually controls the type, the level, and the duration of interaction. The robot must accept the patient's wishes, e.g., when he/she asks to stop an exercise, due to stress or pain).
- *Privacy* (Securing privacy during human-robot interaction is of utmost importance).
- *Autonomy* (A person mentally healthy has the right to make informed decisions about his/her care).
- *Human-human relation* (HHR is a very important ethical issue that has to be considered when using assistive/socialized robots. Typically, robots are used as means of enhancement of therapy given by caregivers, not as a replacement of them).
- *Justice and responsibility* (Here, the standard ethical issues of 'fair distribution of scarce resources', and 'responsibility assignment' should be addressed, especially if the robots are extremely costly).

War Roboethics deals with the ethical issues that stem from the use of robots for conducting wars. Here, the international laws of war must always be adhered. *Just war* theory involves three parts which are known by their Latin names, namely:

- *Jus ad Bellum* which specifies the conditions under which the use of military force must be justified (just cause, right intention, legitimate authority and declaration, last resort, proportionality, chance of success).
- *Jus in Bello* which refers to justice in war, i.e., to conducting a war in an ethical manner. The fundamental principles of the *humanitarian 'just in bello'* law are: (i) Discrimination (not killing civilians/non-combatants), (ii) Proportionality (use only force proportional to the goal

sought), (iii) Benevolent treatment of prisoners of war (captive enemy should be provided with benevolent, not malevolent, quarantine, away from battle zones) (iv) Controlled weapons (soldiers are allowed to use controlled weapons and methods which are 'not evil in themselves'), (v) No retaliation (this occurs when a state A violates *jus in bello* in war in country B, and B retaliates with its own violation of *jus-in-bello*, in order to force A to obey the rules).
- *Jus post Bellum* which refers to justice during the final stage of war, i.e., at war termination, The return to peace is left to moral laws (proportionality, rights vindication, distinction, punishment).

Although fully autonomous robots are not yet operating in war fields the benefits and risks of the use of such lethal machines for fighting in wars are of crucial concern. The use of robots must respect the *just war* principles, and additional ethical rules about:

- *Firing decision* (The decision to use robotic weapons to kill people still lies with the human operator).
- *Discrimination* (Distinction between combatants and civilians, as well as military and civilian objects).
- *Responsibility* (To whom blame and punishment should be assigned for improper fight and unauthorized harms caused by an autonomous robot: to the designers, robot manufacturer, procurement officer, robot controller/supervisor, military commander, a state's prime minister/ president, or to the robot itself?).
- *Proportionality* (Even if a weapon meets the test of distinction, any weapon must also involve evaluation that sets the anticipated military advantage to be gained against the predicted civilian harm (civilian persons or objects)).

The use in war of autonomous robotic weapons is subject to a number of serious objections, the major of which are [29]:

- Inability to program war laws.
- It is wrong *per se* to take human out of the firing loop.
- Autonomous robotic weapons lower the barriers to war.

Automated Car Roboethics deals with the ethical issues rising by the use of self-driving cars. Scientists and engineers are now not debating whether self-driven cars will come to be, but how it will happen and what it will mean, i.e., discussions are dominated not about technology itself, but about the behavioral, legal, ethical, economic, insurance, environmental, and policy implications. The fundamental ethical question here is: '*who will be liable when an automated driverless car crashes?*' It will be the car itself, the driver, the manufacturer, or the people who designed and built the hardware or software? Or the mapping platform or blame another car that sent an erroneous signal on the highway? The answer to this question is not straightforward. It needs careful and deep technical, legal, ethical, and social considerations.

Other questions referring to the use of autonomously driven cars are the following [61]:

- What would be the behavior of people still driving old-fashion cars around autonomous vehicles or a mix of the two on the road?
- What if an autonomous car does not drive like people do? For example, if an autonomous car drives excessively faster than normal drivers do? How will you react when sitting in the front seat of a car that drives faster than you are used to do?
- How people learn to trust automated vehicles?
- If for some reason a vehicle requires you to suddenly take wheel, will you be able to quickly turn your attention away from what you were doing while your car was doing driving for you?
- How will autonomous cars change our travel and consumption pattern?
- Will people want to buy autonomous cars appreciating the benefits involved (e.g., safer driving, congestion reduction, cut down of the amount of source urban land needed for parking, etc.)?
- What will be the proper mode of vehicle-to-vehicle driverless communication that lets cars tell each other what they are doing so they won't collide?
- What will be the proper changes of auto insurance?

A discussion of the implications of self-driving vehicles is presented in [62]. With reference to the last question it is noted that fully autonomous cars and other vehicles are not yet on road. Only partially autonomous driving (also referred to as *assisted driving*) is now really possible. Technologies include lane keeping, adaptive cruise control, and automatic braking. The lane-keeping function steers the car so that it stays in its lane even as the highway changes direction. The adaptive cruise control applies brakes or throttle to maintain speed while keeping the car a safe distance from other cars. Finally, automatic braking applies the brakes if it senses the car tailgating or a collision coming. What does all this mean for auto insurance? How insurers will price risk when fully autonomous cars will be a reality? Currently, auto-insurers base pricing primarily on driver characteristics, but increased automation may shift the emphasis toward vehicle, usage, and geographic characteristics. Given that technology is expected to make further big advancements, it is hard to imagine that vehicle automation will go no further than assisted driving technologies. A time is likely to come that automatic driving will be the standard mode of driving. Does that mean accidents will no longer happen and auto insurance will be obsolete? With high probability, the answer to this question is 'no'. An accident-free world is only possible in a perfect world. Auto insurance in some form is likely here to remain. It is just going to be different.

10.6.3 Professional Codes of Ethics

In conducting their professional work, robotics engineers must apply the professional 'codes for engineers' such as the code of the *National Society of Professional Engineers* **(NSPE)** described in Section 10.3.2, and the **IEEE** and **ASME** codes. A code referred especially to robotic engineers was formulated by **WPI** *(Worchester Polytechnic Institute)* [32] and is the following.

WPI Code of Ethics for Robotics Engineers

'As an ethical robotics engineer, I understand that I have responsibility to keep in mind at all times the well-being of the following communities:

- *Global:* the good of the people and the environment.
- *National:* the good of the people and government of my nation and its allies.
- *Local:* the good of the people and environment of affected communities.
- *Robotics Engineers:* the reputation of the profession and colleagues.
- *Customer and End-Users:* the expectations of the customers and end-users.
- *Employers:* the financial and reputation well-being of the company.

To this end and to the best of my ability I will:

- Act in such a manner that I would be willing to accept responsibility for the actions and uses of anything in which I have a part in streaming.
- Consider and respect people's physical well-being and rights.
- Not knowingly misinform, and if misinformation is spread do my best to correct it.
- Respect and follow local, national, and international laws whenever applicable.
- Recognize and disclose any conflicts of interest.

- Accept and offer constructive criticism.
- Help and assist colleagues in their professional development and in following this code.'

As stated in [32], 'this code was written to address the current state of robotics engineering and cannot be expected to account for all possible future developments in such a rapidly developing field. It will be necessary to review and revise this code as situations not anticipated by this code need to be addressed'. Detailed discussions on robotic ethics and the WPI code of ethics for robotics engineers can be found in [33, 34], and a useful discussion on ethics and modular robotics is provided in [35].

10.7 Ethics of Management Control

Conventional analyses of management control systems may lead to the so-called '*illusion of control*', i.e., managers may wrongly believe that everything can be controlled and monitored. In organizations (as in all social functions or technological systems) the ethical behavior of managers and individual employees is crucial for long-run survival of the organizations. The ethical behavior of an employee depends on several factors such as his/her ethical philosophy, ethical decision ideology, organization/position-related factors, and external environmental factors,

Very broadly, management is the art or skills for guiding human activities and physical resources towards the achievement of given goals. According to *Henry Fayol*, 'To manage is to forecast and plan, to organize, to command, to coordinate, and to control'. *Louis Allen* says that 'Management is what managers do in the organizations'.

Management (classical perspective) has three primary subfields, namely (Figure 10.10):

- *Scientific management* (Emphasizes scientific changes for improving labor productivity).
- *Administrative management* (Looks on overall organization rather than the individual worker and is based on the functions of forecasting, planning, organizing, commanding, coordinating, and controlling as suggested by Henry Fayol).
- *Bureaucratic management* (A general systematic integrated approach developed in Europe that considers the organization as a whole).

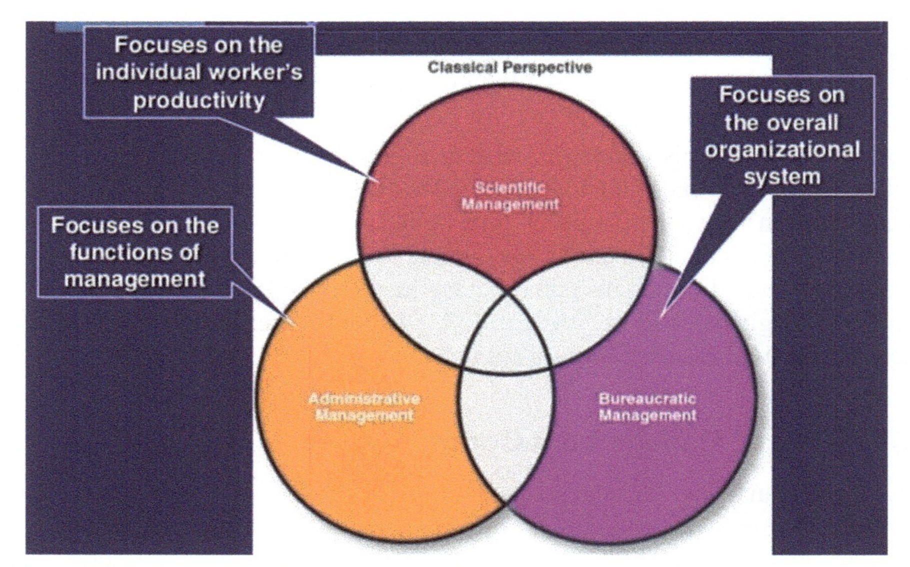

Figure 10.10 The three principal subfields of classical management that emerged during the 19th and 20th Centuries.

Source: www.slideshare.net (Foundations of Management, J.J. Llerin).

Managers attempt to improve their organizations by optimally selecting the following seven aspects (the so-called "*seven S's*") which are fully interconnected as shown in Figure 10.11.

- *Structure*: The framework in which the members of the company are coordinated.
- *Strategy*: The path selected by the company for its future growth.
- *Systems*: The formal and informal processes including management, information systems, innovation systems, capital control systems, etc., that govern every day operations.
- *Style*: The leadership style of the top management, and the overall operating scheme of the company.
- *Skills*: The competencies possessed by the organization, and what it does best.
- *Staff*: The human resources of the company, their past experience and knowledge, and their current motivation, training and integration in the organization.

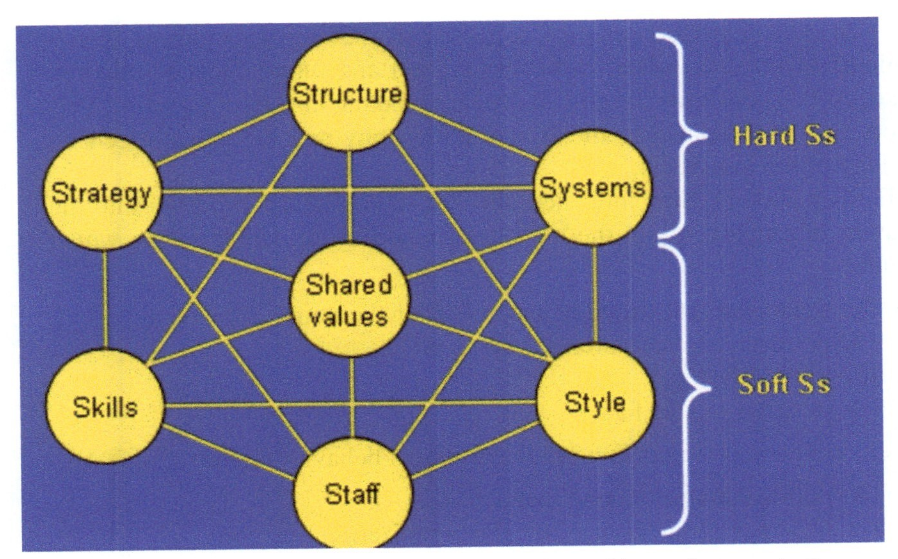

Figure 10.11 The seven S's framework for analyzing and improving organizations.
Source: www.1000ventures.com/business_guide/mgmt_inex_7s.html

- *Shared values*: The ethical values and principles adopted by the organization, and shared by its members, which are often unwritten and go beyond the conventional corporate objectives.

There are three possible different philosophies of individual person's behavior in their organizational work. These are *utilitarianism, individual rights*, and *justice*. Ethical decision performance can be classified on the basis of two dimensions:

- *Idealism* (the belief that behaving ethically ensures positive outcomes).
- *Relativism* (the belief that moral values depend on circumstances).

The *value system* of a person is an important factor that specifies whether he/she will behave ethically or not when faced with an external dilemma. Other particular factors which influence the ethical behavior of a person include:

- The level of his/her moral development.
- The age of the person.
- His/her focus on control.
- His/her ego strength.

The past ethical performance of a person in making decisions is also a dominant factor that influences his/her current and future decision making. Past decisions constitute the ethical decision/making history of a person. This history is tightly related to the ethical philosophy and the ethical decision ideology of the individual.

Further factors that influence the decision-making process and the behavior of employees are the ones determined by the organization which include:

- The organizational structure and operational culture.
- The performance measurement systems.
- The reward systems.
- The position-related factors.

The environmental factors that influence ethical behavior include:

- The political and economic factors.
- The legal environment in which the organization functions.
- The social factors.

The integration of all the above factors (ethical philosophy, ethical decision-making, ideology, decision history, individual factors, organizational factors, and external environmental factors), helps to fully understand the steps involved in ethical decision making and behavior, and provides pointers/guidelines as to how this behavior can be *'controlled' by* managers [30, 31].

The *management control* ethical issues may arise in any department or function of the organization. *Operational function* ethical issues may arise in terms of productivity, quality, and safety. Better quality leads to more efficient utilization of resources, thus enhancing productivity. To regulate ethical conduct, organizations develop their own code of ethics and establish their Ethics Committee which must assure the balance between the ethical issues that are due to the strategic decisions taken at top management level and the ethical problems that the employees face at all levels of functioning.

Figure 10.12 shows the principal characteristics of *ethical leadership*, namely:

- *Respect for others*: Recognize and respect the goals and ambitions of the colleagues and followers.
- *Serve other people*: This can be done through mentoring, building team, empowering, etc.
- *Show justice*: Fairness and justice are high in the ethics priority list.

Figure 10.12 Characteristics of ethical leadership.

Source: https://foramshah/wordpress.com/2014/07/01/ethical-leadership/

- *Honesty*: This ethical competency increases the trust and builds a good relationship between the leader and the follower.
- *Build community*: Leadership aims to influence colleagues and subordinates to achieve organization or team goal.

Other leadership characteristics, very important for the organization, include:

- *Creativity* (Exceed expectations, inspire new ideas).
- *Innovation* (Take challenges, pursue innovative research leading to new concepts and methods).
- *Learning* (Acquire new knowledge, enhance the ability for effective assessment).

Principal features of an ethical organization are:

- Values and integrity.
- Openness and transparency.
- Objectivity and fairness.
- Effective communication upwards and downwards.

Ethical culture of an organization, which typically is unwritten, means the extent to which the organization regards and respects its values, and makes doing what is right a priority. It is based on shared values, principles and traditions of thinking and doing things that influence the ways the organization members behave towards the achievement of its goals.

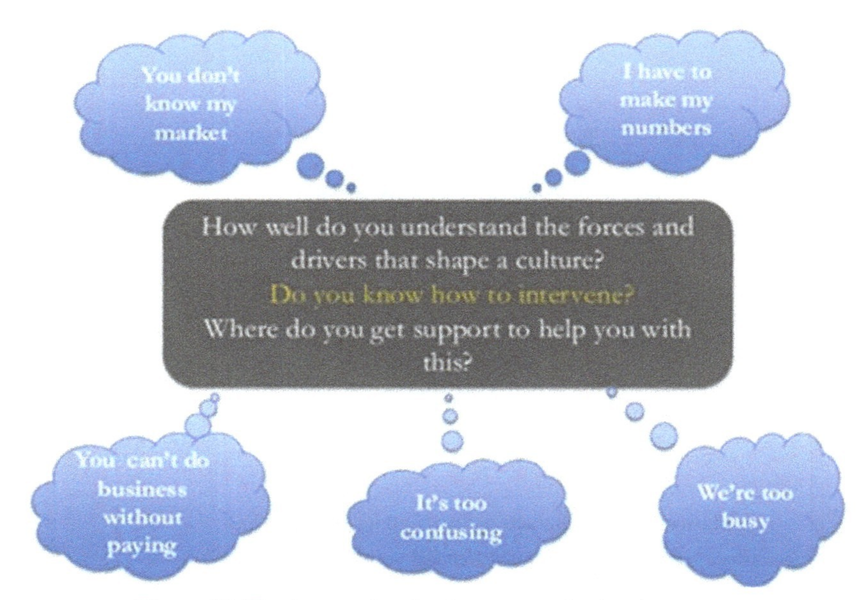

Figure 10.13 Aspects for shaping an organizational structure.

Source: https://www.slideshare.net/the SCCE/scce-chicago-session802-master20141208 [59].

The first important prerequisite in shaping a strong organizational culture (general and ethical) is to understand the forces and drivers that shape it and how to intervene. Particular aspects of this are shown in Figure 10.13.

In particular, to build an ethical organization culture engineers and managers have to do the following [54]:

- Communicate ethical behaviors.
- Provide ethical training.
- Reward ethical acts and punish unethical ones.
- Provide protective measures.
- Show low to moderate aggressiveness.
- Focus on means as well as outcomes.
- Perform the acts shown in Figure 10.14.

The potential benefits/payoffs of the creation of a strong ethical organization culture include [58]:

- Improved reputation.
- Enhanced access to capital.
- Potential avoidance of penalties.
- Stronger employee commitment.

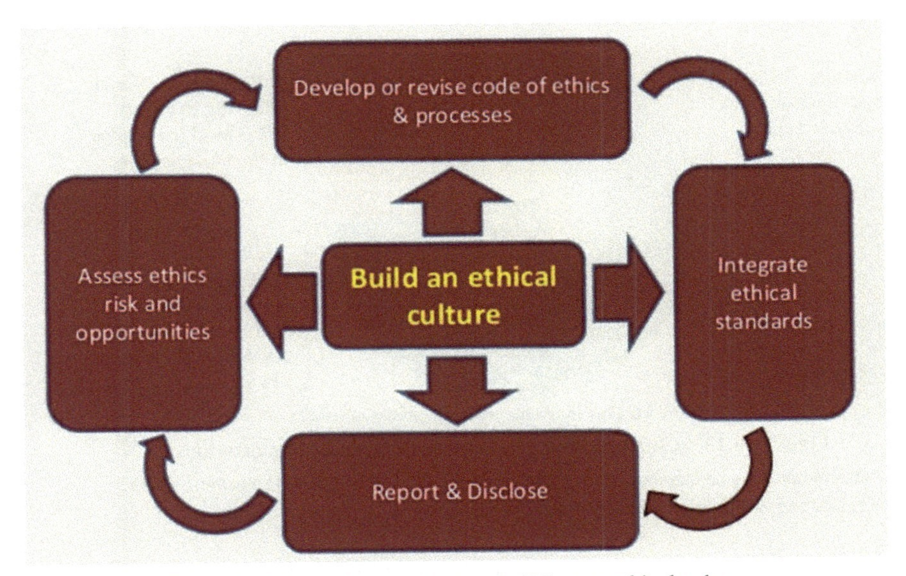

Figure 10.14 Actions needed for building an ethical culture.

Source: www.slideshare.net (M. Mayer, The role of ethics in HR standards, competence and governance, Business, 2015) and www.ethicssa.org

- Increased customer loyalty.
- Decreased vulnerability.

Three ways for managers to accomplish their responsibilities are:

- *Best ratio*: People are normally good and so they create appropriate conditions.
- *Black and white*: Right is right and wrong is wrong. Therefore ethical decisions can be made and materialized.
- *Full potential*: People must know and realize their full potential. Therefore they should make decisions towards achieving this potential.

In organizations, strategic management should be accompanied by ethical management. Strategy is established by knowledge and intelligence, whereas ethics can be set up by adopting moral values and standards. Strategic management helps in achieving the standards rather than defining them, but ethical management assists in setting the standards. Strategies may differ from company to company, but ethical practices are universal and follow the accepted code of ethics.

Some ethical challenges of organizations for resolving common ethical dilemmas are depicted in Figure 10.15.

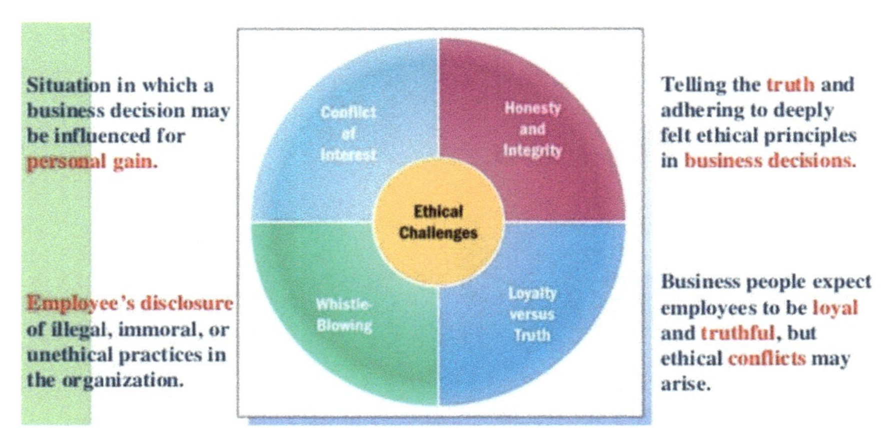

Figure 10.15 Common ethical dilemmas of business organizations.
Source: https://slideshare.net/parabprathamesh/business-social-responsibility (Prathamesh Parab, Business Ethics and Social Responsibility, Business Technology).

An authoritative code for managers is the '**AOM**' Code of Ethics [36] which involves principles that underlay the professional responsibilities and conduct of AOM's members, and enforces ethical standards that apply to members in official AOM roles and those participating in AOM-sponsored activities. AOM's 'Code of Ethics' consists of the following sections.

- Introduction.
- Preamble.
- General principles.
- Professional principles.
- Ethical standards.

The general principles serve as guide to AOM members in determining ethical courses of action in various contexts, and exemplify the highest ideals of professional conduct. The principles are the following:

Responsibility AOM members establish relationships of trust with those with whom they work (students, colleagues, administrators, clients). They are aware of their professional and scientific responsibilities to society and to the specific communities in which they work. AOM members uphold professional standards of conduct, clarify their professional roles and obligations, accept appropriate responsibility for their behavior, and seek to manage conflicts of interest that could lead to exploitation or harm. They are concerned about the ethicality of their colleagues' scientific, educational, and professional conduct. They strive to contribute portions of their professional time for little or no compensation or personal advantage.

Integrity AOM members seek to promote accuracy, honesty, and truthfulness in the science, teaching, and practice of their profession. In these activities AOM members do not steal, cheat, or engage in fraud, subterfuge, or international misrepresentation of fact. They strive to keep their promises, to avoid unwise or unclear commitments, and to reach for excellence in teaching, scholarship, and practice. They treat students, colleagues, research subjects, and clients with respect, dignity, fairness, and caring. They accurately and fairly represent their areas and degrees of expertise.

Respect for People's Rights and Dignity AOM members respect the dignity and worth of all people and the rights of individuals to privacy, confidentiality, and self-determination. AOM members are aware of and respect cultural, individual, and role differences, including those based on age, gender identity, race, ethnicity, culture, national origin, religion, sexual orientation, disability, language, and socioeconomic status, and they consider these factors when working with all people. AOM members try to eliminate the effect on their work of biases based on these factors, and they do not knowingly participate on or condone activities of others based upon such prejudices. The AOM and its members are also committed to providing academic and professional work environments that are free of sexual harassment and all forms of sexual intimidation and exploitation.'

The professional principles are to enhance the learning of students and colleagues, and the effectiveness of organizations through AOM teaching, research and practice of management. AOM has the following five responsibilities:

- To AOM students.
- The advancement of managerial knowledge.
- To the AOM and the larger professional environment.
- To both managers and the practice of management.
- Ascribing to the Code of Ethics.

A detailed description of these responsibilities is provided in the AOM's document [36].

Another Code of Conduct for Managers is provided in Amio [37]. This code is the following:

- **Honesty** Managers in every industry must understand their company's policies and guidelines, as well as its mission, and how they are expected to go about accomplishing their goals. They also need to know and follow the laws of the government, particularly as they pertain to business. Mostly, effective managers must be honest about aspects such as production and profit at all times. While being dishonest isn't always a federal offense, it can result in numerous issues for a company.

- **Accountability** Good managers expect their workers to take responsibility for their actions and overall performance, and demand the same of themselves. That means answering to ownership or executive boards when things don't go right, accepting the blame, and coming up with solutions to avoid future issues.
- **Integrity** Managers who perform their jobs with a high level of integrity are widely the best type of supervisors to work for. That's because managers who possess integrity are often consistent in their decision-making and resolution of issues. These managers also make their goal clear and assist employees when it comes to reaching those goals.
- **Respect** Appropriate behavior is a key factor in a code of conduct for a manager, who must demonstrate acceptable behavior in the workplace. That doesn't mean managers need to act like robots and display little signs of personality. Quite the opposite, as, actually, many managers are expected to be energetic and lead in areas of teamwork and motivation. But they also need to treat staff members, customers and their own supervisors with the same respect they would expect for themselves.
- **Flexibility** While most companies don't expect their managers to display sympathy to employees who aren't meeting expectations, most businesses prefer leaders who are patient and work with those in need of assistance. Good managers show their workers how jobs are best performed, then monitor workers and offer suggestions and tips. After all, the goal of managers in every industry is to make sure workers stay productive and the company stays profitable."

To increase the commitment of the employees toward the ethical programs of the organization, it helps very much if the organization incorporates proper reward systems which promote ethical means of achieving the specified objectives.

The formulation of a procedure for ethical decision making helps essentially to make successful ethical decisions. A potential general ethical decision making procedure, which may be combined with that depicted in Figure 10.2, is the following:

- Define the ethical dilemma or problem.
- Identify the potential issues involved.
- Examine ethical values and principles.
- Apply the ethical principles and codes of ethics.

- Know the applicable laws and regulations.
- Obtain consultation.
- Consider possible courses of action.
- Explore the consequences of various actions.
- Decide on the course of action.

Management control systems need to consider rationality, creativity, morality, and human association. In particular, the design of management control systems that support ethical behavior should involve the following features [60]:

- Control system design must comply with all laws, ethical codes and policies.
- Managers and employees must be sensitized about proper behavior.
- Employee behavior must be audited with all stakeholders.
- Considerable deviations must be reported.
- Violations must be evaluated.
- Regulations must be implemented to ensure ethical behavior.

10.8 Systems Philosophy

10.8.1 What is Systems Philosophy?

Systems philosophy **(SP)** is a philosophical field which attempts to develop a new philosophy (worldview) through the use of systems concepts. SP was coined by Ervin Laszlo in his 1972 book entitled 'Introduction to Systems Philosophy: Toward a New Paradigm of Contemporary Thought [38]. SP has been viewed as the 'reorientation of thought and world view' inspired by von Bertalanffy's 'General Systems Theory' [39]. It is noted that the term *systems philosophy* was agreed by Laszlo and von Bertalanffy before the publication the SP book. The term 'systems philosophy' is used as shorthand to refer to 'the philosophy of systems', but this usage can create confusion. Actually, for von Bertalanffy, philosophy of systems is the element of SP called '*systems ontology*', but for Laslo it is the element called '*systems metaphysics*' [38, 39]. Systems ontology provides a good grounding for '*systems thinking*' but does not include the central focus of SP, which is concerned with the articulation of a worldview based on systems perspectives and humanistic concerns.

Although systems philosophy was only formally presented as a discipline endeavor in 1972, the concepts it involves have a long history in philosophy starting with the Pre-S ocratic philosophers (7th–5th BC), e.g., Heraclitus and Anaximander. Pre-Socratic philosophers are the first philosophers known who attempted to work out the nature and dynamics of the world from the observations and reasoning, rather than just founding their '*world picture*' (ontology, cosmology) on *mythodology*. These earliest philosophical studies were typically formal from a point of view we would today consider as '*systematic*'. For the ancient Greeks the world was '*kosmos' (from the word 'κόσμημα' (kosmima) meaning jewel, ornament),* i.e., something beautifully ordered rather than '*chaos'*, or, otherwise, an 'ordered harmonious structure where everything is a living substance that tries to adjust itself according to its kind and its context. This means that the early Greek philosophers viewed the Universe as a kind of *organism* rather than as a type of great machine. The Pre-Socratic traditions were integrated with the Classical ones by Aristotle resulting in explicit systematic approaches to the organization of knowledge. Aristotle is recognized to be the first who argued that 'a system is more than the sum of its parts' and created the earliest synthesis of all the principal branches of learning, thus producing a taxonomy of categories which later was regarded as a form of *complexity hierarchy* (the so-called '*Three of Porphyry*', 3rd century CE') [40].

During the medieval period these important ideas were diluted with the use of '*scholaticism*' in Europe. A revival of the ambitions of the Pre-Socratics occurred in the 16th century, during which the rise of '*reductionism*' and 'mechanism' has taken place that devalued the wonderful *organismic* features exhibited by Nature which admired the Greek '*systemicists'*. Of course, the power of the reductionist approach is well recognized, but it was later realized that linear reductionism and physicalism mechanism cannot adequately reveal all the principles and real properties of the phenomenal world.

According to von Bertalanffy, system philosophy is one of the three areas studied within the *systemic discipline.* These areas are [39]:

- *Systems Science* (i.e., the scientific exploration and theory of '*systems*' in the various sciences).
- *Systems Technology* (i.e., the problems of modern society that involve computer design and use, automatic control, automation, etc.).
- *Systems Philosophy* (i.e., the new philosophy of '*nature*' which views the world as a great *organization* that is '*organismic*' rather than '*mechanistic*' in nature.).

The principal domains of SP are [39]:

- *Systems ontology* (what is *system*? How systems are realized?)

- *Systems paradigms* (i.e., development of worldviews that consider humankind as a species of an actual system in the natural hierarchy of likewise actual physical, biological, and social systems).
- *Systems axiology* (i.e., development of models and systems that involve humanistic concerns).
- *Applied system philosophy* (i.e., use of the insights from other domains of SP to solve practical social, technological, or philosophical problems).

10.8.2 A Look at the Evolution of Systems Philosophy

Based on Laszlo's and von Bertalanffy's founding work, SP was in subsequent years evolved as follows [56]:

Global Problematique *(Hasan Ozbekham)* The 'global problematique' was originated in the Ozbekham's proposal to the *Club of Rome* which contained 49 Continuous Critical Problems (CCPs). This work was not adopted by the Club as too humanistic. Instead, Forrester's system dynamics was adopted, and as a result the book entitled: 'The Limits of Growth' was published [41]. Later (1995) Ozbekhan in cooperation with Alex Christakis, revised the 49 CCPs following the *Structured Dialogue Design*, and developed an influence map that provided leverage points for surpassing the global Problematique.

Integrated Worldview *(Leo Apostel)* It is a fact that disciplinary views are increasingly becoming fragmented, and so there is a potential for the interdisciplinary and transdisciplinary work in order to solve cultural, social and economic problems of modern society. Recognizing this fact, Leo Apostel has tried to fill in the gap of the non-existence of integrated world views and published a book on this subject entitled: 'From Fragmentation to Integration' [42]. He founded the Worldviews Group and the Leo Apostel Center for Interdisciplinary Studies (at the Free University of Brussels) which is focused on developing integrated systematic models of the nature and structure of worldviews in order to promote research towards an integrated and unified perspective of the world.

Value Realism *(David Rousseau)* Value relativism seems to be problematic for social and individual welfare, and does not address the holistic issues of SP and universalist issues of moral and spiritual intuitions and experiences. Inspired by this fact David Rousseau promoted work toward revealing the ontological foundations of moral values and normative intuitions in order to incorporate values into Laszlo's model of natural systems in a holistic manner. To promote this approach Rousseau established a Center for Systems Philosophy [43, 44].

Systemic Intervention *(Gerald Midgley)* This approach is the result of concerns that developments in philosophy of language, philosophy of science, and philosophy of sociology do not agree in their views about reality modeling. This issue has inspired Midgley to promote practices for systemic interventions that could bypass the above disagreements by focusing on the processes involved in making boundary estimations in real-life situations. This approach which is known as *Critical Systems Thinking*, 'critical' in the sense of 'reflective', is the dominant focus of the Center for System Studies at the University of Hull, founded by Midgley. One of the major influential publications on critical thinking is Midgley's 2000 book [45].

Luhman [46] argued that GST surpasses the traditional epistemological problem of the 'criterion of truth' by replacing the Kantian question of 'how a subject can have objective knowledge?' with the question 'how is organized complexity possible?' He also claimed that his systems theory exceeds the division between *subject* and *object* (which was of concern by both rationalists and positivists), the divisions between *whole* and *parts* (replaced by the distinction between system and environment), and also the divisions between *individuals* and *societies*, and the divisions between the *given* and the *possible* – replaced by the distinction between the *possible* and the *actual* divisions which were the principal considerations of dialectical thinkers.

Epistemological philosophical questions about general systems theory include the following [47]:

- How do we know which definition of GST will comfort us with the following ideal outcome: GST, based on the definition, will offer a solid and unified theoretical foundation for all system analysis methods developed in the different fields?
- What is the difference of the concept 'set' in theories and applications?
- How can the general systems theory be built on the axiomatic set theory?
- Why is mathematics so unreasonably effective in the study of natural science?
- What is the meaning of truth?
- Is GST a language of science and technology?
- How can some results and concepts in system theory be applied to unify different concepts in different fields?

Formal mathematical answers to these questions using the language of set theory were provided in Lin et al. [47], where it is concluded that a completely new understanding about the world could be obtained within GST if for each application a proper revolution is implemented, without resorting to small incremental changes but to complete reassessment of structures, roles, and behaviors [48].

10.9 Control and Cybernetics Philosophy

Initiated in the 19th Century the work of Control theory was followed by the meta-sciences of cybernetics and systems theory. Cybernetics naturally became the subject of philosophical studies performed by leading thinkers in cybernetics and professional philosophers. In the 20th Century *management science* was developed as an area of control theory concerned with the study of practical control in organization/enterprise systems. Management philosophy has emerged from management science by the beginning of the 20th Century, and extensive research was pursued in both *control philosophy* and *management philosophy* by professional thinkers.

One can say that like other philosophies (philosophy of artificial intelligence, philosophy of information systems, etc.) control philosophy is 'a system of generalizing philosophical considerations about the subject and methods of control, the position of control among other scientific branches, etc. More specifically, control philosophy can be defined as 'an area of philosophy concerned with comprehension and interpretation of control processes and control cognition, and exploration of the essence and role of control' [49]. This definition of the term 'control philosophy' has a rich internal structure covering epistemological research of control science, and the analysis of ontological, logical and other aspects of control science, including management philosophy. In Novikov [49] it is stated that cybernetics, is a branch of control science that studies its most theoretical regularities. In other words, it is argued that 'cybernetics must and would play the role of control philosophy', as a branch of control theory, studying its most general regularities.

Figure 10.16 illustrates the connections between *control philosophy*, *cybernetics*, and *management philosophy* interpreted in the widest way [49]. Control includes *methodology*, *control activity*, and *structure of control activity*. We see in Figure 10.16 that 'management philosophy' is a major player (domain) in the triangle "*control philosophy, cybernetics, management philosophy*".

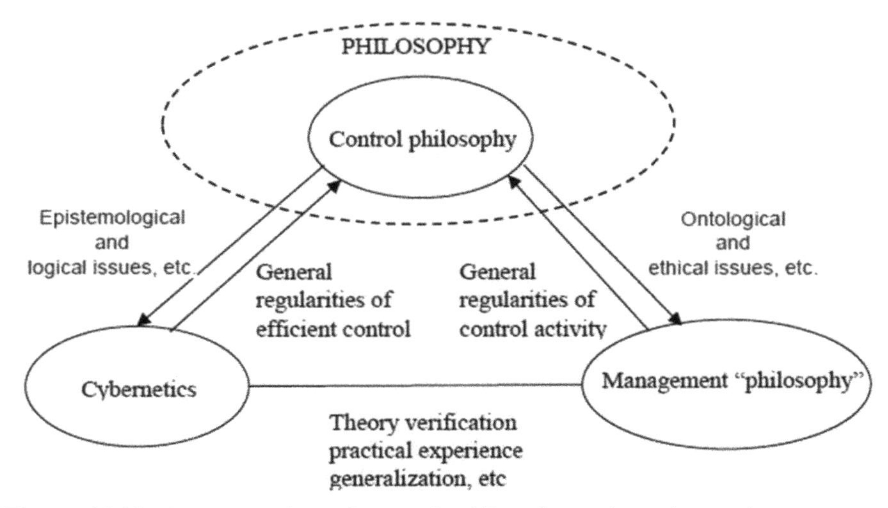

Figure 10.16 Interconnection of control philosophy, cybernetics, and management philosophy.

Like any other philosophy, *control philosophy* is concerned with:

- *Ontological issues* (including logical and other related foundations of control and management practice).
- *Epistemological issues* (including logical and other related foundations of science).

In general, philosophical aspects (components) of control theory that need deep consideration include the following, pictorially shown in Figure 10.17 [49]:

- Control tasks.
- Subjects of control.
- Scheme of control activity.
- Conditions of control.
- Types of control.
- Methods of control.
- Forms of control.
- Control means.
- Control functions.
- Factors having an impact on control efficiency.
- Control principles.
- Control mechanisms.

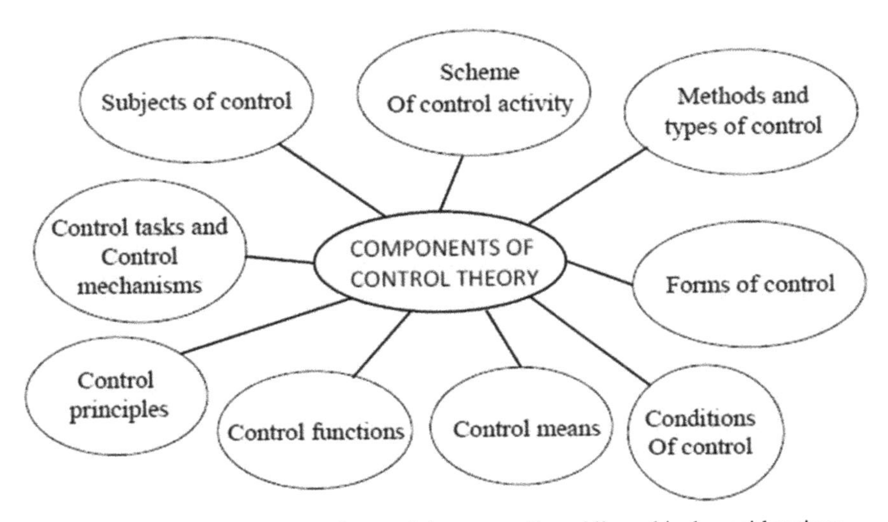

Figure 10.17 Components of control theory needing philosophical considerations.

Control methodology is the theory of activity organization, and *activity* is a purposeful human action. Control activity represents a certain form of practical activity. If a human being is included in a control system, activity becomes activity on activity organization. Actually, control theory investigates the interaction of *control subject* and *controlled object* (which maybe another subject) as shown in Figure 10.18 [49].

Cybernetics is different than *Artificial Intelligence* (**AI**). AI scientists employ computers to build intelligent machines and develop real implementations, i.e., to construct working examples which they consider to be their most important achievement. Cybernetics researchers develop models of organizations, and use feedback, goals, and conversation to understand the capacity and limits of systems of all kinds (biological, man-made, social). They provide powerful descriptions of the models which help to identify their most important features, and exploit them in complex applications. For AI, knowledge is a commodity that can be stored in a computer. As pointed out in [57], the application and use of such stored knowledge in the real world constitutes intelligence. Such a '*realist*' view of the world allowed AI to construct semantic networks and rule-based expert systems as a way to construct realistic intelligent machines. Cybernetics in contrast has evolved from a 'constructivist' view of the world, where objectivity comes from shared agreement about meaning, and where intelligence (or information) is

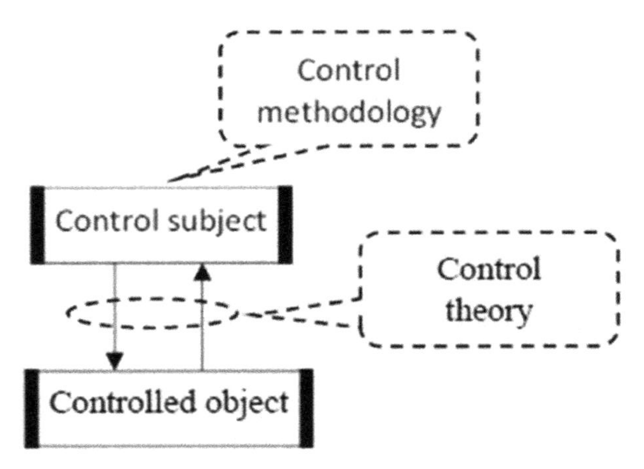

Figure 10.18 Control methodology deals with the control subject and controlled object. Their interaction is studied by control theory.

the outcome of an interaction rather than commodity stored in a computer. These differences determine basically the start, the source and the direction of research carried out from a cybernetic, versus an AI, point of view.

A comprehensive volume on the *philosophy of automation* is the proceedings of the '1978 *Automation in Coloration: The Philosophy of Automation' Conference* [50]'. Two legitimate questions which are still open are:

- 'Is it good sense to seek full sequential automation of any batch process?'
- 'Should one not change the technology rather than carry the process of mechanization in this ultimate stage?'

10.10 Conclusions

Ethics is a dominant issue of society in all its activities and manifestations from ancient times. Many theories and approaches toward assuring the proper behavior of humans in their interrelations and interactions were developed. The design and operation of man-made systems should also follow the ethical rules, Scientists and engineers should be responsible for assuring the design of safe systems as much as possible having always in mind that they are to be used by humans.

In this chapter we have studied the ethical issues raised during the entire life cycle (design, implementation, operation, and use) of four classes of scientific/technological artifacts, namely systems, cybernetics, control, and automation artifacts. We have studied the ethical issues of systems engineering, systems thinking, cybernetics organisms and human implants, and control and automation. An overview of systems and cybernetics philosophy was also included. Intercultural philosophical issues can be found in [26, 51, 52].

References

[1] Dewey, J. (1972). *Theory of Valuation.* Chicago, IL: Chicago University Press.

[2] Jonsen, A., and Toulmin, S. (1990). *The Abuse of Casuistry: A History of Moral Reasoning*. Los Angeles, CA: The University of California Press.

[3] Bromberg, S. E. (2013). *The evolution of ethics: An introduction to Cybernetic ethics*. Available at: http://www.evolutionaryethics.com

[4] Kasser, J. (1995). "Ethics in Systems engineering," in *Proceedings of the 5th Annual International Symposium, National Council on Systems Engineering*, St. Louis, MO.

[5] Gert, B. (1988). *Morality.* Oxford: Oxford University Press.

[6] Brinsjord, S., and Taylor, J. (2011). "The divine command approach to robotic ethics," in *Robot ethics: The ethical and social implementations of robotics*, (Cambridge, MA: The MIT Press).

[7] Rowan, J. R., and Sinaich, S. Jr. (2002) *Ethics for the Professions.* Boston, MA: Cengage Learning.

[8] National Society of Professional Engineers [NSPE] *Code of Ethics for Engineers.* Available at: www.nspe.org/Ethics/CodeofEthics/index.html

[9] IEEE *Code of Ethics and Code of Conduct.* Available at: www.ieee.org/about/corporate/governance/p7.html www.ieee.org/ieee_code_of_conduct.pdf

[10] ASME *Code of Ethics of Engineers.* Available at: https://community.asme.org/colorado_section/w/wiki/8080.code-of-ethics.aspx

[11] Harter, N., Dean, M., and Evanecky, D. (2004). "The ethics of systems thinking," in *Proceedings of 2004 American Society for Engineering Education: Annual Conference and Exposition.*

[12] Blackburn, S. (1999). *Think.* Oxford: Oxford University Press.

[13] Hitt, W. (1990). *Chapter-2 Ethics and Leadership.* Columbus OH: Batelle Press.

[14] Wheatley, M. (2001). *Leadership and the New Science.* San Francisco, CA: Berret-Koehler.

[15] Turkle, S. (1995). *Life on the Screen*. New York, NY: Simon and Schuster.
[16] Introna, L. D. (2001). Proximity and Simulacra: On the possibility of ethics in an electronically mediated world. *Philos. Contemp. World.*
[17] Cohen, R. A. (2000). Ethics and Cybernetics: Levinasian reflections. *Ethics Inf. Technol.* 2, 27–35.
[18] Palese, E. (2012). *Robots and cyborgs to be or to have a body?* Springer. http://www.nebi.nlm.nih.gov/pmc/articles/PMC3368120/ [accessed May 30, 2012].
[19] Sai Kumar, M. (2014). Cyborgs-the future mankind. *Int. J. Sci. Eng. Res.* 5, 414–420.
[20] BMJ (2005). Implant ethics. *J. Med. Ethics* 31, 519–525.
[21] Roskles, A. (2002). Neuroethics for the new millennium. *Neuron* 35, 21–23.
[22] Satava, M. (2003). Biomedical, ethical and moral issues being forced by advanced medical technologies. *Proc. Am. Philos. Soc.* 147, 246–258.
[23] Mother Nature Network [MNN] (2015). *Seven real life human cyborgs.* http://www.mnn.com/leaderboard/stories/7-real-life-human-cyborgs [accessed January 26, 2015].
[24] Brian, A. W. (2009). *The Nature of Technology*. New York, NY: Free Press.
[25] Bunge, M. (1977). Towards a Technoethics. *Monist* 60, 96–107.
[26] Tzafestas, S. G. (2016). *Roboethics: A Navigating Overview*. Berlin: Springer.
[27] Rehabilitation Engineering and Assistive Technology Society [RESNA]. Available at: http://resna.org/certification/RESNA_Code_of_Ethics.pdf
[28] Feid-Seifer, D., and Mataric, M. J. (2011). Ethical principles for socially assistive robotics. *IEEE Robot. Autom. Mag.* 18, 24–31.
[29] HRW-IHRC *Losing Humanity: The Case Against Killer Robots, Human Rights Watch*. Available at: www.hrw.org
[30] Rosanas, J. M., Velilla, M., and Bus, M. J. (2005). The ethics of management control systems: Developing technical and moral values. *Ethics* 57:83.
[31] ICMR (2012). "Chapter-8 IBS Center for Management Research," in Management Control Systems," in *Business Ethics and Management Control: Overview.*
[32] WPI (2010). *Code of Ethics for Robotics Engineers*. Availble at: http://ethics.iit.edu/ecodes/node/4391

[33] Ingram, B., Jones, D., Lewis, D., and Richards, A. (2010). "A code of ethics for robotic engineers," in *Proceedings of 5th ACM/IEE International Conference on Human-Robot Interaction (HRI'2010)*, Osaka.

[34] www.wpi.edu/Pubs/E-project/Available/E-project-0304410-172744/unrestricted/A_Code_of_Ethics_for_Robotics_Engineers.pdf

[35] Smyth, T. *Paper Discussing Ethics and Modular Robotics*. Available at: www.pitt.edu/~tjs79/paper3.docx

[36] AOM *Code-of-Ethics*. Available at: www.aom.org/About-AOM/Code-of-Ethics.aspx

[37] Amio S. *Code of Conduct for Managers*. Available at: http://smallbusiness.chron.com/code-conduct-managers-2733.html

[38] Laszlo, E. (1972). *Introduction to Systems Philosophy: Toward a New Paradigm of Contemporary Thought*. Philadelphia, PA: Gordon and Breach Publishers.

[39] von Bertalanffy, L. (1976). *General Systems Theory*. New York, NY: Brazillier.

[40] Centre for Systems Philosophy [CSP] *The Development and Status of Systems Philosophy*. Availble at: http://systemsphilosophy.org/history-and-development-of-systems-philosophy

[41] Meadows, D. H., and Randers, J. (1972). *The Limits to Growth*. New York, NY: Universe Books.

[42] Aerts, D., Apostel, L., De Moor, B., Hellemans, S., Maex, E., Van Belle, H., and Van der Veken, J. (1994). *Worldviews: From Fragmentation to Integration*. Brussels: VUB Press.

[43] Rousseau, D. (2014). *Systems Philosophy and the Unity of Knowledge. Syst. Res. Behav. Sci.* 31, 146–159.

[44] Rousseau, D. (2012). "Could spiritual intuitions map a scientifically plausible ontology?," in *Proceedings of Joint Conference of the Scientific and Medical Network and the Society for Scientific Explorations; Mapping Time, Mind and Space*, Drogheda, 18–21.

[45] Midgley, G. (2000). *Systemic Intervention: Philosophy, Methodology, and Practice*. Berlin: Springer.

[46] Luhmann, N. (2012). *Introduction to Systems Theory*. New York, NY: J. Wiley.

[47] Lin, Y., Ma, Y., and Port, R. (1990). Several epistemological problems related to concept of systems. *Math. Comput. Model.* 14, 52–57.

[48] Wood-Harper, A. T., and Fitzgerald, G. (1982). A taxonomy of current approaches to systems analysis. *Comput. J.* 25, 12–16.

[49] Novikov, D. A. (2016). "Chapter-2 Cybernetics," in *Springer Series, 'Studies in Systems, Decision and Control'*, ed. J. Kacprzyk, Vol. 47, (Berlin: Springer).

[50] Cowan, I. J. (ed.). (1978). "Automation in Coloration: the Philosophy of Automation," in *Proceedings 1978 Conference*, Manchester.

[51] Wimmer, F. M. (1996). "Is intercultural philosophy a new branch or a new criterion in philosophy?," in *Interculturality of Philosophy and Religion, National Biblical Catechetical and Liturgical Centre*, Bangalore, ed. G. D. Souza (Aachen: Augustinus), 101–118.

[52] Mall, R. A. (2000). *Intercultural Philosophy*. Washington, DC: Rowman & Little Publishers, Inc.

[53] ABET *Code of Ethics*. Available at: www.coursehero.com/file/1128606 4/Ethics-code-ABET/

[54] Kushwaha, M., and Chhabra, P. *Business Technology*. Available at: www.slideshare.net/meghakushwaha90/0rg-culture-33987737

[55] Keeney, R. L. (1994). *Creativity in decision making with value focused thinking, MIT Sloan Management Review*. Available at: www.sloanreview.mit.edu/article/creativity-in-decision-making-with-valuefocused-thinking/

[56] Wikipedia *Systems Philosophy*.

[57] https://www.ukessays.com

[58] Barringer, B., and Ireland, R. D. (2008). *Chapter-7 Entrepreneurship: Successfully Launching New Ventures*. London: Pearson.

[59] Steiholtz, R., and Gee, I. (2014). *The alchemy of ethics: The Organizational Development and the Ethics Agenda*. SCCE Compliance and Ethics Institute.

[60] Das, D. (2013). *Control Systems Make Management of an Organization Possible, Business,* May 10, 2013. Available at: www.slideshare.net

[61] Kemp. D. S. *Autonomous Cars and Surgical Robots: A Discussion of Ethical and Legal Responsibility*. Available at: http://verdict.justin.com 2012/11/19/autonomous-cars-and-surgical-robots

[62] O'Donnell, J., and Mitchell, B. *USA Today*. http://www.ustoday.com/ story/money/cars/2013/06/10/automakers-develop-self-driving-cars/ 2391949

A Selection of Books on Systems, Cybernetics, Control, and Automation

SYSTEMS THEORY

1. **L. VON BERTALANFFY** General System Theory: Foundations, Development, Applications, Penguin, New York, 1969.
2. **L. VON BERTALANFFY** Perspectives on General Systems Theory: Scientific Philosophical Studies, G. Brazilier/Oxford U. Press, Oxford, 1976.
3. **Y. LIN** General Systems Theory: A Mathematical Approach, Springer, 2002.
4. **N. LUHMANN** Introduction to Systems Theory. Polity, Cambridge, UK, 2012.
5. **A. RAPOPORT** General Systems Theory: Essential Concepts and Approach, CRC Press, Boca Raton, 1980.
6. **R. ACKOFF** On Purposeful Systems, Aldine Press, Oxford, 1972.
7. **I. V. BLAUBERG** et al. Systems Theory: Philosophy and Methodological Problems, Progress Publishers, Moscow, 1977.
8. **G. KLIR** An Approach to General Systems Theory, Van Nostrand, New York, 1969.
9. **C. W. CHURCHMAN** Systems Approach, Delta, New York, 1968.
10. **R. LILIENFELD** Rise of Systems Theory: An Ideological Analysis, Wiley, New York, 1977.

SYSTEM DYNAMICS

1. **J. W. FORRESTER** World Dynamics, Wright Allen Press, Laurence, Kansas, 1977.
2. **J. W. FORRESTER** Industrial Dynamics, MIT Press, Cambridge, 1961.

3. **G. ROWELL and G. WORMLEY** System Dynamics, Prentice Hall, Upper Saddle, 1997.
4. **J. RANDERS** Elements of the System Dynamics Method, MIT Press, Cambridge, 1980.
5. **J. AWREJCEWITZ** Dynamical Systems: Theoretical and Experimental Analysis, Springer, Berlin, 2010.
6. **B. K. BALA, F. M. AZHAD and K. M. NOH** System Dynamics: Modeling and Simulation, Springer, Berlin, 2015.
7. **M. M. MEERSCHAERT** Mathematical Modeling, Academic Press/ Elsevier, New York, 1999.
8. **J. STERMAN** Business Dynamics: Systems Thinking and Modeling in a Complex World, Irwin/McGraw-Hill, New York, 2000.
9. **SUSHIL K. GUPTA** Systems Dynamics: A Practical Approach for Managerial Problems, Wiley, New York, 1999.
10. **E. BELTRAMI** Mathematics for Dynamic Modeling, Academic Press, Orlando, 1987.

SYSTEMS THINKING

1. **P. SENGE** The Fifth Discipline: The Art and Practice of the Learning Organization, Double Day, New York, 1990.
2. **P. CHECKLAND** Systems Thinking, Systems Practice, Wiley, New York, 1981.
3. **D. H. MEADOWS and D. WRIGHT** Thinking in Systems: A Primer, Chelsea Green Publishing, White River Junction, VT., 2008.
4. **D. P. STROH** Systems Thinking for Social Change, Chelsea Green Publishing, White River Junction, VT, 2015.
5. **P. MELLA** System Thinking: Intelligence in Action (Perspectives in Business Culture), Springer, Berlin, 2012.
6. **G. M. WEINBERG** An Introduction to System Thinking, Dorset House Publ., New York, 2013.
7. **J. GHARAJEDAGHI** Systems Thinking: Managing Chaos and Complexity, Elsevier/Morgan Kaufmann, New York, 2012.
8. **M. C. JACKSON** Systems Thinking: Creative Holism for Managers, Wiley, New York, 2003.
9. **S. G. HAINES** The Manager's Pocket Guide to Systems Thinking and Learning, Human Resource Development Press, Amherst, MA, 1999.
10. **R. L. ACKOFF and H. J. ADDISON** Systems Thinking for Curious Managers: With 40 New Management f-Laws, Triarchy Press, Axminster Devon, UK, 2010.

CYBERNETICS

1. **N. WIENER**, Cybernetics or Control and Communication in the Animal and the Machine MIT Press, Cambridge, 1948.
2. **H. VON FOERSTER** The Cybernetics of Cybernetics (BCL Report 1975), Future Systems, Inc., Minneapolis, 1995.
3. **R. ASHBY** Introduction to Cybernetics, Methuen, London, 1056.
4. **B. CLEMSON** Cybernetics: A New Management Tool, Abacus Press, Kent, 1984.
5. **R. TRAPPI (Ed.)** Cybernetics: Theory and Applications, Hemisphere, Washington, 1983.
6. **H. VON FOERSTER** Understanding of Understanding: Essays on Cybernetics and Cognition, Springer, Berlin, 2013.
7. **V. M. GLUSHKOV** Introduction to Cybernetics, Academic Press, New York, 1966.
8. **R. M. GLORIOSO** Engineering Cybernetics, Prentice-Hall, Upper Saddle River, 2008.
9. **A. I. LERNER** Fundamentals of Cybernetics, Chapman and Hall, 1977.
10. **H. BRUN and S. SLOAN** Cybernetics of Cybernetics, Future Systems, Inc., Minneapolis, 1995.

CONTROL

1. **J. GIBSON** Nonlinear Automatic Control. McGraw-Hill, New York, 1963.
2. **J. J. SLOTINE AND W. LI** Applied Nonlinear Control, Pearson, New York, 1990.
3. **J. MELSA and D. SCHULTZ** Linear Control Systems, McGraw-Hill, New York, 1969.
4. **L. VINCENT and J. WALTER** Nonlinear and Optimal Control Systems, Dover, 2015.
5. **R. ISERMAN** Digital Control Systems, Springer, Berlin, 1981.
6. **R. DORF and R. BISHOP** Modern Control Engineering, Prentice-Hall, Upper Saddle River, 2001.
7. **B. C. KUO** Automatic Control Systems, Prentice-Hall, Upper Saddle River, 1996.
8. **K. OGATA** Modern Control Engineering, Prentice-Hall, Upper Saddle River, 1997.
9. **W. GRANTHAM and T. L. VINCENT** Modern Control Systems Analysis and Design, Wiley, New York, 1993.

10. **E. JARZEBUSKA** Model Based Tracking Control of Nonlinear Systems, CRC Press, Boca Raton, 2012.

AUTOMATION

1. **J. LOVE** Process Automation Handbook, Springer, Berlin, 2007.
2. **S. Y. NOF** Handbook of Automation, Springer, Berlin, 2009.
3. **H. A. FREEMAN** Tutorial: Office Automation/No Dq 711, IEEE Computer Society, 1986.
4. **N. G. TIAN** Real Time Control Engineering: Systems and Automation, Springer, Berlin, 2016.
5. **M. HORN and D. WATZENING** Automated Driving, Springer, Berlin, 2017.
6. **T. R. KURFERS** Robotics and Automation Handbook, CRC Press, Boca Raton, 2004.
7. **R. KRUTZ** Industrial Automation and Control Security Principles, ISA, 2013.
8. **D. TSICHRITZIS** Office Automation: Concepts and Tools, Springer, Berlin, 1969.
9. **R. REILLY** The Handbook of Office Automation, iUniverse, Inc., New York, 2004.
10. **M. D. ZISMAN** Office Automation: Revolution or Evolution? Sloan School of Management, Cambridge, 1978.

COMPLEXITY AND NONLINEAR SYSTEMS

1. **N. F. JOHNSON** Simply Complexity: A Clear Guided Tour to Complexity Theory, Oneworld Publications, London, 2009.
2. **M. MITCHELL** Complexity: A Guided Tour, Oxford U. Press, Oxford, 2011.
3. **J. H. HOLLAND** Complexity; A Very Short Introduction, Oxford U. Press, 2014.
4. **H. SAYAMA** Introduction to the Modeling and Analysis of Complex Systems, SUNY Press, 2015.
5. **P. C. MOLEMAN and K. M. NEWEL (Eds.)** Applications of Nonlinear Dynamic Development, Psychology Press, Hove, UK, 1998.
6. **T. P. LEUNG and H. S. QIN** Advanced Topics of Nonlinear Systems, World Scientific, Singapore, 2001.

7. **S. ARORA and D. BARAK** Computational Complexity: A Modern Approach, Cambridge U. Press, Cambridge, 2009.
8. **S. NEIL RASBAND** Chaotic Dynamics of Nonlinear Systems, Wiley, New York, 1980.
9. **W. M. HADDAD and V. CHELLABOINA**, Nonlinear Dynamical Systems and Control: A Lyapunov-Based Approach, Princeton U. Press, Princeton/Oxford, 2008.
10. **J. M. LIN and L. S. TSIMRIN** Digital Communications Using Chaos and Nonlinear Systems, Springer, Berlin, 1998.

CHAOS AND FRACTALS

1. **E. LORENZ** The Essence of Chaos, U. Washington Press, Washington, 1995.
2. **L. SMITH** Chaos: A Very Short Introduction, Oxford U. Press, Oxford 2007.
3. **J. GLEICK** Chaos: Making of a New Science, Viking, New York, 1987.
4. **N. LESMOIR-GORDON** Introducing Fractals: A Graphic Guide, ICON Books, London, 2005.
5. **F. FEMAT and R. SOLIS-PERALES** Robust Synchronization of Chaotic Systems via Feedback, Springer, Berlin, 2008.
6. **B. B. MANDELBROT** Fractal Geometry of Nature, W. H. Freeman, San Francisco, 1882.
7. **B. B. MANDELBROT** Fractals and Chaos, Springer, Berlin, 2004.
8. **K. FALCONER** Fractals: A Very Short Introduction, Oxford U. Press, Oxford, 2008.
9. **I. PRIGOZINE et al.** Order out of Chaos, Bantam, New York, 1984.
10. **S. H. STROGATZ** Nonlinear Dynamics and Chaos: With Applications to Physics, Chemistry, Biology, and Engineering, West View Press, Boulder, Colorado, 2014.

ADAPTATION AND SELF-ORGANIZATION

1. **F. CAPRA and P. L. LUISI** The Systems View of Life: A Unifying Vision, Cambridge U. Press, Cambridge, 2014.
2. **A. KAUFFMAN** The Origins of Order: Self-Organization and Selection, Oxford U. Press, Oxford, 1993.
3. **G. NICOLIS and I. PRIGOGINE** Self-Organization in Non-Equilibrium Systems, Wiley, New York, 1977.

4. **S. J. FARLOW** Self-Organizing Methods in Modeling, Marcel Dekker, New York, 1984.
5. **R. D. BRUNNER and G. D. BREWER** Organized Complexity, Free Press, New York, 1971.
6. **J. HOLLAND** Adaptation in Natural and Artificial Systems, U. Michigan Press, Ann Arbor, 1976.
7. **J. MILLER and S. E. PAGE** Complex Adaptive Systems: An Introduction to Computational Models of Social Life, Princeton U. Press, Princeton, 2007.
8. **R. BELLMAN** Adaptive Control Processes: A Guided Tour, Princeton U. Press, Princeton, 1972.
9. **S. JOHNSON** Emergence The Connected lives of Ants, Brains, Cities, and Software, Scribner/Simon and Schuster Books, New York, 2001.
10. **J. HOLLAND** Hidden Order: How Adaptation Builds Order, Helix Books, New York, 1996.

SOCIO-TECHNICAL SYSTEMS

1. **S. BORRAS and J. ELDER** The Governance of Sociotechnical Systems Explaining Change, Edward Elgar Publications, Cheltenham, 2014.
2. **R. E. CAVALLO** Role of Systems Technology in Social Science Research, Martinus Nijhoff, Boston, 1979.
3. **W. B. ROUSE, K. R. BOFF and P. SANDERSON** Complex Socio-Technical Systems: Understanding and Influencing the Causality, IOS Press, Amsterdam, 2012.
4. **G. WALKER, N. A. STANTON and D. P. JENKINS** Command and Control: The Sociotechnical Perspective, CRC Press, Boca Raton, 2009.
5. **A. W. AULIN** Cybernetic Laws of Social Progress, Pergamon Press, Oxford, 1983.
6. **D. FLUELER** Decision Making for Complex Sociotechnical Systems, Springer, Berlin, 2006.
7. **S. STAFFORD** Platform for Change, Wiley, London, 1975.
8. **N. LUHMAN** Social Systems, Stanford U. Press, St. Redwood City, CA,1996.
9. **S. MINUCHIN and H. FISHMAN** Family Therapy Techniques, Harvard U. Press. Cambridge, MA, 1981.
10. **D. G. JOHNSON and J. M. WETMORE** Technology and Society: Building our Sociotechnical Future, MIT Press, Cambridge, MA, 2008.

APPLICATIONS

1. **L. SKYTTER** General Systems Theory: Ideas and Applications, World Scientific, Singapore, 2001.
2. **J.-P. CORRIOU** Process Control: Theory and Applications, Springer, Berlin, 2010.
3. **S. SMITH et al.** Systems Theory in Action: Applications to Individual, Couple, and Family Therapy, Wiley, New York, 2010.
4. **G. MEYER and S. BEIKER (Eds.)** Road Vehicle Automation 3, Springer, Berlin, 2016.
5. **J. H. CHOW, P. V. KOKOTOVIC and R. THOMAS** Systems and Control Theory for Power Systems, Springer, Berlin, 1995.
6. **E. B. ROBERTS** Managerial Applications of System Dynamics, MIT Press, Cambridge, 1978.
7. **C. E. BILLINGS** Human-Centered Aviation Automation Principles and Guidelines, NASA, 2013.
8. **A. Weintrit** Marine Navigation and Safety of Sea transportation, CRC Press, Boca Raton, 2013.
9. **J. STERNESON** Industrial Automation and Process Control, Prentice Hall, Upper Saddle River, NJ, 2002.
10. **B. R. MERTA and Y. J. REDDY** Industrial Process Automation Systems, Elsevier, New York, 2017.

IMPACT ON SOCIETY

1. **N. WIENER** The Human Use of Human Beings: Cybernetics and Society, Perseus Books, New York, 1954.
2. **R. JERVIS** System Effects: Complexity in Political and Social Life, Princeton U. Press, Princeton, 1999.
3. **C. R. DECHERT (Ed.)** The Social Impact of Cybernetics, Notre Dame Press, Notre Dame, Indiana, 1966.
4. **M. FORD** The Lights in the Tunnel: Automation, Accelerating Technology, and the Economy, Create Space Independent Publications Platform, Boulder, CO, 2009.
5. **L. WEATHERBY** The Cybernetic Humanities, John Hopkins U. Press, Baltimore, MD, 2015.
6. **E. BRYNJOLFSSON and A. Mc AFEE** The Second Machine Age, Work, Progress, and Prosperity in a Time of Brilliant Technology, W. W. Norton and Company, New York, 2016.

7. **G. LEONHARD** Technology vs. Humanity: The Coming Clash Between Man and Machine, Fast Future Publishing Ltd., Tonbridge, Kent, 2016.
8. **M. FORD** Rise of the Robots: Technology and the Threat of a Jobless Future, Basic Books, New York, 2016.
9. **O. DALE and R. SMITH** Human Behavior and the Social Environment, Pearson, New York, 2013.
10. **M. KNUDSEA and W. VOGD** Systems Theory and the Sociology of Health and Illness, Routlege, Boca Raton, 2015.

ETHICAL ISSUES

1. **I. Van de POET and L. ROYAKKERS** Ethics, Technology, and Engineering, An Introduction, Wiley- Blackwell, Oxford, 2011.
2. **M. WINSTON and R. EDELBACK** Society, Ethics, and Technology, Wadsworth Publishing, Belmont, CA, 2011.
3. **O. THYSSEN** Business Ethics and Organizational Values: A Systems Theoretical Analysis, Palgrave Macmillan, London. 2009.
4. **S. P. Van RYSEWYK and M. PONTIER (Eds.)** Machine Medical Ethics, Springer, Berlin, 2014.
5. **S. G. TZAFESTAS** Roboethics: A Navigating Overview, Springer, Berlin, 2015.
6. **M. KAPTEIN** Workplace Morality: Behavioral Ethics in Organizations, Emerald Group Publishing, Bingley, UK, 2017.
7. **K. A. MERCHANT and W. A. VAN der STEDE** Management Control Systems: Performance Measurement, Evaluation and Incentives, Prentice Hall, Upper Saddle River, NJ, 2003.
8. **H. T. TAVANI** Ethics and Technology: Controversy, Questions, and Strategies for Ethical Computing, Wiley, New York, 2013.
9. **S. JASANOFF** The Ethics of Invention: Technology and the Human Future, W.W. Norton and Company, New York, 2016.
10. **S. BROMBERG** The Evolution of Ethics: An Introduction to Cybernetic Ethics, Dianic Publications, Luxembourg, 1999.

PHILOSOPHICAL ISSUES

1. **G. MIDGLEY** Systemic Intervention: Philosophy, Methodology, and Practice, Springer, Berlin, 2000.
2. **C. HOOKER** Philosophy of Complex Systems, Elsevier/North Holland, Amsterdam, 2017.

3. **M. BUNGE** Treatise on Basic Philosophy, Vol. 4: A World of Systems, D. Reidel, Boston, 1979.
4. **J. C. PITT** Thinking About Technology: Foundations of the Philosophy of Technology, Seven Bridges Press, New York, 2006.
5. **J. FREDERICK and K. M. SAYER (Eds.)** Philosophy of Cybernetics, Clarion-Simon and Schuster, New York, 1968.
6. **F. H. GEORGE** Philosophical Foundations of Cybernetics, Abacus Press/CRC Press, Boca Raton, 1979.
7. **K. M. SEYE** Cybernetics and the Philosophy of Mind, Routlege, Boca Raton, 1976.
8. **N. CARS** The Glass Cage: Automation and Us (Audiobook), Brilliance Audio, New York, 2014.
9. **J. MINGERS** Systems Thinking: Critical Realism and Philosophy—A Confluence of Ideas, Routlege, New York/Abingdon, 2014.
10. **S. G. TZAFESTAS** An Introduction to Robophilosophy: Cognition, Intelligence, Autonomy, Consciousness, and Conscience, River Publishers, Delft, 2016.

Index

About the Author

Spyros G. Tzafestas was born in Corfu, Greece on December 3, 1939. Degrees: B.Sc. in Physics (1962) and P.G. Dipl. in Communications (1963) from Athens Univ. D.I.C. and M. Sc. (country-region Eng.) in Control from London Univ. (1967). Ph.D. in Systems and Control from Southampton Univ. (March 1969). Positions: Research leader at 'Demokritos' Res. Center in Computer Control (1969–1973). Professor of Control Systems at the E. Eng. Dept. of Patras Univ., Patras, Greece (Oct. 1973–Mar. 1985). Professor in Control and Robotics, Dir. of R&A Lab. at the School of E.& C. Eng. (SECE) of the Natl. Tech. Univ. of Athens (NTUA, Apr. 1985–Aug. 2006), Dir. of NTUA Inst. Comm. & Comp. Syst. (1999–2009). Supervisor of 35 Ph.D., and over 100 Master Theses. Recipient of D.Sc. (Southampton Univ.) and two honorary doctorates in engineering (TU Munich, Germany, and EC Lille, France). Over the years he worked in D. P.S., M.V. and M.D. Systems, Robotic systems, and AI, KB and CI control techniques (over 700 papers). He is a LIFE FELLOW of IEEE and a FELLOW of IET (IEE). Founding editor of J. Int. and Robotic Syst. (1988–2006). Chief Editor of the Springer ISCA book Series. Editor of 40 research books and 25 journal special issues. Organizer and/or chair of many Intl. conferences (IEEE CDC, EUCA, etc.), and coordinator of many national and European projects in IT, CIM, Robotics, Intelligent Systems, and Control. Author of seven international books and seven Greek books, His biography is included in more than 30 intl. biographical volumes. Currently, Dr. Tzafestas continues his scientific work at NTUA as a Professor Emeritus-Senior Research Associate of SECE.

For Product Safety Concerns and Information please contact our EU
representative GPSR@taylorandfrancis.com
Taylor & Francis Verlag GmbH, Kaufingerstraße 24, 80331 München, Germany

www.ingramcontent.com/pod-product-compliance
Ingram Content Group UK Ltd.
Pitfield, Milton Keynes, MK11 3LW, UK
UKHW021055080625
459435UK00001B/5